AIRBORNE PARTICULATES

AIRBORNE PARTICULATES

MING CHENG
AND
WEN LIU
EDITORS

Nova Science Publishers, Inc.
New York

For permission to use material from this book please contact us:
Telephone 631-231-7269; Fax 631-231-8175
Web Site: http://www.novapublishers.com

LIBRARY OF CONGRESS CATALOGING-IN-PUBLICATION DATA

Airborne particulates / [edited by] Ming Cheng and Wen Liu.
p. ; cm.
Includes bibliographical references and index.
ISBN 978-1-60692-907-0 (hardcover)
1. Air--Pollution--Health aspects. 2. Aerosols. I. Cheng, Ming, 1964- II. Liu, Wen, 1966-
[DNLM: 1. Air Pollutants. 2. Particulate Matter. 3. Environmental Exposure. 4. Environmental Health. 5. Environmental Monitoring. WA 754 A2997 2009]
RA576.A47 2009
615.9'02--dc22

2009000968

Published by Nova Science Publishers, Inc. ✢ New York

CONTENTS

PREFACE

This book presents studies of airborne particulates and their role in human health in various geographical regions. Various methodologies are used to monitor the transport and composition of these particulates. The negative impact of antimony and its compounds on the environment and ecosystem is discussed. Furthermore, this book describes, compares and contrasts airborne particulate techniques, both theoretically and in practice. Potential human health risks related to a myriad of inorganic and organic species, especially the polycyclic aromatic hydrocarbons, requires a detailed understanding of pollutant levels both indoors and outdoors. The photochemical degradation of selected polycyclic aromatic hydrocarbons, which is being done using simulated solar UV-irradiation, is examined. Particularly urban air, in high traffic areas, containing matagenic and carcinogenic pollutants. Finally, this book describes the genotoxic effects of carbon black, fine particulates on the murine macrophages. The ultrastructural evidences of damage induced by fine and coarse air particulate matter are also discussed.

Chapter 1 - PM_{10} (particles with aerodynamic diameter less than 10μm) play an important role in human health, numerous studies have related PM (ambient particulate matter) levels with adverse health outcomes, primarily respiratory-related, in the general population. The Health Effect is closely correlated with Properties of PM, Toxicology mechanism is studied on the molecular level by means of molecular biology method mostly, and the vitro plasmid assay together with FESEM (Field emission scanning electron microscopy) and ICP-MS (inductively coupled plasma-mass spectrometry) were employed to investigate the physical-chemical characteristics and the bio-reactivity of PMs in Beijing air. It was found that the TD20 (toxic dose of PM causing 20% of plasmid DNA damage) of Beijing PM can be as low as $28\mu g\cdot ml^{-1}$ and as high as $>1000\mu g\cdot ml^{-1}$, Comparison of the physical properties indicates that $PM_{2.5}$ (particles with aerodynamic diameter less than 2.5μm)is more toxiferous than PM_{10}, the oxidative capacity of PM_{10}s of dust storms is much lower than that of the non-dust storm PM_{10}s in spite of the mass of PM_{10} during dust storms is commonly 5 times higher than that during non-dust storm period and water water-soluble fractions and bulk solutions of Beijing PMs produce similar plasmid assay results, indicated that the bioreactivity of Beijing PMs is mainly sourced from the water-soluble fraction and water-soluble zinc is probably the major element responsible for the plasmid DNA damage.

Chapter 2 - The Canary Islands are located on the western dominant transport path of atmospheric dust plumes generated in the Saharan desert. The Saharan region is widely regarded as the world's greatest source of natural mineral dust, much of which is delivered to the ocean and, from time to time, reaches Southern Europe and the Americas. The fact that the Canary Island archipelago is well placed to receive advected Saharan dust plumes was registered by the State Agency of Meteorology of Spain as long ago as the 1950s. Most studies focus on the observation and monitoring of the dust transport, by analyzing satellite images provided by various agencies (AERONET; MPLNET; EARLINET). Recent research on particulate matter has expanded in recent years to include particle size measurements, aggregation, mineralogy, geochemistry, and particle reactivity, as well as considering the impact on human health and global climate.

This chapter will provide a review of previous research on Saharan dust and other aerosols undertaken on the Canary Islands, research groups and agencies involved and their key findings, as well as the main thrust of current research and prospects for future work. It will consider accumulation of Saharan dust both at present and in the past as well as its textural and mineralogical characteristics in deposits found on the islands. Particular attention will be afforded to the impact on the land surface and human population. It will be shown that for the period 2005-2007 the deposition rates on the island of Gran Canaria were of the order of 20 ± 11 g m^{-2} y^{-1}. A high percentage of aggregate in the deposited dust (> 60%) has been determined using Image Analysis of SEM-EDX. Results from these studies show that particle aggregates have lower bulk densities than individual particles, a characteristic with relevance to some existing dust plume transportation models.

On the basis of geomorphologic mapping, GIS-calculations, luminescence dating and sedimentologic-pedologic analyses of sediments trapped in volcanically-dammed valleys on the Canary Island of Lanzarote, estimates of soil moisture fluctuations in deposited Saharan dust and the hydrologic budget of the island are put forward for the Late Quaternary. These deposits provide information on anthropogenic influence on the dust source area of the nearby Saharan Desert during the Holocene and constitute a basis for quantitative estimates of dust input to this island during a large part of the Quaternary.

Chapter 3 - Antimony (Sb) and its compounds have distinguished physicochemical properties. Hence, they have been widely used as functional materials; moreover, awareness regarding its negative impact on the environment and ecosystem has increased considerably. This chapter aims at providing an overview of the available studies on the recent enrichment of Sb in airborne particulates and its potential sources. The analyses of samples of environmental archives such as peat and ice cores have revealed marked changes in the concentrations of various atmospheric pollutants, probably due to the changes in human activities after the Second Industrial Revolution. In fact, historical studies have revealed that the temporal trend of the anthropogenic impact of Sb on the environment between the mid- and late 20th century was strongly dominated by two emission sources: coal combustion and metallurgical processes. Interestingly, another remarkable enrichment of Sb has been observed during the last several decades, implying the emergence of some new anthropogenic Sb sources. This emergence is most likely to be related with the recent expansion of Sb applications as functional materials, such as a flame retardant for plastics and chemical fablics and as a lubricant for automotive brake friction materials. In this chapter, first, the latest knowledge on the fates of Sb through mining, fabricating, usage, and emission to the atmosphere and the toxicities of various Sb compounds are briefly summarized with the aim of providing an in-depth understanding of this topic. Then, the temporal trends of Sb in the environmental samples are reviewed for the purpose of discussing the relationship between atmospheric Sb behaviors and human activities. Subsequently, current studies on airborne Sb, which are carried out in various regions of the world, are overviewed, and the potential sources of Sb are characterized. Moreover, abrasion dust from automotive brakes is highlighted to be an important Sb source that potentially affects highly-trafficked areas. Finally, we focus on the outlook on the demand for Sb compounds and subsequent changes in the Sb concentration in airborne particulates.

Chapter 4 - Airborne particulate matter (PM) from various combustion emission sources is considered as the most important pollutant for adverse health effects in the respiratory system. There is a large number of toxic and carcinogenic substances from airborne particulate matter entering the lung's alveoli which cause oxidative stress, inflammation and cytotoxic damage. These conditions lead progressively to increased morbidity and mortality as well as to increased risk to lung cancer. Studies investigated their size and Physicochemical characteristics, focusing especially to fine particulate matter with aerodynamic diameter below 10 μm (PM_{10} and $PM_{2,5}$), their transition metals, polycyclic aromatic hydrocarbons (PAH), stable quinoids and carcinogenic nitropyrenes. Fine particles enter the lungs and because of their small size are trapped in the alveoli and pulmonary tissues, where toxic substances in the presence of oxygen and redox reactions promote the production of reactive oxygen species (ROS) especially hydroxyl radicals, which are linked to the pathogenesis of pulmonary and cardiovascular adverse effects causing lipid peroxidation, damages to peptides and cellular DNA. In this study we investigated the problem of ROS generation from airborne PM, exhaust soot (diesel and gasoline) and soot from burned wood. Additionally, using Electron Paramagnetic Resonance (EPR) we examined the persistent free radicals in various samples, characteristic to a mixture of semiquinone radicals. Also, we studied the direct production of superoxide anion ($O_2^{\bullet-}$), hydrogen peroxide (H_2O_2) and the damaging hydroxyl radicals ($HO^{\bullet}$) by various extracts of particulate matter. Finally, we examined by EPR the formation of oxidative damage to guanosine nucleobase by PM in aqueous phosphate buffer (pH 7.4). From these results, it was suggested that the cytotoxic and carcinogenic potential of PM is the result of redox cycling of persistent quinoid radicals and transition metal ions on the surface of particles in the generation of ROS with potential for oxidant injurious effects in pulmonary tissues.

Chapter 5 - Airborne particulates are an important public health concern. Because of toxic effects of airborne particulates, they are monitored regularly for total loading and chemical constituents. Monitoring includes various techniques to measure airborne particulates. The primary objective of this paper is to describe, compare, and contrast airborne particulate techniques, both theoretically and in practice. Many of these techniques have been applied recently in a single location–Fallon, Nevada–where a cluster of childhood leukemia has been ongoing since 1997. This offers an opportunity to assess advantages and disadvantages of airborne particulate techniques.

Techniques: Total suspended particulate (TSP) chemistry is the filtering of air to capture airborne dust for measurement and interpretation of elements. Surface dust chemistry is the measurement and interpretation of elements in fine sediments that accumulate on outdoor surfaces. Lichen chemistry is the measurement and interpretation of elements in or on lichens. Leaf chemistry is the measurement and interpretation of elements in or on leaves of trees and other plants. Bark chemistry is the measurement and interpretation of elements on the surface of bark of trees and other plants. Dendrochemistry is the measurement and interpretation of chemical elements in tree rings.

Fallon: TSP chemistry showed elevated tungsten and cobalt in airborne particulates of Fallon compared to other towns of west central Nevada. Morphological and chemical testing of Fallon airborne tungsten particles showed them to be anthropogenic in origin, not natural. Surface dust chemistry and leaf surface chemistry showed peaks of tungsten and cobalt just northwest of the center of town. Lichen tissue chemistry showed elevated airborne tungsten and cobalt within Fallon compared to outlying desert. Dendrochemistry showed environmental availability of tungsten increasing in Fallon during the mid 1990s, coinciding roughly with the onset of the cluster of childhood leukemia in that town.

No single technique for measuring airborne particulates is perfect. Consequently, it would be best to apply multiple techniques in order to capitalize on advantages and offset disadvantages, as illustrated by this comprehensive study of airborne particulates in Fallon.

Chapter 6 - Airborne particle matter contains a myriad of inorganic and organic species, many of which can adversely affect human health. Of these constituents, the class of polycyclic aromatic hydrocarbons (PAHs) is noteworthy because it is ubiquitous and includes several potent carcinogens. Most PAHs emitted to the ambient air are formed during the incomplete combustion of organic materials; that is, anthropogenic activities such as motor vehicle use and industrial activities become major sources of atmospheric PAHs associated with particles. PAH profiles based on specific PAH concentration showed differential behaviors for certain relatively heavy molecular weight PAHs, suggesting that the ratios give valuable information regarding the stability and emission sources of the PAHs. Such outdoor PAHs pollution is considerably impacted on the indoors; indoor PAH concentrations are lower than, and significantly correlated to outdoor PAH concentrations. Humans spend most of their lives indoors. Therefore any evaluation of the human health risks caused by exposure to air pollution requires a detailed understanding of pollutant levels not only indoors but also outdoors. Furthermore, as a series of structurally related aromatics, halogenated aromatic hydrocarbons (HAHs) such as polychlorinated dibenzo-*p*-dioxins/dibenzofurans and polychlorinated biphenyls have been also widespread in the environment, which have prompted great concern about their adverse human effects. Recently, chlorinated PAHs (ClPAHs), structural hybrids of PAHs and HAHs, have been found in the atmospheric particles in an urban city, Japan; 1-chloropyrene was detected at the highest concentration (7.5 pg/m^3 in annual mean), followed by 6-chlorobenzo[*a*]pyrene (5.6 pg/m^3) and 9,10-dichlorophenanthrene (5.1 pg/m^3). Those ClPAHs have been shown to elicit dioxin-like toxicity and mutagenic properties, especially the toxicities of exposure to ClPAHs based on aryl hydrocarbon receptor-mediated activity were estimated to be 30~50 times higher than that of dioxins. The estimation raises the problem that the adverse effects of exposure to ClPAHs deserve further attention.

Chapter 7 - Particulate matter (PM) is a major factor which affects the air quality of ambient environment in Taiwan region. The consecutive study of ambient air pollutants for different particulate seizes and their chemical compositions were conducted. Several character sampling sites were selected in this study. They are Taichung Harbor (TH) and WuChi traffic (WT) and Taichung airport (TA) sampling sites. As for TH and WT sampling sites, measured the concentrations of $PM_{2.5}$ (PM with aerodynamic diameter <2.5 μm, Fine particulate), $PM_{2.5-10}$ (PM with aerodynamic diameter 2.5-10 μm, Coarse particulate) and total suspended particulates (TSP) and analyzed their chemical compositions during March 2004 to January 2005. The measurement results obtained that average TSP concentrations (μg/m^3) were 149.50±29.71 and 106.67±32.04 at TH and WT sampling sites, respectively.

Average concentrations ($\mu g/m^3$) of coarse particulate were 28.42±8.67 and 22.31±5.46 at TH and WT sampling sites, respectively, and average concentrations ($\mu g/m^3$) of fine particulate were 44.77±9.20 and 36.81±5.61 $\mu g/m^3$ at TH and WT sampling sites, respectively, during sampling periods. Additionally, the two primary components of metallic elements on TSP particulates were Fe (1151±236 ng/m^3) and Zn (793±104 ng/m^3), and Fe (1303±298 ng/m^3) and Zn (1081±277 $\mu g/m^3$) at the TH and WT sampling sites, respectively. The two primary components of metallic elements in coarse particulates were Fe (825±155 ng/m^3) and Mg (363±66 ng/m^3), and Fe (839±188 ng/m^3) and Mg (294±96 ng/m^3) at the TH and WT sampling sites, respectively. Furthermore, the two primary components of metallic elements in fine particulates were Fe (253±67 ng/m^3) and Mg (167±44 ng/m^3), and Zn (491±133 ng/m^3) and Fe (382±103 ng/m^3) at TH and WT sampling sites, respectively, during sampling periods. As for Taichung TA sampling site, measured the concentrations of different particulate sizes ($PM_{2.5}$, $PM_{2.5-10}$ and TSP) and analyzed their compositions during September to December of 2005. The measurement results obtained that average TSP, fine and coarse particulates were 120.6, 38.13 and 24.45 $\mu g/m^3$, respectively. Besides, two main compositions of average metallic element concentrations in atmospheric particulates were as follows: Fe, 301.67±89.68 ng/m^3 and Mg, 110.22±29.95 ng/m^3 in fine particulate; Fe, 67.07±29.74 ng/m^3 and Mg, 59.67±22.92 ng/m^3 in coarse particulate; Fe, 385.78±117.84 ng/m3 and Mg, 119.21±39.81 ng/m^3 in fine particulate.

Chapter 8 - Among common ambient air pollutants, particulate matter (PM) is currently under intensive epidemiological and toxicological investigation. Epidemiological evidence suggests a strong relationship between elevated urban environmental levels of PM_{10} and increased incidence of morbidity and mortality due to respiratory and/or cardiovascular diseases in the general population. This association seems to be more closely related to fine particles ($PM_{2.5}$). The consistency between the findings of different epidemiological studies argues for a causal association, but the attribution of acute health effects to concentration levels appears very difficult to explain in the light of the current knowledge of the toxicological mechanisms. The mechanisms underlying these adverse effects are not well understood and major questions concerning the specific size fraction, chemical composition, and causative mechanisms leading to the observed health effects remain. The objective of this in-vitro approach was to correlate the composition of PM with cytotoxicity in the lung. Under this framework, we determined the physical and chemical characterization of the $PM_{2.5}$ fraction from a sample collected in Dunkerque, a French seaside City heavily industrialized. Accordingly, we found that 92.15% of PM were equal or smaller than 2.5 μm and their specific surface area was 1 m^2/g. Inorganic (i.e. Fe, Al, Ca, Na, K, Mg, Pb, etc.) and organic (i.e. volatile organic compounds, polycyclic aromatic hydrocarbons, etc.) chemicals, usually associated with natural environment, but also so-called anthropogenic elements, were found in PM, suggesting that much of PM derived from wind-borne dust from the industrial complex, and the heavy motor vehicle traffic. Thereafter, we focused our attention on the capacity of ambient air pollution $PM_{2.5}$ to induce cytotoxicity, oxidative stress, inflammatory, and pro-apoptotic responses in human epithelial lung cells (L132) in culture. We notably reported that short-term in-vitro exposure to collected $PM_{2.5}$ induced oxidative stress conditions (i.e malondialdehyde production; superoxide dismutase activity; 8-hydroxy-2'-deoxyguanosine formation; poly(ADP-ribose) polymerase activity), thereby activating the nuclear factor-kappa B/inhibitory kappa B (NF-κB/IκB) complex, as showed by the concentration- and/or time-dependent increases in the nuclear p65 and cytoplasmic inhibitory kappa B phosphorylation, and the nuclear p65 and p50 DNA binding. Moreover, PM-induced-activation of NF-κB/IκB complex resulted in the excessive mRNA expression and/or protein secretion of various inflammatory mediators (i.e *nitric oxide/inducible nitric oxide synthase, tumor necrosis factor-alpha, interleukine-1 beta , interleukine-8, granulocyte macrophage-colony stimulating factor, interleukine-6, transforming growth factor-beta 1*) in L132 cells. Thereafter, we observed that $PM_{2.5}$-derived reactive oxygen intermediates mediated apoptotic events by activating not only the tumor necrosis factor-alpha-induced pathway (i.e. caspase-8 and -3 activation), but also the mitochondrial pathway (i.e. cytochrome c release from mitochondria, caspase-9 and -3 activation), thereby leading to subsequent changes in the transcription rates of *P53*, *BCL-2*, and *BAX* genes, on the one hand, and DNA fragmentation, on the other hand. Taken together, these results showed that the occurrence of oxidative stress conditions, the subsequent activation of NF-κB/IκB complex, and, therefore, the production of various inflammatory mediators and the development of pro-apoptotic events, that aggravate according to PM concentration and/or incubation time, proceeded cytotoxicity in ambient air pollution $PM_{2.5}$-exposed target L132 cells in culture.

Chapter 9 - The objective of this paper is to provide a documented discussion on modelling of radionuclides interaction with suspended particulate matter in the marine environment. Aquatic environments occupy a major portion of the earth's surface and, therefore, an understanding of the radionuclide pathways is essential for radioecology purposes. Many radionuclides within such systems exhibit non-conservative behaviour, i.e., they undergo a change in the distribution between the particulate and dissolved phases, due to sorption reactions. The first reactive transport models for non-conservative radionuclides were based upon the equilibrium distribution coefficient, kd. However, recent models are based on kinetic rates of uptake/release of radionuclides between waters and the solid phases when short timescales are involved. Recently, much experimental and modelling effort has been focused on determining those factors which affect the kinetics and the final equilibrium conditions for the uptake of pollutants in aqueous suspensions under dynamic or static situations. Some of the obtained results appear to be either surprising or contradictory and introduce some uncertainty on which parameter values are most appropriate for environmental modelling.

The kinetic box models are widely used in modelling of radionuclides dispersion in aquatic systems, and are based on reversible reactions of constant coefficients. Indeed, several models have been proposed in the literature taking account of parallel, consecutive reactions and some combinations of both depending on the uptake experimental data. Research in this field of radioecology has been stimulated by the high level of public concern over health aspects of the presence of contaminant radionuclides in the environment, which has prompted a generally anti-nuclear posture. In this paper we present a review of the most relevant mathematical models developed so far and a brief revision of the historic developments of theories on surface-electrolyte interactions. Laboratory tracing experiments to study the uptake of ^{239}Pu, ^{241}Am, ^{133}Ba and ^{85}Sr in natural aqueous suspensions from several aquatic systems (reservoir, river, estuary and sea) are also presented to illustrate the application of box models covering a large number of environmental situations and different reaction channels.

Chapter 10 - Suspended particles, particularly fine particles of less than 2.5 μm in aerodynamic diameter ($PM_{2.5}$) in the ambient air, has been reported to increase risks of cardiopulmonary diseases on daily death as well as deaths in long-term followed-up cohort studies. We conducted a cross-sectional study in which vital statistics in Japan were used to correlate disease-specific mortalities with air pollution data, including suspended particulate matter (SPM, less than 10 μm in aerodynamic diameter), nitrogen dioxide (NO_2), nitrogen oxide (NO), sulfur dioxide (SO_2) and photochemical oxidants (Ox). A single-pollutant regression analysis revealed a strong correlation of SPM with ischemic heart disease, hypertensive heart disease in both genders, and pneumonia, asthma, chronic bronchitis / emphysema, lung cancer and also breast, uterine and ovarian cancers in females. Cerebrovascular disease, lung infarction and most of cancers did not show any correlation in either gender. NO_2 also showed a high statistically significant correlation with these diseases in a similar manner to SPM, whereas Ox had no association with any of these diseases. Results from the multi-pollutant Poisson regression analysis showed that cardiac diseases remained to correlate with SPM as well as NO_2, while pulmonary diseases correlated only with SPM. The unexpected correlation of female reproductive organ cancer, observed with SPM or NO_2 should further be studied in the future as possible endocrine disrupting-like effects.

Short Communication A - Polycyclic aromatic hydrocarbons (PAHs) and their nitrated/oxygenated derivatives (NPAHs, OPAHs) are atmospheric pollutants which exhibit various biological activities such as mutagenic and carcinogenic activities. These compounds are generally generated via incomplete combustion of organic materials or chemical conversion of the parent compounds and released into the atmosphere with airborne particles. In this study, photolysis rate constants and quantum yields of seven OPAHs and eighteen NPAHs including 3-nitrobenzanthrone (3-NBA), a nitrated aromatic ketone showing extremely strong mutagenicity were newly determined to simulate the atmospheric fate of the compounds on airborne particles under UV irradiation. NPAHs were more easily photodecomposed than the corresponding oxygenated derivatives. 9-Nitroanthracene had the largest photolysis rate constant, while 4-nitropyrene and 3-NBA were found to be the most photoresistant of the nitrated aromatic compounds tested in this study under UV irradiation. Benzo[*c*]phenanthrene-5,6-quinone which has photolysis rate constant similar to that of 9-nitrophenanthrene degraded the fastest of the seven OPAHs. 1,2-Benzanthraquinone was the most photoresistant of all the substituted PAHs examined in this study. Atmospheric lifetime of the compounds due to photolysis was successfully estimated using the actinic flux on the earth's surface, photolysis rate constants, and quantum yields obtained in this study.

Short Communication B - Urban air, particularly in high traffic areas, contains mutagenic and carcinogenic pollutants, including both organic compounds, such as benzene and polycyclic aromatic hydrocarbons, and inorganic substances, as for instance nickel, chromium and other heavy metals. Extracts of urban air particulates induce cancer in animals and are mutagenic in bacteria and mammalian cells. In recent years, increasing levels of PM_{10} (coarse) and $PM_{2.5}$ (fine) particulate matter, i.e. particles with equivalent diameter less than, respectively, 10 μm and 2.5 μm, have been associated with an increment in cardiorespiratory mortality and respiratory-related emergency department visits. Furthermore, a correlation between mutagenitICY data, airborne micropollutants and meteorological parameters has been established. In this work we have evaluated: a) the genotoxic effects of carbon black, coarse, fine airborne particles and particulate ranging from and 0.65-0.43 μm collected in L'Aquila (a town in Central Italy) on the murine macrophage cell line RAW 264.7; b) the ultrastructural evidences of damage induced by fine and coarse air particulate matter. The results obtained give evidence of higher cytotoxicity of fine PM on cultured murine macrophages, as compared with coarse PM.

RESEARCH AND REVIEW STUDIES

In: Airborne Particulates
Editor: Ming Cheng and Wen Liu

ISBN: 978-1-60692-907-0

Chapter 1

A STUDY ON BIOREACTIVITY AND HEALTH EFFECT OF INHALABLE PARTICULATES IN BEIJING CHINA

***Longyi Shao**[*a], **Lin Zhou**[a], **Xiaoyan Song**[a], **Shushen Yang**[a], **Tim Jones**[b] and **Keith Sexton**[c]*

[a] State Key Laboratory of Coal Resources and Safe Mining and the Department of Resources and Earth Sciences,
China University of Mining and Technology, D11 Xueyuan Road,
Beijing 100083, PR China

[b] School of Earth, Ocean and Planetary Sciences, Cardiff University,
Park Place, Cardiff CF1 3YE, UK

[c] School of Biosciences, Cardiff University, Museum Avenue,
Cardiff CF10 3US, UK

ABSTRACT

PM_{10} (particles with aerodynamic diameter less than 10μm) play an important role in human health, numerous studies have related PM (ambient particulate matter) levels with adverse health outcomes, primarily respiratory-related, in the general population. The Health Effect is closely correlated with Properties of PM, Toxicology mechanism is studied on the molecular level by means of molecular biology method mostly, and the vitro plasmid assay together with FESEM (Field emission scanning electron microscopy) and ICP-MS (inductively coupled plasma-mass spectrometry) were employed to investigate the physical-chemical characteristics and the bio-reactivity of PMs in Beijing air. It was found that the TD20 (toxic dose of PM causing 20% of plasmid DNA damage) of Beijing PM can be as low as $28\mu g\cdot ml^{-1}$ and as high as $>1000\mu g\cdot ml^{-1}$, Comparison of the physical properties indicates that $PM_{2.5}$ (particles with aerodynamic diameter less than 2.5μm)is more toxiferous than PM_{10}, the oxidative capacity of PM_{10}s of dust storms is

* Corresponding. Tel. /fax: +86 10 62331248x8523
E-mail address: shaoL@cumtb.edu.cn

much lower than that of the non-dust storm PM_{10}s in spite of the mass of PM_{10} during dust storms is commonly 5 times higher than that during non-dust storm period and water water-soluble fractions and bulk solutions of Beijing PMs produce similar plasmid assay results, indicated that the bioreactivity of Beijing PMs is mainly sourced from the water-soluble fraction and water-soluble zinc is probably the major element responsible for the plasmid DNA damage.

1. INTRODUCTION

1.1 General Properties Of Particulate Matter

Atmosphere particulate matter includes all solid and liquid particles that are suspended in the ambient atmosphere, These atmospheric particles can be sourced from nature or anthropogenic sources. Particles can be described by size, formation process, chemical composition, origin and atmospheric behavior.

Particle size is commonly measured by aerodynamic diameter, which is the diameter of a uniform sphere of unit density that would attain the same terminal settling velocity as the particle of interest. This measure facilitates size comparison among irregularly shaped particles and refers to the physical behavior of the particles rather than the actual size. The diameters of atmospheric particles span orders of magnitude, commonly varying from 0.001 to 100 μm. The particles with aerodynamic diameter less than 100μm are generally named as total suspend particles (TSP). Among them, the particles whose aerodynamic diameter less than 10μm is PM_{10}, also called inhalable particular matter. Those with aerodynamic diameter less than 2.5μm called fine particle or $PM_{2.5}$. Since the particles with aerodynamic diameter more than 10 μm can sink to the ground in a short time because of its gravitational settlement, they can be called dust fall or coarse particles. Fine particles and coarse particles are distinct entities with fundamentally different sources and formation processes, chemical composition, atmospheric residence times and behaviors, and transport distances. These differences also play a role in how PM affects human health.

PM can be emitted from primary or secondary sources. Coarse or fine particle are often the primary sources, are generally anthropogenic in nature. While the finer PM tends to come from secondary sources, and formed in the atmosphere through physical and chemical conversion of gaseous such as nitrogen dioxide (NO_2), sulfur dioxide (SO_2), and volatile organic compounds (VOCs). Typically, there are two broad sets of primary source categories in urban environments: the first is combustion sources, including vehicles sources, industrial sources and power plants, the second source is the mechanical forces, including wind, and vehicle traffic and other activities. The smaller particles result from combustion, whereas the larger particles tend to come from mechanical forces, such as dust storm. PM and other major air pollutants in urban air have some sources in common and exist as a complex mixture. For example, sulfur dioxide. PM can also come from the secondary aerosol formation. For example, the gaseous pollutants nitrogen dioxide (NO_2) and SO_2 contribute to PM formation, and they are generated by the combustion of sulfur-containing fuels, particularly coal. They are a variety of environmental and anthropogenic factors contribute to ambient PM concentrations. Coarse particles can result from mechanical disruption of large particles by crushing or grinding, from evaporation of sprays, or from dust suspension. In general, the construction and destroy activities such as mining and mineral processing, sea spray, dust, and suspension of settled biological material from soil surfaces and roads are the sources of coarse particles. By contrast, high-energy processes, such as combustion, atmospheric photochemistry, high-pressure grinding and long-term weathering, produce free-mode particles (Source: www.epa.gov).

The size of particles predominately control on their behavior in the respiratory system, involving the degree of the particles penetrating, the place of they depositing, and the effectiveness of the body's clearance mechanisms to remove them. The particle properties, such as surface area, chemical composition, water solubility, and ambient concentration are varying with particle size, therefore, their effects on body's health are associated with diverse size.. All the properties mentioned above are related to the formation processes and emissions sources. In addition, one of the most important parameters in determining the residence time of particles in ambient air is particle size. And it is essential to assess exposure. (USEPA, 2003).

The chemistry composition analysis of the atmosphere particles is one of the most pop research fields from 1960's to now. The composition of PM differs by geographic area and can vary with season, source, and meteorology (USEPA, 2002 and USEPA, 2002a).The known PM_{10} chemistry composition currently includes inorganic composition, organic composition(e.g., pollen, spores, and plant and animal debris), trace heavy metals element, element carbon. The organic composition are found predominantly in coarse particles, while some components, such as potassium and nitrate, may be present in both fine and coarse particles. The secondary organic compounds, and primary organic species from cooking and combustion; and certain metals, primarily from combustion processes, are found predominately in fine particles. The particles with different chemistry composition have different influence on climate, health and visibility of atmosphere, and relate to their internally and externally mixing in particles (Haywood& Shine, 1995; Fuller et al., 1999). In addition, the particles from different sources have the different chemistry constitutions; so the chemistry constitutions of particles can be used to identify the source of particles. Potassium in coarse particles comes almost entirely from soil while nitrate comes primarily from the reaction of nitric acid with pre-existing coarse particles (USEPA, 2003). Potassium in free particles comes mainly from emissions of burning wood or cooking meat, and nitrate comes primarily from the reaction of nitric acid and ammonia to form ammonium nitrate particles.

Fine and coarse mode particles typically exhibit different behaviors in the atmosphere. These differences may affect several exposure-related considerations. The residence time of atmospheric particles varies with size. Fine particles remain suspended in the atmosphere longer than coarse particles and can be removed by diffusion with falling raindrops. In their suspended state, particles can be transported thousands of kilometers and remain in the atmosphere for days to weeks. Fine particles also serve as condensation nuclei for cloud droplet formation and eventually are removed from the atmosphere by precipitation or gravitational settling and impaction on surfaces. By contrast, coarse-mode particles tend to settle more rapidly from the atmosphere and are not readily transported across urban or broader areas because they are generally too large to follow air streams. They tend to be easily removed by gravitational settling and by impaction on surfaces. Coarse-mode particles also are readily removed by precipitation (USEPA, 2003).

1.2 The Heath Effect of Particulate Matter

Although the atmospheric particles take up a very small proportion of the atmosphere, they are paid much attention due to their important influence on environment, climate, especially human health. The heath effect of particulate matter is becoming one of the pop fields the world over. Numerous studies have related ambient particulate matter (PM) levels with adverse health outcomes, primarily respiratory-related, in the general population. This group includes the very young, the elderly, and other sensitive or susceptible subgroups.

Epidemiology suggests an association between ambient PM exposures and various acute health outcomes. The acute and chronic health effects of airborne particles have been vigorously investigated for five decades since the London episode (USEPA, 1999 and Pope and Dockery, 1999). PM has been linked to numerous adverse health effects including increased hospital admissions and emergency room visits, respiratory symptoms, exacerbation of chronic respiratory and cardiovascular diseases, decreased lung function (USEPA, 1999). Epidemiologic survey indicates that there is a link between increase of PM_{10} mass levels and adult morbidity and mortality (Schwartz, 1994; Pope III etc., 1995; Chapman etc., 1997 ; Ostro etc., 1999 ; Maynard, 2001). Scientific studies have linked PM, especially fine particles, with a series of significant health problems, including: Premature death; Respiratory related hospital admissions and emergency room visits; Aggravated asthma; Acute respiratory symptoms; Chronic bronchitis; Decreased lung function that can be experienced as shortness of breath; and Work and school absences, The potential health outcomes organized by system effects are summarized in Table 1.

Table 1 summary of Potential Health Outcomes Based on Emerging Toxicological Evidence

Effect	Health Outcomes
Direct Pulmonary Effects	● Lung injury and inflammation ● Increased susceptibility to respiratory infections ● Increased airway reactivity and exacerbation of asthma
Systemic Effects Secondary to Lung Injury	● Lung injury leading to impaired heart function from lowered blood oxygen levets and increased difficulty in breathing ● Lung inflammation and cytokine production leading to adverse systemic hemodynamic effects (e.g., arrhythmia) ● Lung inflammation leading to increased risk of heart attacks and strokesdue to increased blood coagulability ● PM/lung interactions negatively affecting blood cell formation
Direct Effects on the Heart	● Uptake of particles and/or distribution of soluble components from the lungs into the systemic circulation ● Adverse effects on the autonomic control of the heart and ● Cardiovascular system

Source: USEPA, 2003c.

Evidence from a limited number of toxicological studies suggest that exposure to particles can also increase risk of hospitalization for pneumonia and respiratory infection (Yin et al, 2002; Ohtsuka et al, 2000 a,b; and USEPA, 2003). Some toxicological evidence also indicates that PM can aggravate asthmatic symptoms or increase airway reactivity (USEPA,

2003), while a number of epidemiologic studies have reported increased risk of exacerbation of asthma with ambient PM exposure (Vedal et al, 1998; Guillen et al, 1997; Peters et al, 1997; and Delfino et al, 1998). The more demonstrative and sometimes controversial evidence comes from a number of recent epidemiological studies. Many of these studies have linked short-term increases in particulate levels, such as the ones that occur during pollution episodes, with immediate (within 24 hours) increases in mortality. During major pollution events, such as those involving a 200μg increase in particulate levels, an expert panel at the WHO estimated that daily mortality rates could increase as much as 20 percent (USEPA, 2002).

The portal of entry for PM air pollution is the lung; and PM interactions with respiratory tissue likely mediate a wide range of effects. These include respiratory as well as systemic and cardiovascular effects. The recent literature contains the results of several studies that evaluate the effects of long-term PM exposure on respiratory illness, pulmonary function, cardiovascular morbidity, and mortality and cancer rates. In general, chronic respiratory illness and pulmonary function decrement studies are less numerous than acute studies and the findings are inconclusive and, in some cases, inconsistent. Some studies show effects for some health endpoints, but other studies fail to find the same effects. For example, chronic pulmonary studies by several groups of researchers showed no effect for children from airborne particle pollution (Ware et al, 1986; Dockery et al, 1989; and Neas et al,1994). In contrast, another group of researchers studying US and Canadian children found significant associations between pulmonary function and PM levels (Raizenne et al, 1996) Results of multi-city studies provided data on the positive relationship between chronic respiratory diseases and elevated PM levels. Results of the studies suggest a potential long-term PM exposure effect on chronic respiratory disease (Schwartz et al,1996). Other studies report associations between PM exposures and bronchitis rates and/or lung function decrements or slowed lung function growth in children (USEPA,2002; and McConnell et al, 1999) Numerous epidemiology studies have implicated ambient PM levels as a likely contributor to mortality and morbidity effects. For example, of particular interest with regard to PM-related effects on cause-specific mortality is a growing body of evidence linking long-term PM exposure with increased risk of lung cancer, Historical evidence has included studies of lung cancer trends, studies of occupational groups, comparisons of urban and rural populations, and case-control and cohort studies using diverse exposure metrics (Cohen and Pope, 1995). These studies of have generally indicated a somewhat elevated risk for lung cancer relative to living in urban areas with higher PM exposures (Cohen, 2000). Other recent findings suggest that living in a city with higher exposures (Cohen, 2000). Other recent findings suggest that living in a city with higher PM levels may be associated with an elevated risk of lung cancer amounting to an increase of about 10 to 15% above the lung cancer risk in a cleaner city (Pope et al, 2002; Abbey et al, t999; and Beeson et al, 1998)

PM may stimulate airway sensory nerves; leading to changes in lung function and autonomic tone, which influence cardiac function. Ultra fine particles by virtue of their extremely small size may enter pulmonary capillary blood and be rapidly transported to tissues such as liver, bone marrow, and heart, with either direct or indirect effects on organ function (Lippman et al,2003).

An investigation on the 29 countries in Europe Shows that case-fatality rate of cardiovascular diseases will increase 0.76%, mortality of respiratory apparatus will increase 0.58% accompany with PM_{10} increasing $10\mu g/m^3$.

The acute and chronic health effects of airborne particles have been vigorously investigated for five decades since the London episode (USEPA, 1999 and Pope and Dockery, 1999). PM has been linked to numerous adverse health effects including increased hospital admissions and emergency room visits, respiratory symptoms, exacerbation of chronic respiratory and cardiovascular diseases, decreased lung function, and premature mortality (USEPA, 1999). Specific attention is being paid to the particle size fraction capable of being inhaled and reaching the gas exchange region of the lungs. Particles of about 10 micrometers or less in aerodynamic equivalent diameter (PM_{10}) are capable of reaching the alveoli (the gas exchange region of the lungs) (Rabovsky, 1997), and those measuring between 0.5 and 1.0 micrometers aerodynamic diameter have the highest possibility of being deposited and retained in the alveoli (Parkes, 1983). Particles also may carry other harmful substances with them to these regions, with the smaller particles having the greatest surface area available for such transport (USEPA, 2003).

Based on studies of human populations exposed to high concentrations of particles and laboratory studies of animals and humans, there are major effects of concern for human health. These include effects on breathing and respiratory symptoms, aggravation of existing respiratory and cardiovascular disease, alterations in the body's defense systems against foreign materials, damage to lung tissue, carcinogenesis and premature death. The major subgroups of the population that appear to be most sensitive to the effects of PM include individuals with chronic obstructive pulmonary or cardiovascular disease or influenza, asthmatics, the elderly and children CUSEPA, 2004).

Health effects are not restricted to occasional episodes when pollutant levels are particularly high. Numerous studies suggest that health effects can occur at particulate levels that are at or below the levels permitted under national and international air quality standards. According to the World Health Organization (WHO) and other organizations, as noted in USEPA (2002), no evidence exists that indicates that there is a threshold below which particle pollution does not induce some adverse health effects, especially for the more susceptible populations (e.g., the very young, the elderly, the infirmed, and susceptible populations). The portal of entry for PM air pollution is the lung; and PM interactions with respiratory tissue likely mediate a wide range of effects. These include respiratory as well as systemic and cardiovascular effects. PM may stimulate airway sensory nerves; leading to changes in lung function and autonomic tone, which influence cardiac function. Ultra fine particles by virtue of their extremely small size may enter pulmonary capillary blood and be rapidly transported to extmpulmonary tissues such as liver, bone marrow, and heart, with either direct or indirect effects on organ function (Lippman et al,2003).

1.3 The Main Influential Factors of Heath Effect

The heath effect depends on the physical and chemical properties of PM. The physical properties include the mass concentration, number concentration, size distribution, surface area, aggregating status of single particles and so on, and also include the adsorption, hygroscopic behavior and the absorption and scattering of the light etc. The chemistry properties include inorganic and organic chemical compositions and their distribution, the dissolubility of the chemistry composition, heterogeneous reaction on surface of aerosol particles, and the mineral compositions, etc.

The composition of PM differs by geographic area and can vary with season, source, and meteorology (USEPA, 2002 and USEPA, 2002a).Sulfate, ammonium, and hydrogen ions; black carbon, secondary organic compounds, and primary organic species from cooking and combustion; and certain metals, primarily from combustion processes, are found predominately in fine particles. Sand-related materials (e.g., calcium, aluminum, silicon, magnesium, and iron), and primary organic materials (e.g., pollen, spores, and plant and animal debris) are found predominantly in coarse particles. Some components, such as potassium and nitrate, may be present in both fine and coarse particles. Potassium in coarse particles comes almost entirely from soil while nitrate comes primarily from the reaction of nitric acid with pre-existing coarse particles (USEPA, 2003). Potassium in free particles comes mainly from emissions of burning wood or cooking meat, and nitrate comes primarily from the reaction of nitric acid and ammonia to form ammonium nitrate particles. In addition, the mineral dust surface can also adsorb the volatility organic matter (Goss, 1994, 1996, 1997), and these compounds are cancer-causing inducements. Someone thinks that inhalable particles matter itself do not cause disease; the reason of causing disease is their chemistry compositions or their synergic action (Linn et al., 1999). But there are some experiments confirming that the quartz with extrication SiO_2 can contact the DNA directly and cause pathological changes of DNA, then cause the high blood fat of the lung cell. The asbestos mineral also can cause lose of lung viscosity and appear non- viscosity lung disease and non-viscosity pleural disease. (Guthric et al., 1993) The example of a great deal of researches shows that the particles in the atmosphere are the major carrier of the poisonous and harmful elements and hydrocarbon compounds. They mainly lodge at the outside of the particles surface (Ghio et al., 1999; Bonvallot, 2001). The poly-annulus hydrocarbon (PAHs) in atmosphere mainly distributes among fine particles range, and the content of PAHs within course particles is few. About 70%~90% of PAHs adhere on inhalable matters less than 5μm, among them the BaP , the cancer-causing matter, adhere the particles less than 1.1μm (Sun,1997).The chemical elements with the great danger to the human body healthily in atmosphere , such as the Cr, Ni, As, Pb, V, Cu, Zn, Mn etc, mainly enrich on the surfaces of particles less than the 2μm .They are easy to sink to the area in the lung after breathed into the human body (Yang et al., 1993,1994). In the report of Paoletti et al. (1991), the types of metal elements and minerals in the lung organization depend on the compositions of the particles in atmospheres. Metals elements checked with high frequency in the city air can always check in the person's lung organization .The international standardize organization (ISO) put forward the " the high risk particles", which are easy to cause the children and adult lung disease, are the particles smaller than 2.4 μm. It is very near to the PM2.5 (Yang et al., 2000)

Particle properties and their associated health effects differ by size. The diameters of atmospheric particles span orders of magnitude, commonly varying from 0.001 to 100 μm. The size of particles is a major control on their behavior in the respiratory system, including how far the particles penetrate, where they deposit, and how effectively the body's clearance mechanisms remove them. The particles whose mean aerodynamic diameter less than 2.5μm can enter the air exchange system of the lung. The finer particles have more surface areas, and would absorb more harmful materials; therefore, the particle size is one of the most important parameters in determining the residence time of particles in ambient air, a key consideration in assessing exposure. Particle surface area, chemical composition, water solubility, and ambient concentration vary with particle size and are a function of the formation processes and emissions sources (USEPA, 2003).

The mass and quantities of the particles can be used to evaluate how much particulate matter inhaled by human will make the person feel ill. The size of the particles decides the position that they enter the human body. The size of particles varies by several orders of magnitude, the total mass concentrations can't characterize the properties of the particles efficiently, the particulate matters with different size distributions would display different physical and chemical properties. In fact, all physical and chemical properties of particles are closely related to their size distribution. The mass and quantities of the particles can be used to evaluate how much particulate matter inhaled by human will make the person feel ill. The size of the particles decides the position that they enters the human body, and the particles whose mean aerodynamic diameter less than 2.5μm can enter the air exchange system of the lung. The finer particles have more surface areas, and would absorb more harmful materials. The size of aerosol particles vary by several orders of magnitude, the total mass concentrations can't characterize the properties of the particles efficiently; the particulate matters with different size distributions would display different physical and chemical properties. In fact, all physical and chemical properties of particles are closely related to their size distribution; therefore, this is one of most important attributes of the aerosol. There are different pollution sources in different region, so the size distributions are different, and the ratios of $PM_{2.5}$ in the PM_{10} are also different (Meszaros, 1999).

The adsorption of fine particles due to the very strong adsorptive properties of surface of agglomerate nucleus because of the finer size, so they can adsorb harmful gas, metals ion and strong cancer-causing organic matters such as BaP, as a result, the fine particles enrich great quantities harmful heavy metals element (Hg, As, Pb etc)(Artaxo et al., 1992). The fine particle is also the carrier of germs and viruses. The organic dust provides the essential nourishment and develop place for the germs and virus in air. The particle surface still has the catalysis function; it can promote the atmosphere chemical reaction to carry on (Dai, 2001). The hygroscopic behavior of aerosol particle depends on the properties of water-love or water-hate. The size distributions of particle are different under different relative humidity (RH); their optical properties are also influenced. The most important result of the hygroscopic behavior of aerosol particle is the formation of cloud and rain. The direct climate forcing of aerosol depends on the optical property. The particles interact with light in two different ways: scattering and absorption (Horvath, 1992). Its efficiency primarily depends on the size and chemistry composition of particles, for example, water or liquid droplets are purely scattering, whereas black carbon particles constitute the most important absorbing materials in atmosphere. The sum of scattering and absorption is called extinction or attenuation. The atmosphere visibility relates to the extinction of aerosol particles too (Haywood& Shine, 1995; Fuller et al., 1999).

1.4 Research Methods of Particles Toxicology Mechanism

The research of particles toxicology mechanism is a pop field at present. The domestic and international scholars study the particles poison mechanism on the molecular level by means of molecular biology method mostly (Hu, et al., 1998). These methods include in vivo method and in vitro method (Kadiiska et al., 1997). The in vivo method mainly indicate the bronchus lavement method (BAL) (Kadiiska et al., 1997;Ghio et al., 1999;Rice et al.,

2001;Schins et al., 2004). This method mixes the atmosphere particles to certain concentration, laves them into the body of the experiment animal (such as big rat), then flush, and separate target cells from bronchus lavement liquid (BALF), observing the changes of the target cells, and studying the lung tissue (Adamson et al., 1999). Otherwise, the in vitro method usually is used to evaluate the toxin effects of PM (Eisenbrand et al., 2002). The in vitro method brings up the target cells such as the lung cells of the human body, the epithelium cells etc., then acts with certainly concentration PM in different time (6,12,24 hours, et al).It observes the reaction of the terminal cells (the end points), and study relation between dose and reaction or time and effect. The damages of PM to the cells, especially a carcinogen that has the heredity toxicity usually effect DNA in the cell directly, so the relative simple experiment can be used to carry on the toxicology research (Kramer et al., 1998; The Ding et al., 1999). The in vitro method in common use includes dissolves the blood activity experiment method, Ames experiment (Yan et al., 2001), the Micronuclei Test, the chromosome mutation test, the single cell gel electrophoresis method (SCGE, the comet experiment) (Sardas et al., 1995;Devi et al., 2000) and the Plasmid assay method (Donaldson et al., 1997;Greenwell et al., 2002). The in vitro experiments need to choose the certain terminal, and the toxicology research usually chooses the Biomarker, such as IL-6 (interleukin-6) or IL-8 etc. The toxicology effect of PM can be provided by studying the Biomarker, and can also use DNA as the Biomarker directly. The relation between the biology activity and toxicities dose and effect is the foreland of the study of toxicology (Hu et al., 1998). At present, most methods can only study qualitatively, but the single cell gel electrophoresis method and the Plasmid assay method can process quantificational study. Using the Plasmid assay, Donaldson et al (1997) confirms that Fe ion in particles can catalyze oxidize to produce the free radicals of OH. There are more amount of Fe (III) in particles, but not of Fe(II). The research of the EHC-93 by Adamson et al. (2000) showed that its toxicity may come from the level of water-soluble Zn in the particles. The research of the British Port Talbot by Moreno et al. (2004) also showed that the metal's elements in inhalable particles, particularly Zn, may be the main reason to cause DNA damage. In China, Shi & Shao et al. (2004), Shao et al. (2005) receives the same conclusion by the study oxidize damages of inhalable particles in Beijing.

The research of particles toxicology mechanism is studied on the molecular level by means of molecular biology method mostly (Hu, et al., 1998). These methods include in vivo method and in vitro method (Kadiiska et al., 1997). The in vivo method mainly indicate the bronchus lavement method (BAL) (Kadiiska et al., 1997;Ghio et al., 1999;Rice et al., 2001;Schins et al., 2004). This method mixes the atmosphere particles to certain concentration, laves them into the body of the experiment animal (such as big rat), then flush, and separate target cells from bronchus lavement liquid (BALF), observing the changes of the target cells, and studying the lung tissue (Adamson et al., 1999). Otherwise, the in vitro method usually is used to evaluate the toxin effects of PM (Eisenbrand et al., 2002). The in vitro method brings up the target cells such as the lung cells of the human body, the epithelium cells etc., then acts with certainly concentration PM in different time (6,12,24 hours, et al).It observes the reaction of the terminal cells (the end points), and study relation between dose and reaction or time and effect. The damages of PM to the cells, especially a carcinogen that has the heredity toxicity usually affect DNA in the cell directly, so the relative simple experiment can be used to carry on the toxicology research (Kramer et al., 1998; The

Ding et al., 1999). The in vitro method in common use includes dissolves the blood activity experiment method, Ames experiment (Yan et al., 2001), the Micronuclei Test, the chromosome mutation test, the single cell gel electrophoresis method (SCGE, the comet experiment) (Sardas et al., 1995;Devi et al., 2000) and the Plasmid assay method (Donaldson et al., 1997;Greenwell et al., 2002). The in vitro experiments need to choose the certain terminal, and the toxicology research usually chooses the Biomarker, such as IL-6 (interleukin-6) or IL-8 etc. The toxicology effect of PM can be provided by studying the Biomarker, and can also use DNA as the Biomarker directly. The relation between the biology activity and toxicities dose and effect is the foreland of the study of toxicology (Hu et al., 1998). At present, most methods can only study qualitatively, but the single cell gel electrophoresis method and the Plasmid assay method can process quantificational study. Using the Plasmid assay, Donaldson et al (1997) confirms that Fe ion in particles can catalyze oxidize to produce the free radicals of OH. There is more amount of Fe (III) in particles, but not of Fe (II). The research of the EHC-93 by Adamson et al. (2000) showed that its toxicity may come from the level of water-soluble Zn in the particles. The research of the British Port Talbot by Moreno et al. (2004) also showed that the metal's elements in inhalable particles, particularly Zn, may be the main reason to cause DNA damage. In China, Shi & Shao et al. (2004), Shao et al. (2005) receives the same conclusion by the study oxidize damages of inhalable particles in Beijing.

1.4.1The Application of Individual Particle Analysis in Healthy Effect Evaluation

The atmosphere chemistry behavior of the atmosphere particles has been the stress field in research of atmosphere environment. The analysis of samples focused on the bulk chemistry analysis, the process and products of the chemical reaction before 1970's (Seinfeld et al., 1975; Tang, 1990). The individual particle analysis of the atmosphere particle carried on the analysis to the coarse particles with the help of optical microscope. With the development of nuclear physics technology, the nuclear physics approaches were used to study the individual aerosol particles and the accuracy of analysis were developed greatly. These approaches include instruments for morphology, such as scanning electronic microscope (SEM), transmission electronic microscope (TEM), the atom force microscope (AFM) etc.; the instruments for compositional information, such as electron probe microanalysis (EPMA), includes energy-dispersive X-ray spectrometry (EDX), wavelength-dispersive X-ray spectrometry (WDX), laser microanalysis mass spectrometry (LAMMS), proton induced X-ray emission (PIXE), Micro- PIXE, nuclear microprobe (NM) and time of flight- secondary ionic mass spectrometry (TOF- SIMS) etc.(Mamane et al.,1985;Okada et al.,1987;Van Borm et al.1989;Artaxo et al.,1993;Orlic,1995;Junno et al.,1996; Buseck et al.,1999,2000;Ghermandi et al,1999;Shi and Shao et al.,2003). The principles of these methods mainly are descending the particles surface by high energy particle beam (such as photons, electrons, protons, ions etc.), collecting the physics signal (such as characteristic X-ray, secondary electrons, transmission electrons etc.), exciting by the interaction between high energy particle beam and particles surface, analyzing the signal to obtain the information of morphology, composition, or microstructure, and acquire the relative information of individual particles, such as morphology, size, chemistry composition, structure, and mineral constitution etc.. In above-mentioned methods, electronic microscope with energy-dispersive X-ray spectrometry (SEM- EDX and TEM- EDX) is the most simple and quick technology for charactering the aerosol particles, Russell et al. (1973) studied the atmosphere particles of Denver downtown by means of SEM-EDX to acquire the half quantitative information of the coarse particle. Ramsden (1968) and Shibaoka et al. (1977) studied the formation of soot during the coal combustion by TEM. Binafede et al. (1973), Carpenter et al. (1980) and Ramsden et al. (1982) studied the morphology and composition of soot by electronic microscope and electronic probe microanalysis (EPMA) (Qi et al., 1999). After 1990's, the paper about the relationship between the mineralogy, morphology, chemistry composition of individual particles and the various effects of particles flowed out in great quantities from 1990. (Van Malderen et al.1992; Katrinak et al, 1995;Hinz et al, 1996;Grime, 1998;Kasparian et al1998; Ganor et al, 1998;Haapala et al., 1998; Kasahara et al., 2002; Aragon et al, 2002; Silva et al., 2000;Li et al, 2003a, b; P óSfai et al, 2003,2004; Liu, et al.1994; Wang et al., 1996;Zhang et al., 1998;Xu et al., 1998,2002;Shao et al., 2000,2003,2004,2005;Chen et al, 2001;Qiu et al., 2001;Dong et al., 2001;Wang et al., 2001).

In recent years, the individual particle analysis technique has a new achievement in the technical development and combination of above-mentioned methods. Ma et al. (2001) analyzed the Asia sandstorm particle samples collected in Japan to acquire the diagram of chemical element distribution by SEM- EDX, PIXE and micro- PIXE. Liang et al. (2001) and Nair et al. (2004) employ the time of flight- secondary ionic mass spectrometry (TOF- SIMS) to investigate aerosol particles. During analysis, Nair et al. (2004) freezed the samples under -140 ℃ quickly by putting they into the liquid nitride to limit the vaporize of the volatility matter, then to make the half volatility matter or volatility matter on the particles surface keep down. Wittmaack et al. (2002) tried to collect particles samples by silicon slice when using SEM-EDX, and found that the silicon slice was better than filter film. Martin et al. (2000) explored the change and formation process of aerosol particles by stimulation to assess the chemistry compositions and size distributions, and compared with the results analyzing by TEM in the every process. Masatoshi et al. (2004) combines the synchronous radiation X-ray fluorescence (SR- XRF) with the SEM- EDX to analysis individual particles. Because the SR- XRF can examine the trace and micro amount of chemical elements, and be sensitive to the heavy metals element, as well as SEM- EDX can get the morphology of the individual particles, examine the lighter chemical element, and their advantage get full exertive.

1.4.2 The Plasmid Assay

In toxicology research of atmospheric particulate matter, the plasmid DNA assay is an in vitro method to detect the oxidative damage on plasmid DNA by free radicals. This method has been used to evaluate the free radical activities of fine and ultra fine TiO_2 (Donaldson et al.,1996),fibers(Gilmour etc) carbon black (Li et al.,) and urban airborne particles (Donaldson et al., 1996; Li et al., 1997; Stene et al., 1998) Greenwell et al. (2002, 2003) have optimized and sensitized the plasmid assay, and then used this method to evaluate the bioreactivity of diesel exhaust particles and urban airborne particles in Cardiff, UK.

The principle of the plasmid assay is that any free radicals on the particle surface could cause oxidative stress oil super-coiled DNA (Donaldson et al,, 1996), The preliminary oxidative damage causes the super-coiled DNA to become relaxed (Fig.1), Further damage results in linearization of the DNA (Fig.1), followed by complete fragmentation, The different forms of DNA are separated on agarose gel by electro-phoretic mobility. Using sensitive densitometry, the different forms of DNA are quantified (Fig.1), Bioreactivity is referred as a parentage of linearized and relaxed DNA accounting for total DNA, For this study, airborne PM samples, including dust storm particles, were collected in Beijing. The oxidative capacity of these samples was measured using plasmid DNA assay. In order to study physicochemical properties that may affect the bioreactivity of PM samples, high resolution Field Emission Scanning Emission Microscopy (FESEM), coupled with an image analysis system, was employed to study the microscopic characteristics and size distribution of these samples. Inductive Coupled Plasma-Mass Spectrometry (ICP-MS) was used to measure the water-soluble and whole fraction metal concentrations in the PM samples.

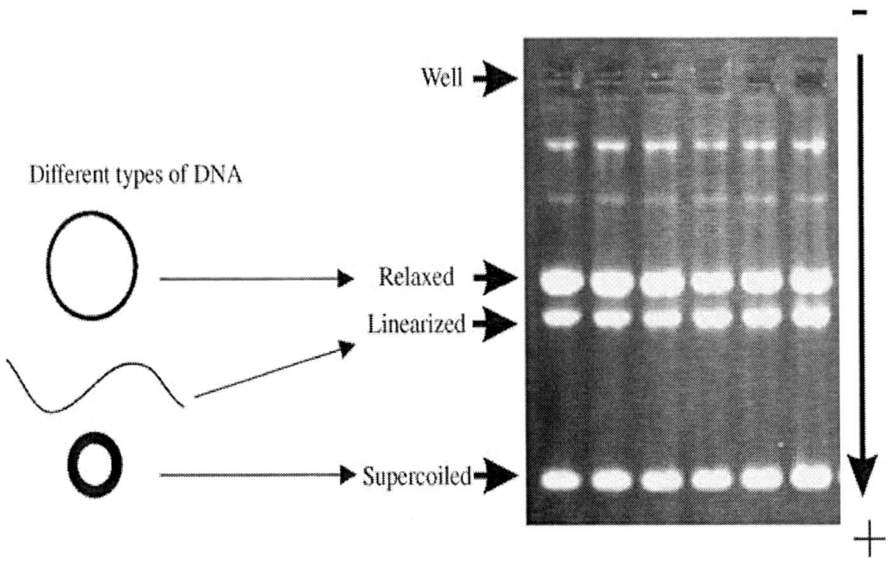

Figure 1. Identification of different types of plasmid φX174-RF DNA (Promega, London, UK) in gel.

2. Materials and Measurement

2.1 Sampling

There are three outdoor collection sites: the campus of the China University of Mining and Technology (simplified as CUMTb, an urban Beijing site near the famous Zhong-guancun Poly-technology Village), Ming Tombs Reservoir (MTR, a relatively clean air site near the Ming Tombs that has no nearby large-scale industrial pollution sources), and Nankou (a satellite city in suburban Beijing). All the samples were collected for 12 h via PM selected inlet heads at a flow rate of $30L\cdot min^{-1}$, using Negretti selective head (UK) with poly-carbonate filters (Millipore, USA, pore size 0.6 μm) at the same time, the meteorological parameters including temperature, humidity, wind direction/speed and weather were recorded.

After considering the meteorological characterization and average mass concentrations in each season, 18 samples were selected for the plasmid assay, including sandstorm particles collected in the spring of 2001. Indoor PM_{10} samples were collected in the living room (n = 2) of a smoker on the campus of China University of Mining and Technology (CUMTB) and a non-smoker's home (living room (n = 2) and kitchen (n = 1)) around the Baishiqiao area in Beijing, China, July 2003. The volunteer homes used gas central heating and gas cookers. PM_{10} was collected for 8 h (on dry days only) via PM_{10}-selective inlet heads at a flow rate of 30 $Lmin^{-1}$, using polycarbonate filters (47mm diameter, 0.67 mm pore size, Millipore, UK).

2.2. The Plasmid Assay

The procedure for the plasmid scission assay was carried out in accordance with the methods used in Merolla and Richards (2005). Control samples were incubated in water and oxidative controls were incubated in a 0.04% hydrogen peroxide solution. PM samples were spiked with a stock particle solution at different concentrations. All incubations were carried out to a final volume of 20 μl, each containing 200ng φ–X174-RF DNA (Promega, London, UK). Incubations were gently agitated (to ensure maximum mixing of the sample and avoid sedimentation) for 6h at room temperature. Bromophenol blue/glycerol loading dye (3μl) was added to each sample before loading. The gels comprised 0.6% agarose and 0.25% ethidium bromide and were run in a 30 V electrophoretic current for 16 h at room temperature in 1% Tris-Borate-EDTA (TBE) buffer. The Syngene Genesnap program (Synoptics Ltd., Cambridge, UK) was used to image gels and to quantify different forms of DNA. Six replicates of each lane were quantified in this way and the means were plotted against particle concentration. Subtracting the damage caused by the negative control (water), the DNA damage induced by airborne particles could be calculated. Using a linear regression, the TD20 (toxic dose causing 20% damage to DNA) was calculated.

2.3 FESEM and Image Analysis

A 4 mm (internal diameter) copper washer was fixed on the standard 12.5 mm aluminum SEM stub using epoxy resin (Araldite). Approximately one eighth of the polycarbonate filter was cut and mounted onto the copper washer using epoxy resin. The stubs were gold-coated to a thickness of 20 nm using a sputter coater (Cressington 208HR). (Fig.2) The specimens were analyzed using a XL30 Field Emission Scanning Electron Microscope (FESEM, Philips Electron Optics, NL) at the Natural History Museum (London, UK). Image analysis was conducted on a customized Quantement 500C system.

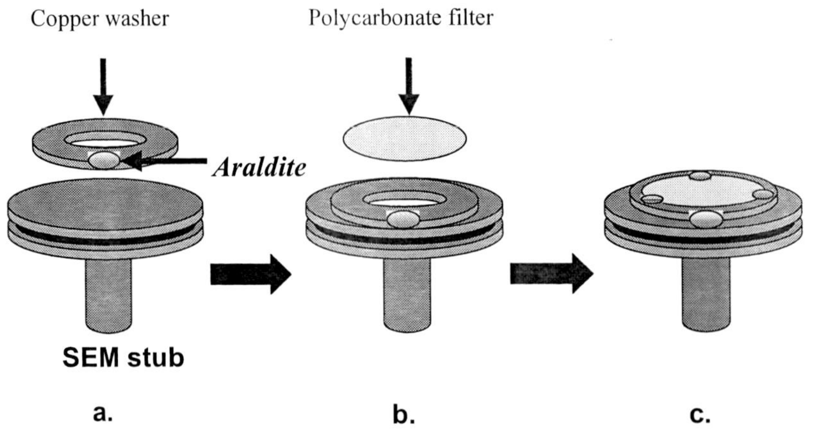

Figure 2. Schematic diagram for FESEM sample preparation. (a) fixing washer to SEM stub;(b) araldite filter sample onto washer; (c) completed sample 'drum' (adapted from Jones,etc, 2000, 2001)

2.4. ICP–MS Analysis

Half of the polycarbonate filters with airborne PM samples were cut and weighted to calculate the mass of the samples. The cut filters were put into a small bottle filled with 10 ml HPLC grade water, and installed on a vortex mixer (Scientific Industries, Vortex Genie 2) set at level 3 for 16 h. Then the solutions were filtered with 0.2 μm pore size polycarbonate (Millipore) filters. 1 ml of the solutions was combined with 0.5 ml rhodium (50 ppb) standard solution and 5 ml HNO_3 (2%). The prepared solution was analyzed with ICP–MS (Perkin Elmer Elan 5000).

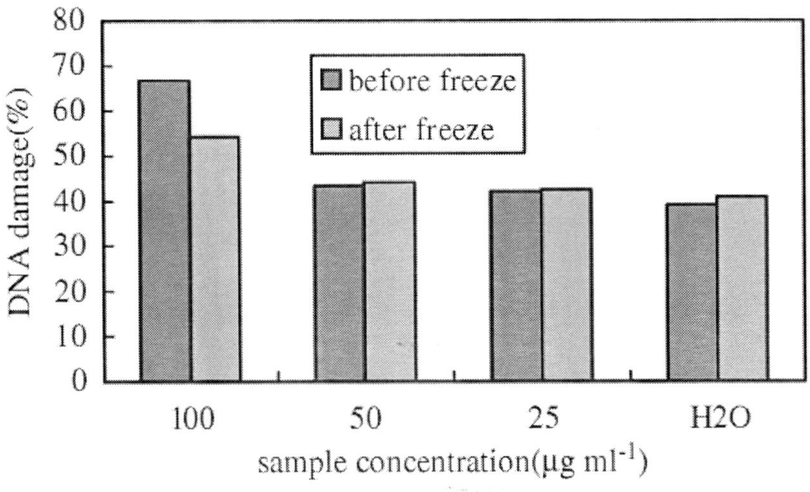

Figure 3. Influence of freeze-drying on the oxidative capacity of PM samples

3. RESULTS

3.1. Oxidative DNA Damage by Beijing Outdoor Airborne PM_{10} and $PM_{2.5}$

The oxidative capacity of Beijing PM_{10} and $PM_{2.5}$ were examined with the plasmid assay and their TD20s were calculated using linear regression. The sample information is shown in Table 2. The gel images and the semi-quantitative analysis of oxidative stress of air-borne particles are shown in Fig. 3 and Table 2, respectively.

The spring PM_{10} samples collected at the urban CUMTb site produce relatively weak DNA damage even under a relatively high dose (for example, 400μg·ml^{-1}; sample Nos. 3, 4 and 5, Table 3). Both No.3 and No.4 samples have very high TD20 values, being >1000 and the TD20 value of the No. 5 sample is 823μg·ml^{-1}. In contrast, PM_{10}s collected at the beginning of November 2000 (Beijing city normally began to start domestic heating on November 15), have demonstrated relatively strong oxidative stress on DNA. One of these PM_{10} samples causes 53% DNA damage at 500μg·ml^{-1} (No. 7, Table 3), with the TD20 being 714μg·ml^{-1}. Another PM_{10} streaks all super-coiled DNA at 400μg·ml^{-1}and 78% at 200μg·ml^{-1} dose (control damage is 42%) (No. 8, Table 3) with the TD20 value of 137μg·ml^{-1}. Two PM_{10} samples collected in the fall of 2001 impose weak oxidative damage to DNA at 100 and 50μg·ml^{-1}, respectively. The TD20 values of these two samples are 619 and 183 μg·ml^{-1} (No. 9 and No. 10, Table3), respectively.

Four PM samples collected at the MTR (background site) were studied by plasmid assay. The $PM_{2.5}$ collected during the heating period (No. 11, Table 3) does not show obvious DNA damage at 50 μg·ml^{-1}, while causing 54% DNA damage at 100 μg·ml^{-1} after freeze-drying (No. 12, Table 3). Their TD20s were calculated to be 165 and 150 μg·ml^{-1}, respectively. The TD20 of the PM_{10} collected during the heating period is 480 μg·ml^{-1} (No. 13, Table 3). The PM_{10} collected during non-heating period causes 50% DNA damage around 200 μg·ml^{-1} and its TD20 was calculated to be 353 μg·ml^{-1} (No. 14, Table 3). The $PM_{2.5}$ collected during the non-heating period causes 70% DNA damage at relatively low concentrations (50 μg·ml^{-1}, No. 15, Table 3). Its TD20 is very low at 28 μg·ml^{-1}. This is much lower than that of the PM_{10} collected during the same season (No. 14) and also lower than that of the $PM_{2.5}$ collected during the heating period (No. 11 and No. 12).

The oxidative capacity of three PM samples collected at Nankou (a suburban satellite city) was also determined. The TD20 of the PM_{10} sample (No. 16, Table 3) was 320 μg·ml^{-1}. This sample caused 100%DNA damage at 750 μg·ml^{-1}, with the TD20 at 260 μg·ml^{-1}. The two $PM_{2.5}$ samples caused 76% and 85%DNA damage at 100 μg·ml^{-1}dose and 43% and 48% DNA damage at 50 μg·ml^{-1}dose (No. 17 and No. 18, Table 3). Thus, their TD20s are 53 and 38 μg·ml^{-1}.Compared with PM samples collected in the urban Beijing area and the background site, the TD20s of PM samples collected in the satellite city are relatively lower, suggesting a higher bioreactivity. $PM_{2.5}$ has a stronger oxidative potential than the PM_{10} in the Nankou area.

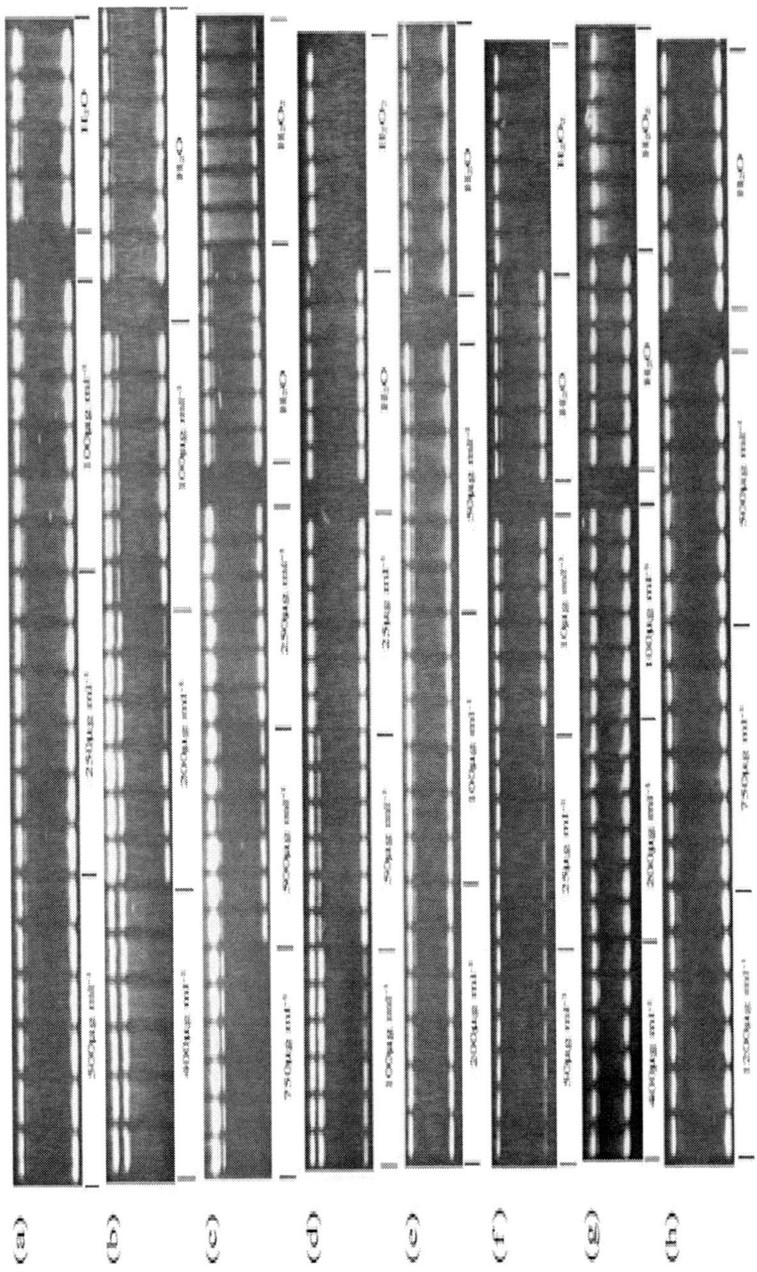

Figure 4. Gel image showing oxidative damage on super-coiled DNA induced by (a) PM_{10} collected at the urban site in spring (No. 6); (b) PM_{10}collected at the urban site in autumn (No. 7); (c) PM_{10} collected at Nankou during heating period (No. 16); (d) $PM_{2.5}$ collected at Nankou during heating period (No. 18); (e) PM_{10} collected at the MTR site during heating period (No. 14); (f) $PM_{2.5}$ collected at the MTR site during non-heating period (No. 15); (g) dust storm sample under low concentration (No. 1); (h) dust storm sample under higher concentration (No. 1). H_2O and H_2O_2 are used as the negative and positive controls.

3.2. Oxidative DNA Damage by Sandstorm Particles

The PM_{10} samples were collected during several dust storm episodes in the spring of 2001, and their oxidative capacities were examined using the plasmid assay. It was found that the three samples caused negligible damage to DNA (for example, Fig.3g and h) compared to control (HPLC grade water) samples, even at relatively high doses (1200 $\mu g \cdot ml^{-1}$). One sample (No.1) was concentrated to 1200 $\mu g \cdot ml^{-1}$ by freeze drying, but did not show DNA damage. Similarly, another sample (No.2) was concentrated up to 800$\mu g \cdot ml^{-1}$, by freeze-drying, but still did not cause obvious DNA damage. Therefore, the plasmid assay results show that dust storm particles have relatively low oxidative capacity.

Table 2 Sample information

Sample No.	Sampling site	Sampling date	Sample type	Collection code
No.1	CUMTb (Urban Beijing)	03/2001	PM_{10}	A26*(Dust storm)
No.2		03/2000	PM_{10}	A32*(Dust storm)
No.3		04/2000	PM_{10}	D1C2
No.4		04/2000	PM_{10}	D1C1
No.5		04/2000	PM_{10}	D2C2
No.6		07/2000	PM_{10}	C8
No.7		11/2000	PM_{10}	A9
No.8		11/2000	PM_{10}	A10
No.9		09/2001	PM_{10}	A61
No.10		09/2001	PM_{10}	A62
No.11	MTR (Background site)	03/2001	$PM_{2.5}$	A48
No.12		03/2001	$PM_{2.5}$	A48*
No.13		03/2001	PM_{10}	A50
No.14		07/2001	PM_{10}	A51
No.15		07/2001	$PM_{2.5}$	A53
No.16	Nankou (Satellite city)	03/2001	PM_{10}	A27
No.17		03/2001	$PM_{2.5}$	A41
No.18		03/2001	$PM_{2.5}$	A44

* Concentrated by freeze-drying.

Table 3 Quantification of oxidative DNA damage induced by airborne particles collected in Beijing urban area, a satellite city and a clean air area

	Sample conc. μg·ml^{-1}	Damage rate ±tdev%	Sample conc. μg·ml^{-1}	Damage rate ±Stdev%	Sample conc. μg·ml^{-1}	Damage rate ±tdev%	Sample conc. μg·ml^{-1}	Damage rate ±Stdev%	Sample conc. μg·ml^{-1}	Damage rate ±Stdev%	Sample conc. μg·ml^{-1}	Damage rate ±Stdev%
	No.1		No.2		No.3		No.4		No.5		No.6	
	1200	39	800	39	400	43±2	400	32±4	400	32±4	500	37
	750	39	400	39	200	39±1	200	36±5	200	36±5	250	44
	500	36	100	40	100	39±2	100	33±7	100	33±7	100	42
	H_2O	40	H_2O	42	H_2O	39±2	H_2O	20±2	H_2O	20±2	H_2O	43
TD20(μg·ml^{-1})	>1000		>1000		>1000		>1000		823		>1000	
R^2							0.8929		0.3469			
	No.7		No.8		No.9		No.10		No.11		No.12	
	500	53±1	400	100±1	100	44±3	50	44±2	50	41±2	100	54±2
	250	46±2	200	78±2	50	40±1	25	39±1	25	41±2	50	45±4
	100	38±3	100	57±4	25	41±1	10	38±1	10	39±1	25	47±2
	H_2O	41±4	H_2O	43±3	H_2O	41±2	H_2O	39±2	H_2O	38±2	H_2O	39±2
TD20(μg·ml^{-1})	714		137		619		183		165		150	
R^2	0.8633		0.9831		0.9423		0.7697		0.9944		0.8521	
	No.13		No.14		No.15		No.16		No.17		No.18	
	200	50±3	200	50±3	50	70±3	750	100±0	100	76±10	100	42±5
	100	40±2	100	40±2	25	82±3	500	65±6	50	43±1	50	14±1
	50	37±1	50	37±1	10	43±1	250	53±4	25	42±1	25	15±3
	H_2O	40±1	H_2O	40±1	H_2O	38±2	H_2O	40±2	H_2O	39±2	H_2O	11±8
TD20(μg·ml^{-1})	480		353		28		260		53		64	
R^2	0.8506		0.7236		0.5547		0.9248		0.8445		0.8385	

3.3 Oxidative DNA Damage by Beijing Indoor Airborne PM10

The oxidative capacity of indoor and outdoor PM10 was examined with the plasmid scission assay and their TD50 values were calculated using linear regression. The sample information is shown in table4. The gel images and the semi-quantitative analysis of oxidative stress of airborne particles are shown in Fig.5 and Table 5, respectively. It can be seen that the PM10 from the non-smoking kitchen (No. 1) generated the strongest oxidative damage of all samples; the value of TD50 being as low as 45μg·ml−1, For this sample, under a dosage of 75μg·ml−1, 97.8%, DNA was damaged, and when the concentration of sample dosage increased, the damage percentage remained relatively constant(Fig 5 Table5) The PM10 from the smoker's home produced stronger DNA damage than those from non-smoker's homes, as well as outdoors.PM10 from a smoker's home (Nos2 and 3) had lower TD50 values, been being 100 and 263μg·ml−1, respectively. The No.4 sample from the no-smoker's home caused 17% and 60% DNA damage suggesting an overall TD50 of 216μg·ml−1. For No.5 sample, even at the concentration of as high as 500μg·ml−1, it cause only 38% DNA damage, suggesting a lower bioreactivity.

Table 4. Sample information for indoor and outdoor PM_{10} collected in Beijing: China in 2003

	Sample No.				
	No.1	No.2	No.3	No.4	No.5
Sampling conditions	Indoor kitchen	Indoor smoker's	Indoor smoker's	Indoor non-smoker's	Indoor non-smoker's
Sampling date	November	July	November	July	July
Mass (μgm^{-3})	205.9	51.1	611.1	29.3	91.2

Moreover, dust storm particles with relatively low oxidative capacity are mainly composed of relatively coarse irregular minerals with few fine soot aggregates and coal fly ash (Fig. 4f).Three random samples (Nos. 10, 15 and 17) have been selected to illustrate the relationship of the compositional constituents with the bioreactivity. FESEM observations coupled with image analyses have revealed the relative percentages of three major constituents; minerals, fly ashes and soot aggregates, as shown in Fig.6. The TD20 values are also shown in this figure. It can be seen that number percentages of mineral and fly ash particles have a good positive correlation with TD20s while the soot aggregates show an obvious negative correlation with TD20s. It is clear that the highest soot number percentage of the No. 15 sample is correspondingly associated with the lowest TD20 values and thus the highest bioreactivity. In contrast, the number percentage of mineral particles in this sample is the lowest and this means mineral particles may be less bioreactive compared with soot. Hung and Wang (2001) demonstrated that reactive oxygen species concentration in airborne particles shows a strong relationship with photochemical reaction strength. It is reported by Zhang et al. (1998a,b) that the photochemical smog in Beijing may increase with the rapid growth of the number of vehicles in Beijing. This suggests that concentration of reactive oxygen species will increase and the corresponding bioreactivity of Beijing airborne particles will increase in the future.

Table 5 Semi-quantification of oxidative DNA damage induced by Beijing indoor PM_{10}

	Sample Dosage (μg·ml−1)	Damage rate±SD%	Sample Dosage (μg·ml−1)	Damage rate±SD%	Sample Dosage (μg·ml−1)	Damage rate±SD%	Sample Dosage (μg·ml−1)	Damage rate±SD%	Sample Dosage (μg·ml−1)	Damage rate±SD%
	No.1		No.2		No.3		No.4		No.5	
	600	97±0.18	500	100±2.56	100±2.56	69±4.77	500	80±0.55	500	38±0.38
	300	97±0.99	250	99±0.95	99±0.95	35±0.89	250	75±1.52	250	27±0.40
	150	95±0.90	125	97±0.85	97±0.85	35±3.03	125	32±0.85	125	27±0.49
	75	98±0.33	75	35±0.25	35±0.25	23±0.07	75	28±0.28	75	26±0.53
	37.5	34±4.15	H2O	14±1.09	14±1.09	22±0.21	H2O	15±1.21	H2O	21±0.85
TD50	45		100		263		216		>500	
R2	0.9535		0.6901		0.7508		0.9494		0.8333	

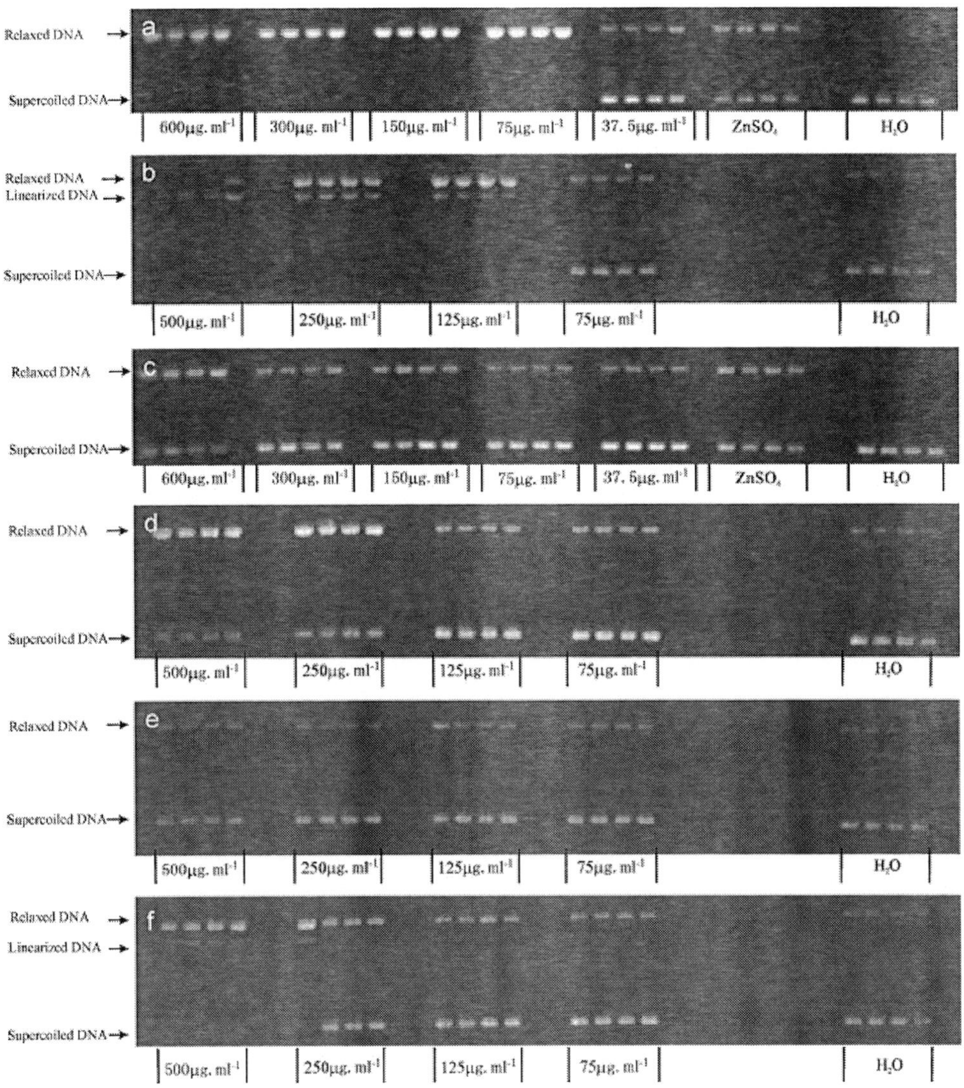

Figure 5. Gel images showing oxidative damage to supercoiled DNA induced by PM l0s. (a) from kitchen (No. 1); (b) from smoker's home (No. 2); (c) from smoker's home (No. 3); (d) from non-smoker's home (No. 4) (e) from non-smoker's home (No. 5)

In this study,the indoor PM_{10} samples were analyzed. It is found that four types of particles were classified in num-per–size distributions: (1) soot aggregates, (2) coal fly ash, (3) minerals (mainly gypsum (calcium sulphate)) and (4) unknown fine particles (Fig.4).The size distributions for all samples are shown inFig.5. It is clear that all samples have bimodal distributions in their number–size patterns; one is at 0.1–0.2 mm and another at 1.0–2.5 mm. The No. 1sample from the kitchen, which induced the strongest oxidative stress (i.e. lowest TD50), revealed a different number–size pattern from all other PM_{10} samples, and its 0.1–0.2 mm peak is clearly higher and the 1.0–2.5 mm is less evident. The relationship between the TD50 values and relative number percentages of the PM_{10} constituents is shown in Fig.6. The number percentages of fly ash and minerals have positive correlations with the TD50 (Fig. 6a and b). The negative correlation between the soot aggregates and TD50 (Fig. 6c) demonstrated a close relationship between oxidative capacity and soot aggregates. The sum of the soot aggregates and unknown fine particles revealed a negative relationship with TD50 (Fig. 6d).

4.2. Comparison of Oxidative Capacity of the Water-Soluble and Whole Fraction of PM Samples

It is widely considered that lung toxicity of atmospheric particles mainly comes from water-soluble fraction (Richards et al., 1989; Adamson et al., 2000; Donaldson et al., 1997; Valavanidis et al., 2000; Ambroz et al., 2001), although Ghio et al. (1999), Imrich et al. (2000) and Soukup and Becker (2001) showed that it is the insoluble fraction in atmospheric particles that causes the biological responses of alveolus

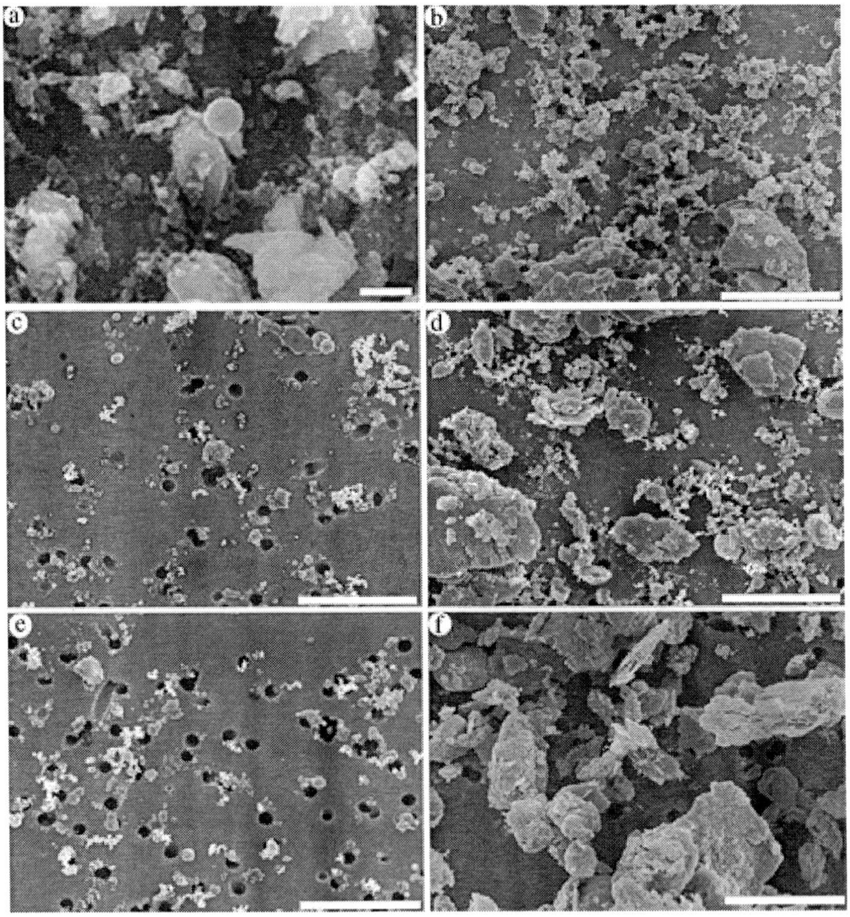

Figure 6. FESEM of Beijing airborne particles. Scale bar 5μm. (a) PM_{10} collected at the urban site in autumn (No. 8); (b) PM_{10} collected at Nankou during heating period (No. 16); (c) $PM_{2.5}$ collected at Nankou during heating period (No. 18); (d) PM_{10} collected at the MTR site during heating period (No. 13); (e) $PM_{2.5}$ collected at the MTR site during non-heating period; (f) dust storm PM_{10} sample (No. 1).

To study the source of bioreactivity of Beijing airborne particles, we have separated the soluble fractions from intact whole solutions of Beijing PM Samples. Solutions before and after centrifugal separation were simultaneously tested with the plasmid assay. The results show that the difference between oxidative damage to super-coiled DNA by the water-soluble fraction and the whole sample are insignificant (Fig.7). The oxidative DNA damage of one sample was reduced from 84% to 60% (Fig.7) at the highest dose (100 μg·ml^{-1}), and that of another sample from 100% to 85% at the highest dose (400 μg·ml^{-1}) after being centrifuged. The Nos. 8 and 18 samples, as well as an extra four sets of samples have been assayed with this method and the results display the similar TD20 values for both insoluble fractions and intact whole solutions (Fig.8). From these results, it is deduced that the oxidative capacity of airborne particulates is mainly derived from the water-soluble fraction.

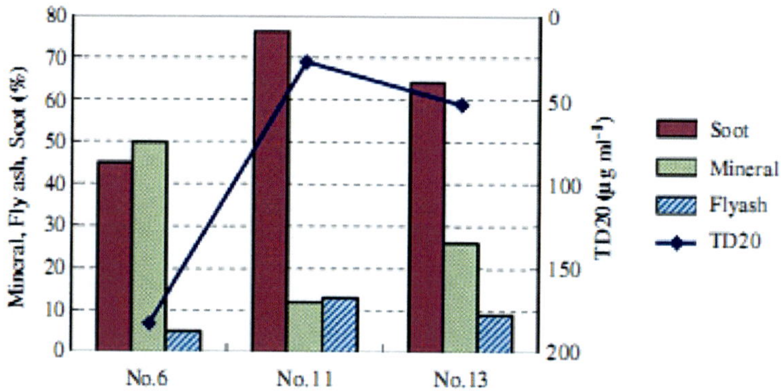

Figure 7. Relationship of the TD20 values with the number percentages of soot aggregates, fly ashes, minerals in three Beijing PM samples.

4.3. Soluble Trace Metal Concentration and Bioreactivity

The trace elements, including V, Cr, Mn, Co, Ni, Cu, Zn, As and Pb, were analysed in the water-soluble fractions of a limited number of Beijing PM samples in order to examine the most probable sources of the oxidative capacities of these particulates. The water-soluble elements, V, Cr, Mn, Co, Ni, Cu, Zn, As and Pb, and the TD20 values of Beijing PM samples are shown in Table 6.

The TD20s of PM samples correspond negatively to total analyzed water-soluble metal concentrations (Fig.9), indicating that the oxidative capacities of the PM samples were probably derived from these metals. It can also be seen that the two $PM_{2.5}$ samples (Nos. 15 and 18) had higher total soluble elements and much lower TD20 values than the PM_{10}s(Table 6). In addition, compared to other samples, the dust storm sample (No. 1) and some spring PM_{10} samples (Nos. 3 and 4) had very high TD20s (N1000), but their total soluble metal concentrations are relatively low (Table 6). Over all the higher the total water-soluble trace element concentration, the lower the TD20 and higher the oxidative capacity.

Table 6 Trace element concentrations and TD20 values of Beijing PM samples

Sample No.	No. 1	No. 4	No. 3	No. 7	No. 8	No. 10	No. 13	No. 15	No. 16	No. 18
Trace element concentrations (*ppm*)										
As	9	25	67	67	131	278	152	153	278	152
Co	14	2	4	3	8	0	0	45	17	18
Cr	0	0	0	6	28	80	31	252	36	57
Cu	76	2	50	70	218	323	135	1226	317	676
Fe	1484	474	385	0	0	0	0	0	0	0
Mn	225	323	582	805	933	1084	647	981	975	814
Ni	0	0	0	33	48	189	0	2685	538	1547
Trace element concentrations (*ppm*)										
Pb	0	29	113	92	262	396	74	338	328	853
V	8	9	12	10	16	22	15	12	41	27
Zn	0	0	512	871	3581	6256	1007	7763	5473	5812
Sum	1816	864	1725	1957	5225	8631	2061	13455	7939	10112
Bioreactivity										
TD20(μgml^{-1})	>1000	>1000	>1000	714	137	183	480	28	260	64

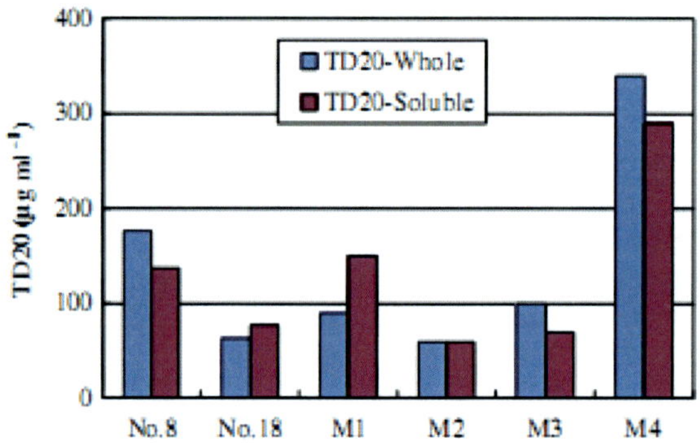

Figure 8. A comparison of TD20s between water soluble fractionand intact whole solution of different PM_{10} samples.

The water-soluble zinc concentration and TD20s of PM samples show a particularly strong negative relationship (Fig.9). For example, water-soluble zinc concentrations in the two $PM_{2.5}$ samples are relatively high (7763 ppm for No. 15 and 5812 ppm for No.18) compared with PM_{10}s (871~ 5473 ppm) and dust storm sample (below detection, 0 ppm) (Table 6). It is known that the TD20s of the $PM_{2.5}$s are much lower than those of the storm dust and PM_{10}s. Therefore, the higher the zinc concentration in PM, the lower the TD20. This is consistent with the results from Richards et al. (1989) and Adamson et al. (2000) who regard zinc to be an important element when considering the oxidative capacity of PMs. The relationship between the water-soluble Cu concentration in PM samples and their TD20s is similar to that of zinc, however, the correlation is not so strong, and this could be explained by the Cu redox cycle in Singh et al. (2001).Iron concentration in PM samples is often considered to be responsible for oxidative capacity (Donaldson et al., 1997; Valavanidis et al., 2000; Ambroz et al., 2001). According to the SEM–EDX results.

The atomic weight of iron in the Beijing dust storm samples is as high as 2.5%. The ICP–MS analysis also shows that the iron concentration in the dust storm samples is much higher than that in non-dust storm PM samples. However, the oxidative capacity of the dust storm samples are smaller compared with non-dust storm PM samples. Thus, iron concentration in Beijing PM samples does not correspond to oxidative potentials, suggesting that oxidative capacity of Beijing airborne particles is not due to iron.

In this investigation, the kitchen PM_{10} sample exhibited the highest TD50, as determined by a plasmid scission assay, as well as the largest numbers of ultrafine particles (IA revealed a dominate peak at 0.1–0.2 mm that corresponded to soot aggregates and unknown fine particles). With regard to the chemical composition of the various PM_{10} samples, there was a positive correlation between the TD50 values and the total analyzed elements in whole samples. In contrast, the TD50 of PM_{10} samples corresponded negatively to the total analyzed water-soluble metal concentrations, inferring that the oxidative capacity of the samples was mainly derived from their water-soluble fractions. Most notable was water-soluble Zn, which revealed a negative correlation with its TD50 values (i.e. high water-soluble Zn concentrations were associated with low TD50 values, and thus, higher oxidative capacity). In the five analyzed samples, PM_{10} derived from the kitchen had the highest level of water-soluble Zn and the lowest TD50, followed by the smoker's home sample, which had a high concentration of water-soluble Zn and a low TD50. These endings were consistent with other investigators (e.g. Richards et al., 1989; Adamson et al., 2000), who regarded zinc to be an important element when considering the oxidative capacity of PM_{10}. The source of this airborne zinc is unclear, however, clearly relates to human activities. Since the highest levels are found in the kitchen, we can speculate that the levels could be the result of cooking Zn-containing foodstuffs at high temperatures with liquids such as vinegar. The amount of bioavailability zinc found in vegetables (the most common constituent of Chinese cooking) is dependant on the soil in which they grew. Cooking the plants in an open, very hot system generating an aerosol would seem to be a plausible pathway to increase the proportion of bioavailable Zn in the local atmosphere. Previous work by BeruBe et al.(2004) has shown that indoor-generated PM_{10} rapidly distributes itself around a home, and is not confined to the room in which it was generated. Other bioavailable metals, especially iron, have been linked to the oxidative damage caused by particles (Donaldson et al., 1997; Wang et al., 1999;Valavanidis et al., 2000; Ambroz et al., 2000; Kimet al., 2001; Han et al., 2001). SEM-EDX revealed the percentage of Fe in the Beijing PM_{10} samples to be as high as 2.5%. The ICP-MS analysis also showed that the Fe concentration in the whole particle samples was much higher than that in the soluble fractions of the particle samples; the concentration of soluble Fe was near zero or below detection. In addition, the relationship between Fe and the bioreactivity of PM_{10} was insignificant, and hence, it is unlikely to be associated with oxidative potential. Epidemiological studies have shown an increased risk of respiratory tract cancer among cooks and bakers (Lund and Borgan, 1987). The metabolic extract of oil fumes, to which the effect has been attributed, can reduce the activity of anti-oxidative enzymes and promote the formation of mutagenic aldehydes such as trans-trans-2, 4-decadienal, which apparently lead to cytotoxicity and oxidative DNA damage (Wu and Yen, 2004). In addition, the components of heated cooking oil in the kitchen,such as the smoke-condensed matter, residual oil, particulate matters and volatile organics, can induce oxidative DNA damage in vitro (Xi et al., 2003). These further complications must be noted when evaluating the causes of oxidative stress in PM_{10} samples collected from kitchens.Environmental tobacco smoke contains nearly 5000 chemical species, including high concentrations of tar and organic compounds. It is worth while to make note that the tar content in cigarette smoke is dominated by trace metals, especially Zn (Landsberger and Wu, 1995). Consequently, the relationship(s) between the contents of the tar and organic compounds, as well as the size and soluble trace metal composition of PM_{10} requires further investigations when considering the driving forces behind the oxidative capacity of environmental tobacco smoke.

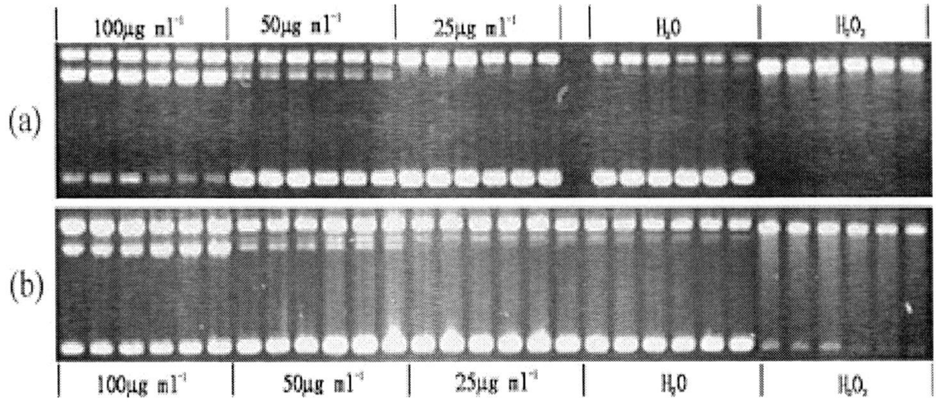

Figure.9. Comparison of the oxidative damage of super-coiled DNA induced by whole and water soluble fraction particles for the No. 18 sample. (a)Whole particles; (b) Water soluble fraction.

4.4 Indoor Air Pollution and Health

The PM generated in kitchens appears to be the more toxic. The gases or particles released by indoor pollution sources should be responsible for this.

There are many sources of indoor air pollution in any home. These include combustion sources such as oil, gas, kerosene, coal, wood, and tobacco products; building materials and furnishings as diverse as deteriorated, asbestos-containing insulation, wet or damp carpet, and cabinetry or furniture made of certain pressed wood products; products for household cleaning and maintenance, personal care, or hobbies; central heating and cooling systems and humidification devices; and outdoor sources such as radon, pesticides, and outdoor air pollution. The relative importance of any single source depends on how much of a given pollutant it emits and how hazardous those emissions are. In some cases, factors such as how old the source is and whether it is properly maintained are significant. For example, an improperly adjusted gas stove can emit significantly more carbon monoxide than one that is properly adjusted. Some sources, such as building materials, furnishings, and household products like air fresheners, release pollutants more or less continuously. Other sources, related to activities carried out in the home, release pollutants intermittently. These include smoking, the use of unvested or malfunctioning stoves, furnaces, or space heaters, the use of solvents in cleaning and hobby activities, the use of paint strippers in redecorating activities, and the use of cleaning products and pesticides in house-keeping. High pollutant concentrations can remain in the air for long periods after some of these activities.

4.5. Mineral Components in Beijing Aerosol and Bioreactivity

Information on the bioreactivity of the urban PM using methods such as the plasmid assay is very limited. Compared with UK Cardiff city (mean TD50 is about 10 $\mu g \cdot ml^{-1}$) as reported by Greenwell et al. (2002, 2003), the bioreactivity of Beijing air-borne PM is clearly lower. According to He et al. (2001),$PM_{2.5-10}$ accounts for 54% of PM_{10} mass. Shao and Shi (2003) reported that although the number percentage of mineral particles in Beijing air is small, the mass percentage could be higher than 50%. The FESEM and image analyses revealed that $PM_{2.5-10}$ was mainly composed of mineral particles. As previously described, dust storm particles are mainly composed of coarse mineral grains and thus have a lower oxidative capacity. Therefore, the probable reason for the lower oxidative capacity of the Beijing PM is the relatively high percentage of mineral particles.

5. Conclusions

As a result of the research the following conclusions are presented:

1) Beijing airborne particles with different Properties damage super-coiled DNA at various degrees, and the TD20 of which vary from 28$\mu g \cdot ml^{-1}$ to more than 1000$\mu g \cdot ml^{-1}$,as the oxidative capacity, Beijing PM2.5 is higher than that of PM_{10}. and samples collected during the heating period are higher than those of the non-heating period. The higher number percentages of soot aggregates and lower number percentages of mineral and fly ashes are associated with the lower TD20s and thus the higher oxidative capacity.

2) Results indicate that dust storm particles have a lower oxidative capacity compared to non-dust storm particles. In this study, neither the particles at 400$\mu g \cdot ml^{-1}$ nor that at 1200$\mu g \cdot ml^{-1}$caused much DNA damage.

3) The water-soluble components are believed to be the main source of bioreactivity of Beijing airborne particles. It is found that the lower TD20s and thus the higher bioreactivity of PM are commonly associated with the higher total analyzed water-soluble V, Cr, Mn, Co, Ni, Cu, Zn, As and Pb concentrations. The obvious negative correlation between the water-soluble trace elements and the TD20 values indicates that the water-soluble zinc of the particles is probably the most important element that is responsible for the plasmid DNA damage.

4) Aerosols generated in kitchens appear to be the most toxic.

References

Adamson I Y.R., Prieditis H, Hedgecock C, Vincent R ,2000 .Zinc is the toxic factor in the lung response to an atmospheric particle sample. *Toxicology and Applied pharmacology,* 166:111~119

Adamson I. Y.R., Prieditis H., Vincent R. 1999, Pulmonary toxicity of an atmospheric particulate sample is due to the soluble fraction. *Toxicology and applied pharmacology,* 157: 43~50.

Ambroz HB, Bradshaw TK, Kemp TJ, Kornacka EM, Przybytniak GK. Role of iron ions in damage to DNA: influence of ionizing radiation, UV light and H2O2. J Photochem *Photobiol A Chem* 2001;142:9-18.

Analitis A, Katsouyanni K, Dimakopoulou K, et al. Short-Term Effects of Ambient Particles on Cardiovascular and Respiratory Mortality. *Epidemiology*, 2006, 17(2): 230-233.

Anderson, J, Aggett, F, Buseck, P, Germani, M, Shattuck, T, 1988. Chemistry of individual aerosol particles from Chandler, Arizona, an arid urban environment. *Environmental Science and Technology* 22, 811~818.

Aragon Pina A, Torres G, Santiago P, et.al, 2002, Scanning and transmission electron microscope of suspended lead-rich particles in the air of San Luis Potosi, Mexico. *Atmospheric Environment*, 36: 5235-5243

Artaxo P, Marta L C, Rabello et al, 1993, Nuclear microprobe analysis and source apportionment of individual atmospheric aerosol particles, Nuclear Instruments and *Methods in Physics Research*, B75: 521~525

Artaxo, P, Rabello, M, Maenhaut, W, Grieken, R, 1992. Trace elements and individual particle analysis of atmospheric aerosols from the Antarctic peninsula. *Tellus 44B*, 331~334.

Bai Y, Suzuki AK, Sagai M. The cytotoxic effects of diesel exhaust particles on human pulmonary artery endothelial cells in vitro: role of active oxygen species. *Free Radic Biol Med* 2001;30 (5):555–62.

Be′ ruBe′ , K.A., Sexton, K.J., Jones, T.P., Moreno, T., Anderson,S., Richards, R.J., 2004. The spatial and temporal variations in PM_{10} mass from six UK Homes. *Science of the Total Environment* 324, 41–53.

BéruBé, K.A., Jones, T.P., Williamson, B.J., Winters, C., Morgan, A.J. and Richards, R.J., 1999. Physicochemical characterization of diesel exhaust particles: factors for assessing biological activity. *Atmospheric Environment* 33, 1599~1614.

Bonvallot, V ,2001.Organic compounds from diesel exhaust particles elicit a proinflammatory response in human airway epithelial cells and induce cytochrome expression. *American journal of respiratory cell and molecular biology*. 25:515~521

Buseck, P, Jacob, D, Pósfai, M., Li, J. & Anderson, J, 2000,, Minerals in the air; An environmental perspective. *Int. Geol. Rev.*.

Buseck, P, Pósfai, M, 1999. Airborne minerals and related aerosol particles: Effects on climate and the environment. *Proc. Natl. Acad. Sci.* USA 96, 3372~3379.

Chen, Tianhu, Feng, Huijun, Zhang, yu, et al.,2001, The atmosphere particles constitute and its environment designation meaning of Hefei, *rock mineralogy journal*(in chinese),20:(4)433~436

Chow J C , 1999,Welcome to a special issue on PM2.5:A Fine Particle Standard, J. Air & *Waste Manage. Assoc.*, 49:1

Clauin, T, Schulz, M, Balkanski, Y & Boucher, O, 1998, Uncertainties in assessing radioactive forcing by mineral dust. *Tellus,* 50B: 491~505.

Conner T L, Norris G A, Landis M S, et al, 2001,Individual particle analysis of indoor, outdoor, and community samples from the 1998 Baltimore particulate matter study, *Atmospheric Environment*, 35:3935-3946

Costa DL, Dreher KL. Bioavailable transition metals in particulate matter mediate cardiopulmonary injury in healthy and compro-mised animal models. *Environ Health Perspect* 1997;105 (S5):1053–60.

Dai, Shugui, 2001, Environment chemistry (in chinese, higher education press:75~84

Dentener F J, Carmichael G R, Zhang Y, et al, 1996, Role of mineral aerosol as a reactive surface in the global troposphere, *J. Geophys. Res.*, 101:22869~22889

Devi K.D., Banu B.S., Grover P., Jamil K. 2000, Genotoxic effect of lead nitrate on mice using SCGE(comet assay).*Toxicology*, 145: 195~201.

Ding, Guowu, Wang, Xiaoyun, 1999, The mutation research of suspends particles organic of atmosphere in Lanzhou, *environment healthy magazine* (in chinese). 16 (2) : 96~97

Donaldson K, Beswick PH, Gilmour PS. Free radical activity associated with the surface of the particles: a unifying factor in determining biological activity? *Toxicol Lett* 1996;88:293–8.

Donaldson K, Li XY, MacNee W. Ultrafine (Nanometer) particle mediated lung injury. J *Aerosol Sci* 1998;29:553–60.

Donaldson K., Brown D.M., Mitchell C., Dineva M., Beswick P.H., Gilmour P., MacNee W , 1997,Free radical activity of PM_{10}: iron-mediated generation of hydroxyl radicals. *Environmental health perspectives*, 105(suppl 5): 1285~1289.

Dong, Shuping, Liu, Tao, Sun, Dayong, et al, 2001, Identify the main type of atmosphere particles by scanning electronic microscope , *rock mineral test*(in chinese)20:(3)202~207

Eisenbrand G., Pool-Zobel B., Baker V., Balls M., Blaauboer B.J., Boobis A., Carere A., Kevekordes S., Lhuguenot J.C., Pieters R., Kleiner J. 2002, Methods of in vitro toxicology. *Food and chemical toxicology*.40: 193~236.

FullerK.,Malm,W.& Kreidenweis,S.1999.effects of mixing on extinction by carbonaceous particles. *J Geophys. Res.*,104(15) : 941~954

Ganor E, Levin Z, Van Grieken, 1998, Composition of individual aerosol particles above the Israelian Mediterranean coast during the summer time, *Atmos. Environ*, 32:1631~1642

Ghermandi, G, et al, 1999, Elemental and mineral characterization of Coastal Antarctic Aerosols in snow using PIXE and SEM-EDAX, Nuclear Instruments and Methods in *Physics Research*, B150: 392~397

Ghio A.J., Stonehuerner J., Dailey L.A., Carter J.D. 1999, Metals associated with both the water-soluble and insoluble fractions of ambient air pollution particle catalyze an oxidative stress. *Inhalation toxicology*, 11: 37~49.

Ghio,A, Samet, J. 1999a.Metals and air pollution particles, *Air pollution and health*, 635~652

Gilmour, P., Beswick, P.H., Donaldson, K., 1994. Effects of Kim, B.Y., Han, M.J., Chung, A.S., 2001. Effects of reactive oxygen species on proliferation of Chinese hamster lung fibroblast (V79) cells. *Free Radical Biology and Medicine* 30,686–698.

Goodman A L, Underwood G M, Grassian V H, 2000, A laboratory study of the heterogeneous reaction of nitric acid on calcium carbonate particles, *J. Geophys. Res.*, 105(D23): 29053~29064

Gorden G E, 1980, Techniques for treating multi-element particulate data to obtain information on sources: *Overview, NYAS*, 93

Goss K U, 1994, Adsorption of organic vapors on polar mineral surfaces and on a bulk water surface – Development of an empirical predictive model, *Environ. Sci. Technol.*, 28:640-645

Goss K U, Eisenreich S J, 1996, Adsorption of VOCs from the gas phase to different minerals and a mineral mixture, *Environ. Sci. Technol.*, 30:2135~2142

Goss K U, Eisenreich S J, 1997, Sorption of volatile organic compounds to particles from a combustion source at different temperatures and relative humidities, *Atmos. Environ.*, 31:2827~2834

Greenwell L.L., Moreno T., Jones T.P., Richards R.J. 2002, Particle-induced oxidative damage is ameliorated by pulmonary antioxidants. *Free radical &medicine*, 32(9): 898~905.

Greenwell LL, Moreno T, Richards RJ. Pulmonary antioxidants exer differential protective effects against urban and industrial particu-late matter. *J Biosci* 2003;28(1):101–7.

Grime G W. 1998, Analysis of individual environmental particles using micro-PIXE and nuclear microscopy, *X-Ray Spectrometry*, 27: 221~231

Guthric G D, et al, 1993,Health effects of mineral dusts, in Reviews in Mineralogy, 28:1-584, edited by *Mineralogical Society of America*, Washington, D C, USA

Haapala, H, 1998,The use of SEM/EDX for studying the distribution of air pollutants in the surroundings of the emission source, *Environmental Pollution*, 9:361~363

Han JY, Takeshita K, Utsumi H. Noninvasive detection of hydroxy radical generation in lung by diesel exhaust particles. *Free Radic Biol Med* 2001;30:516–25.

Hasegawa, S, Ohta, S, 2002. Some measurements of the mixing state of soot-containing particles at urban and non-urban sites. *Atmospheric Environment* 36, 3899~3908.

Haywood,J & Shine,K, 1995, The effect of anthropogenic sulfate and soot aerosol on the clear sky planetary radiation budget, *Geophys.Res Lett.*, 22:603~606

He K, Yang F, Ma Y, Zhang Q, Yao X, Chan CK, et al. The characteristics of PM2.5 in Beijing, China. *Atmos Environ* 2001;35:4954–70.

Held, A., Hinz, K, Trimborn, A, Spengler, B, Klemm, O, 2002. Chemical classes of atmospheric aerosol particles at a rural site in Central Europe during winter. *Aerosol Science* 33, 581~594.

Hinz ,K, Kaufmann ,R, Spengler, B, 1996. On-line Measurement and Characterization of Single Particles Using Laser Mass Spectrometry and Multivariate Data Analysis. *J.Aerosol Sci.*, Vol.27, Suppl. pp. S171~S172.

Hinz K P, Greweling M, Iflseder, et al, 1998,Source identification of single particles by on-line laser mass spectrometry, *Journal of Aerosol Science*, 9:1253~1254

Horvath ,H , 1992.Effects on virility, weather and climate, In Radiogenic, M.& Harrison ,R(eds.):atmospheric acidity .sources, consequences and abatement. *London : Elsevier*, 435~466.

Houghton J T ed., Climate Change 1994, Radiative forcing of climate change and an evaluation of the IPCC emission sciences, *Cambridge University Press*, 1995, 339

Hu, Yan, Zhao, Shuyou, 1998. The healthy influence of air pollution and the research of relevant method, *The environment monitor management and techniques*(in chinese).10:(6)11~13

Hung, H.F., Wang, C.S., 2001. Experimental determination of reactive oxygen species in Taipei aerosols. *Journal of Aerosol Science* 32, 1201–1211.

Imrich A, Ning YY, Kobzik L. Insoluble components of concentrated air particles mediated alveolar macrophage responses in vitro *Toxicol Appl Pharmacol* 2000;167:140–50.

Jiang, Hongmei, Wang, Dingyong, 2001, Research advance of atmosphere inhalable particles, *environmental science trend*(in chinese),(1)11~15

Jones, T.P., Moreno, T., Be′ ruBe′ , K.A., Richards, R.J., 2006.Physicochemical characterisation of microscopic airborne particles from south Wales: a review of the locations and methodologies. *Science of the Total Environment* 360, 43–59.

Junno, T., Anand, S., Deppert, K., Samuelson, L., 1996. Mapping of ultra fine particles with the atomic force microscope. *J. Aerosol Science, Suppl.*1.pp. S173~174.

Kadiiska M.B., Mason R.P., Dreher K.L., Costa D., Ghio A.J. 1997,In vivo evidence of free radical formation in the rat lung after exposure to an emission source air pollution particle. *Chem. Res. Toxicol*, 10:1104~1108.

Kaegi R, Holzer L, 2003, Transfer of single particles for combined ESEM and TEM analysis, *Atmospheric Environment*, 37:4353~4359

Kamens R M, Guo Z, Fulcher J N, et al, 1988,Influence of humidity, sunlight and temperature on the daytime decay of polyaromatic hydrocarbons on atmospheric soot particle, *Environ. Sci. Technol.*, 22:103~108

Kasahara M, Akashi S, Ma C J, Tohno S, 2003,Fixation and chemical analysis of single fog and rain droplets, *Atmospheric Rsearch*, 65:251~259

Kasahara, M, Holler,R, Tohno,S, Onishi,Y, Ma, C, 2002. Application of PIXE technique to studies on global warming/cooling effect of atmospheric aerosols. *Nuclear Instruments and Methods in Physics Research B*, 189, 204~208.

Kasparian J, Frejafon E, Rambaldi P, et al, 1998, Characterization of urban aerosols using SEM-microscopy, X-ray analysis and lidar mearsurement, *Atmospheric Environment*, 32(17): 2957~2967

Katrinak, K, Anderson, J, Buseck, P, 1995. Individual Particle Types in the Aerosol of Phoenix, Arizona. *Environmental Science and Technology* 29, 321~329.

Kim BY, Han MJ, Chung AS. Effects of reactive oxygen species on proliferation of Chinese hamster lung fibroblast (V79) cells. Free Radic Biol Med 2001;30:686–98.

Kramer, P.J. 1998, Genetic toxicology. *Journal of Pharmacy and Pharmacology*, 50: 395~405.

Landsberger, S., Wu, D., 1995. The impact of heavy metals from environmental tobacco smoke on indoor air quality as determined by Compton suppression neutron activation analysis. *The Science of the Total Environment* 173/174,323–337.

Levin Z, Ganor E, Van Gladstein, 1996,The effects of desert particles coated with sulfate on rain formation in the eastern Mediterranean, *J. Appl. Meteorol.*, 35:1511~1523

Li XY, Gilmour PS, Donaldson K, MacNee W. In Vivo and in Vitro proinflammatory effects of particulate air pollution (PM_{10}) *Environ Health Perspect* 1997;105(S5):1279–83.

Li, J, Anderson, J and Buseck, P , 2003a. TEM study of aerosol particles from clean and polluted marine boundary layers over the North Atlantic, *J. Geophys. Res.*, 108(D6), 4189, doi:10.1029/2002JD002106.

Liang, Handong, Yu, Chunhai, Liu, Xiande, Zhou, Qiang, Li, Yanfang, Feng, Mingde, 2001, The research of secondary ionic mass spectrum on the inorganic element of atmosphere single particles .*The Journal of China University of mining and Technology*(in chinese) ,30:(5)442~445

Linn W S, Gong H, Clark K W, et al, 1999, Day to day particulate exposures and health changes in Los Angeles area residents with severe lung disease, *J. Air & Waste Manage. Assoc.*, 49:108

Liousse C, Cachier H, Jennings S G, 1993,Optical and thermal measurements of black carbon aerosol content in different environments: variation of the specific attenuation cross-section, sigma(s), *Atmospheric Environment*, 27(A):1203

Liu , Xiande, Jia Hong, Qi,Jianbing,1994, The scanning electron microscope research and identifying of pollution source of Qingdao atmosphere particles, *environmental science study*(in chinese),7:(3)10~17

Lund, E., Borgan, J., 1987. Increased lung cancer mortalityamong Norwegian cooks. Scandinavian *Journal of Work Environment and Health* 13, 156–161.

Luo Yunfeng, Zhou Xiuji, Li, Weiliang, 1998, The research status of radiation force and climate effect of atmosphere aerosol, *geosciences' science progress*(in chinese),13:(6)572~581

Ma C J, Kasahara M, Holler R, 2001,Characteristics of single particles sampled in Japan during the Asian dust-storm period, *Atmospheric Environment*, 35: 2707~2714

Mamane Y, Noll K E, 1985,Characterization of large particles at a rural site in the eastern united states; Mass distribution and individual particle analysis, *Atmospheric Environment,* 19(4):811

Martin, S.Yu, J, Han, J, Verdier, M., Li, J and Buseck, P, 2000. An aerosol chemical reactor for coating metal oxide particles with $(NH_4)_2SO_4$-H_2SO_4-H_2O-part2: manipulation of the metal oxide core. *J. Aerosol Sci.* Vol. 31, No. 11, pp. 1283~1298.

Masatoshi Toyoda, Kazuki Kaibuchi, Mitsuru Nagasono, Yasuko Terada, Teruo Tanabe, Shinjiro Hayakawa, Jun Kawai. 2004, X-ray analysis of a single aerosol particle with combination of scanning electron microscope and synchrotron radiation X-ray microscope .*Spectrochimica Acta Part* B, 59 : 1311 ~1315

Maynard RL. Particulate air pollution. In: Brimblecombe P, Maynard RL, editors. The urban atmosphere and its effects. London: *Imperial College Press*; 2001, 190–1.

McNeilly, J.D., Heal, M.R., Beverland, I.J., Howe, A., Gibson,M.D., Hibbs, L.R., MacNee, W., Donaldson, K., 2004.Soluble transition metals cause the pro-inflammatory effects of welding fumes in vitro. *Toxicology and Applied Pharmacology*196, 95–107.

Merolla, L., Richards, R.J., 2005. In vitro effects of water-soluble metals present in UK particulate matter. *Experimental Lung Research* 31 (7), 671–683.

Meszaros E, Barcza, T, Gelencser, A, Hlavay, J, Kiss, G, Krivacay, Z, Molnar, A&Polyak,K. 1999. Size distributions of inorganic and organic species in the atmospheric aerosol in Hungary, *J. Aerosol Sci.*28:1163~1175

Mikkanen, P, Jokiniemi, J, Kauppinen, E, Vakkilainen, E, 2001. Coarse ash particle characteristics in a pulp and paper industry chemical recovery boiler. *Fuel* 80, 987~999.

Moreno, T., Merolla, L., Gibbons, W., Greenwell, L., Jones, T.,Richards, R., 2004. Variations in the source, metal content and bioreactivity of technogenic aerosols: a case study from Port Talbot, Wales, UK. *Science of the Total Environment333*, 59–73.

Mou Chenggang, 1999, The environment mineralogy studies to the human healthily affect of the mineral in atmosphere dust, *Guizhou geology*(in chinese),16:(1)73~77

Nair, A, Tyler, B., Peterson, R., 2004. Application of a cryo-stage in the TOF-SIMS analysis of atmospheric aerosol surfaces. Applied Surface Science 231–232: 538~542.

Neas LM. Fine particulate matter and cardiovascular disease.*Fuel Process Technol* 2000;65–66:55–67.

Oberdo¨ rster, G., Ferin, J., Lehnert, B.E., 1994. Correlation between particle size, in vivo particle persistence and lung injury. *Environmental Health Perspectives* 102 (S5), 173–179.

Okada K, Kobayashi A, Iwasaka, Y et al, 1987, Features of individual Asian dust storm particles collected at Nagoya, *Japan, Meteorol. Soc.* Jpn., 67:515~521

Orlic I. 1995, The nuclear microscope: A review of applications in the environmental sciences. Nuclear Instruments and Methods in Physics Research, B104: 602~611

Paoletti L, Falche M, Batisti D, et al, 1991, Minoral lung burden of an urban population, *Atmos. Environ.*, 25B:381

Paoletti, L., Diociaiuti, M. Berardis, B., Santucci, S., Lozzi, L., Picozzi, P., 1999. Characterisation of aerosol individual particles in a controlled underground area. Atmospheric Environment 33, 3603~3611.

Park EM, Park YM, Gwak YS. Oxidative damage in tissues of rats exposed to cigarette smoke. *Free Radic Biol Med* 1998;25:79–86.

Pope III CA, Dockery DW, Schwartz J. Review of epidemiological evidence of health effects of particulate air pollution. *Inhal Toxicol* 1995;7:1-18.

Pope III, C.A., Burnett, R.T., Thun, M.J., Calle, E.E., Krewski,D., Ito, K., Thurston, G.D., 2002. Lung cancer, cardiopulmonary mortality, and long-term exposure to fine particulate air pollution. *Journal of American Medical Association* 287(9), 1132–1141.

Pósfai, M., Gelencsér, A. Simonics, R. Arató, K. Li, J. Hobbs, P and Buseck , P. 2004, Atmospheric tar balls: Particles from biomass and biofuel burning, J. *Geophys. Res.,* 109, D06213, doi:10.1029/2003JD004169

Pósfai, M., Simonics, R. Li, J. Hobbs, P.and Buseck , P , 2003, Individual aerosol particles from biomass burning in southern Africa: 1. Compositions and size distributions of carbonaceous particles, J. *Geophys. Res*., 108(D13), 8483, doi:10.1029 /2002JD002291.

Posfai,M, Molnar,A. 2000,Aerosol particles in the troposphere :A mineralological introduction, *EMU Notes Mineral*.2,197~252

Pritchard RJ, Ghio AJ, Lehmann JR,Winsett DW, Tepper JS, Park P, et al. Oxidant generation and lung injury after particulate air pollutantexposure increase with the concentrations of associated metals.*Inhal Toxicol* 1996;8:457–77.

Prospero, J, 1999, Long-range transport of mineral dust in the global atmosphere: Impact of African dust on the environment of the southeastern Unites States. *Proc. Natl. Acad. Sci. USA,* 96: 3396~3403.

Pryor , S, Barthelmie, R , 2000.REVEAL II: Seasonality and spatial variability of particle and visibility conditions in the Fraser Valley。 *The Science of the Total Environment* (257) : 95~110

Puledda, S., Paoletti, L., Ferdinandi, M., 1999. Airborne quartz concentration in an urban site. *Environmental Pollution* 104, 441~448.

Qi, Shihua, Fu, Jiamo, Sheng, Guoying, et al. ,1999, The research progress of the source of atmosphere aerosol matter, *environmental science progress*(in chinese),7:(6)26~31

Qiu, Zhijun, Guo, Panlin, Wang, Jiqing,et al,2001, The atmosphere aerosol single particle source apportionment based on the proton microprobe research, *environmental science*(in chinese),22(2):50~54

Rahman I, Macnee W. Role of oxidants/antioxidants in smoking-induced lung diseases. Free *Radic Biol Med* 1996;21:669–81.

Rice T.M., Clarke R.W., Godleski J.J., Al-Mutairi E., Jiang N.F., Hauser R., Paulauskis J.D. 2001, Differential ability of transition metals to induce pulmonary inflammation. *Toxicology and applied pharmacology*, 177: 46~53.

Richards, R.J., Atkins, J., Marrs, T.C., Brown, R.F.R., Masek,L.C., 1989. The biochemical and pathological changes produced by the intratracheal instillation of certain components of zinc-hexachloroethane smoke. *Toxicology Letters* 54,79–88.

Russell P A, Ruud C O, 1973, An analysis of particulates from the Denver urban plume using scanning electron microscopy and energy dispersive X-ray spectrometry, *In Denver Air Pollution Study* –Vol. I , U.S.EPA Report 600/9-76-007a, 165~233

Sardas S., Walker D., D.Akyol, Karakaya A.E. 1995, Assessment of smoking-induced DNA damage in lymphocytes of smoking mothers of newborn infants using the alkaline single-cell gel electrophoresis technique. *Mutation Research*, 335: 213~217.

Schins R.P.F., Lightbody J.H., Borm P.J.A., Shi T.M., Donaldson K., Stone V. 2004.Inflammatory effects of coarse and fine particulate matter in relation to chemical and biological constituents. *Toxicology and applied pharmacology*, 195: 1~11.

Schwartz J, Neas L. Fine particles are more strongly associated than coarse particles with acute respiratory health effects in school-children. *Epidemiology* 2000;11:6-10.

Schwartz J. Air pollution and daily mortality: a review and meta analysis. *Environ Res* 1994;64:36–52.

Schwartz S E, 1996, The Whitehouse effect shortwave radioactive forcing of climate by anthropogenic aerosols: an overview, *Journal of Aerosol Science*, 27(3): 359~382

Seinfeld J H, 1975, Air Pollution – *Physical and Chemical Fundamentals*, New York

Shao L, Shi Z. Microscopic characteristics and size distribution of summer PM_{10} in the air of northwestern urban site and a clean air site in Beijing. *Environ Sci* 2003;24(5):11–6 [In Chinese with English abstract].

Shao, Longyi, Shi, Zongbo, Huang, Qin, 2000, Study of inhalable particles in city atmosphere environment, *environmental protection*,(1):24~26

Shao, Longyi, Shi, Zongbo,2003, The morphology characteristic and size distribution of PM 10 in northwest city area in Beijing and clear site in summer .*Environmental science*(in chinese),24(5), 11~16.

Shi Z, Shao L, Jones T P, et al, 2003, Characterization of airborne individual particles collected in an urban area, a satellite city and a clean air area in Beijing, 2001, *Atmospheric Environment*, 37:4097~4108

Shi, Zongbo, Shao, Longyi, Jones, T, Whttaker, A, Richards, R, Zhang Pengfei.2004, The oxidize harm to DNA by city atmosphere inhalable particles. *Science bullets* (in chinese), 49(7): 673~678

Silva P J, Carlin R A, Prather K A, 2000, Single particle analysis of suspended soil dust from Southern California, *Atmospheric Environment*, 34:1811~1820

Singh S, Asad SF, Ahmad A, Khan NU, Hadi SM. Oxidative DNA damage by capsaicin and dihydrocapsaicin in the presence of Cu (II*). Cancer Lett* 2001;169:139–46.

Sokolik, I N, Toon O B, 1999, Incorporation of mineralogical composition into models of the radiative properties of mineral aerosol from UV to IR wavelengths, J. *Geophys. Res*., 104:9423~9444

Sokolik, I N, Toon O B, Bergstrom R W, 1998, modeling the radioactive Characteristics of airborne mineral aerosol at infrared wavelengths, J. Geophys. Res., 103:8813~8826

Sokolik,I., Andronova,A. & Johnson, T, 1993, Complex refractive index of atmospheric dust aerosols. *Atmos. Environ.*, 27A: 2495~2502.

Song Yu, Tang Xiaoyan, Zhang,Yuanhang, Hu,min, et al.2003 b, The regulation of atmosphere visibility of Beijing and its reduce reason.. *environmental science study*(in chinese).16 (02):10~12

Song Yu, Tang Xiaoyan, Fang,Chen, Zhang,Yuanhang, Hu,Min, et al. . 2003a, The relation between the reduce of visibility of Beijing and the particles polluting .*Environmental science journal* (in chinese).23(04):468~471

Soukup JM, Becker S. Human alveolar macrophage responses to air pollution particulates are associated with insoluble components of coarse material, including particulate endotoxin. *Toxicol Appl Pharmacol* 2001;171:20–6.

Squadrito GL, Cueto R, Dellinger B, PryorWA. Quinoid redox cycling as a mechanism for sustained free radical generation by inhaledairborne particulate matter. *Free Radic Biol Med* 2001;31 (9):1132–8.

Stone V, Shaw J, Brown DM, Macnee W, Faux SP, Donaldson K. The role of oxidative stress in the prolonged inhibitory effect of ultrafine carbon black on epithelial cell function. *Toxicol in Vitro* 1998;12:649–59.

Sun , Ren, Zhu Tan, Bai,Zhipeng ,1997, The progress of the identifies and source apportionment of poly annulus hydrocarbon of atmosphere particles , *city environment and city ecosystem(*in chinese) ,10(3):27~31

Tang, Xiaoyan ,1990, Atmosphere environment chemistry(in chinese), *higher education press*

Valavanidis A, Salika A, Theodoropoulou A. Generation of hydroxyl radicals by urban suspended particulate air matter. The role of iron ions. *Atmos Environ* 2000;34:2379–86.

Van Borm W A, Adams F C, et al, 1988,Cluster analysis of electron microprobe analysis data of individual particles for source apportionment of air particulate matter, *Atmospheric Environment,* 22(10):2297~2307

Van Borm W A, Adams F C, Mainmast W. 1989. Characterization of individual particles in the Antwerp aerosol. *Atmos. Environ*, 23(5): 1139

Van Malderen H, Rojas C, Van Grieken R, 1992,Characterization of individual giant aerosol particles above the North Sea, *Environ Sci Technol*, 26(4):750

Van Malderen H, Van Grieken R, Bufetov N, et al, 1996. Chemical characterization of individual aerosol particles in Central Siberia, Environ. Sci. Technol., 30:312~321

Volz, F, 1983, Infrated optical constants of aerosols at some locations. *Appl. Opt.*, 22:3690~3700.

Wang, Anpu, 1999. The new trend of atmosphere aerosol research (in chinese). *Environment chemistry* 1, 10~15.

Wang, Anpu,Yang, Shulan, Sha,Yin,1996, The chemistry character of the single particles of atmosphere aerosol of Beijing, *environment chemistry*(in chinese),15:(6)488~495

Wang, Mingxing, 2000, *The aerosol and climate, climate and environment research*(in chinese),5:(1)1~5

Wang, Mingxing, Zhang Renjian, 2001. *The forehand problem of the atmosphere aerosol research.climate and environments study(*in chinese). 6:119~124.

Wang, Wei, Pan, Zhi, Liu,Hongjie, et al., , 2001. The size distribution of the transportation particles and its relation with visibility. *Environmental science research*(in chinese).14, 17~22.

Wei F, Hu W, Wu G, Teng E, Zhang J, Chapman RS. Analyses of effects of air pollution on the indexes of children's lung function. *Environ Monit China* 2001;17(S):61–5.

Whittaker, A., Jones, T.P., Be′ ruBe′ , K.A., Maynard, R.,Richards, R.J., 2004. Killer smogs of the 1950s London:composition of the particles and their bioactivity. *Science of the Total Environment* 334–335, 435–445.

Wilson, M.R., Lightbody, J.H., Donaldson, K., Sales, J., Stone,V., 2002. Interactions between ultrafine particles and transition metals in vivo and in vitro. *Toxicology and Applied Pharmacology* 184, 172–179.

Wittmaack, K., 2002.Impact and growth phenomena observed with sub-micrometer atmospheric aerosol particles collected on polished silicon at low coverage. *Atmospheric Environment* 36, 3963~3971

Wu, S.C., Yen, G.C., 2004. Effects of cooking oil fumes on the genotoxicity and oxidative stress in human lung carcinoma (A-549) cells. *Toxicology in Vitro* 18, 571–580.

Wurzler S, Reisin T G, Levin Z, 2000, Modification of mineral dust particles by cloud processing and subsequent effects on drop size distribution, *J. Geophys. Res.*, 105:4501~4512

Xhoffer C, Bernard P, Van Grieken F, et al, 1991,Chemical characterization of source apportionment of individual aerosol particles over the North Sea and the English Channel using multivariate techniques, *Environ. Sci. Technol.*, 25(8):1470

Xi, Z.G., Li, G.X., Sun, Y.M., Chao, F.H., 2003. Oxidative damage of DNA and formation of its biomarker 8-hydroxydeoxyuanosine induced by heated cooking oil vapors. *Journal of Environmental Health* 20 (5), 259–262 (in Chinese with English abstract).

Xu X, Christinani DC, Li B, Huang H. Association of air pollution with hospital outpatient visits in Beijing. *Archives of Environ-mental Health*, vol. 50; 1995. 214–20.

Xu, Li, Zhang, Peng et al.,2002,The exploration of physics and chemistry characteristic of aerosol during spring end and the beginning of autumn of Beijing , *Atmospheric Sciences(*in chinese),26:(3)401~411

Xu, X., Dockery, D.W., Christiani, D.C., Li, B., Huang, H.,1995. Association of air pollution with hospital outpatient visits in Beijing. *Archives of Environmental* Health 50,214–220.

Xu, X., Gao, J., Dockery, D.W., Chen, Y., 1994. Air pollution and daily mortality in residential areas of Beijing, *China.Archives of Environmental Health* 49, 216–222.

Yan,Yuejin, Wu, Canze, Gong,Xiaoli, 2001, The toxicology experiment method of assessing of harmful degree of dust. *Geology explore safety*(in chinese). 1:47-51

Yang, Fumo, Ma, Yongliang, He, Kebin,2000, The fine atmosphere particles PM2.5 and it's study situation, *environment of world*(in chinese),(4)32~34

Yang, Wenmin, Wu, Bingyao, Wu, Zhonhcheng,1993, The accumulation research of eight metal elements of atmosphere *environment and healthy magazine*(in chinese),10:197

Yang, Wenmin, Wu, Bingyao, Yang, Jianjun, et al.,1994, Surface characteristic analysis of atmosphere particles , *Chinese public hygiene journal*,13:(5)275~277

Ye, Ruqiu, Li, Yongquan, Wang, Jiazhen,1992, The study of characteristic of aerosol particles in Guangzhou by scanning electron microscope and assemble analysis, *environment chemistry*(in chinese),11(6):14-20

Ye, Ruqiu, Wang, Jiazhen, Ren, Yuzheng,1987, The study of characteristic of aerosol particles in Shenyang by scanning electron microscope , *environment chemistry*(in chinese),6(2):1-7

Zhang J, Chen Z, Wang W. Source apportionment on fine particulates in atmosphere in *Beijing. Acta Sci Circumst* 1998b;18(1):32–7 [In Chinese with English abstract].

Zhang Y, Shao K, Tang X, Li J. The study of urban photochemical smog pollution in China. *Acta Sci Nat Univ Pekinensis* 1998a;34:392–400.

Zhang,Daizhou,Zhao,Chunsheng, Qin,Yu,1998,Morphology and composition analysis of sand dust particles, *environment science journal*(in Chinese),18 (5) : 450~456

Zhang,Hong, Hou,Tao,Fan,Wenbiao,2000,The electron microscope analysis and identify of source of atmosphere particles of Jincheng,Journal of Shanxi university(*natural science issue*)(in chinese), 23 (2) : 182~185

Zhou G, Tazaki K, 1996, Seasonal variation of gypsum in aerosol and its effect on the acidity of wet precipitation on the Japan Sea side of Japan, *Atmospheric Environment*, 30(19): 3301~3308

In: Airborne Particulates
Editor: Ming Cheng and Wen Liu

ISBN: 978-1-60692-907-0

Chapter 2

SAHARAN DUST AND THE AEROSOLS ON THE CANARY ISLANDS: PAST AND PRESENT

Inmaculada Menendez[a], Edward Derbyshire[b], Johann Engelbrecht[c], Hans von Suchodoletz[d], Ludwig Zöller[d], Pedro Dorta[e], Teresa Carrillo[f] and Felipe Rodríguez de Castro[f]

[a]Department of Physics, University of Las Palmas de Gran Canaria (Spain)
[b]Royal Holloway, University of London (UK)
[c]Desert Research Institute, Reno (USA)
[d]LS Geomorphologie University of Bayreuth (Germany)
[e]Department of Geography, University of La Laguna, Tenerife (Spain)
[f]Hospital Dr. Negrín, University of Las Palmas de Gran Canaria (Spain)

ABSTRACT

The Canary Islands are located on the western dominant transport path of atmospheric dust plumes generated in the Saharan desert. The Saharan region is widely regarded as the world's greatest source of natural mineral dust, much of which is delivered to the ocean and, from time to time, reaches Southern Europe and the Americas. The fact that the Canary Island archipelago is well placed to receive advected Saharan dust plumes was registered by the State Agency of Meteorology of Spain as long ago as the 1950s. Most studies focus on the observation and monitoring of the dust transport, by analyzing satellite images provided by various agencies (AERONET; MPLNET; EARLINET). Recent research on particulate matter has expanded in recent years to include particle size measurements, aggregation, mineralogy, geochemistry, and particle reactivity, as well as considering the impact on human health and global climate.

This chapter will provide a review of previous research on Saharan dust and other aerosols undertaken on the Canary Islands, research groups and agencies involved and their key findings, as well as the main thrust of current research and prospects for future work. It will consider accumulation of Saharan dust both at present and in the past as well as its textural and mineralogical characteristics in deposits found on the islands. Particular attention will be afforded to the impact on the land surface and human population. It will be shown that for the period 2005-2007 the deposition rates on the island of Gran Canaria were of the order of 20 ± 11 g m^{-2} y^{-1}. A high percentage of aggregate in the deposited dust (> 60%) has been determined using Image Analysis of SEM-EDX. Results from these studies show that particle aggregates have lower bulk densities than individual particles, a characteristic with relevance to some existing dust plume transportation models.

On the basis of geomorphologic mapping, GIS-calculations, luminescence dating and sedimentologic-pedologic analyses of sediments trapped in volcanically-dammed valleys on the Canary Island of Lanzarote, estimates of soil moisture fluctuations in deposited Saharan dust and the hydrologic budget of the island are put forward for the Late Quaternary. These deposits provide information on anthropogenic influence on the dust source area of the nearby Saharan Desert during the Holocene and constitute a basis for quantitative estimates of dust input to this island during a large part of the Quaternary.

1. Introduction

Advected Saharan air reaches the Canary Islands with relative frequency. Different studies carried out to date indicate that Saharan air masses are present over the islands between 20 and 25% of the days of the year [1, 2, 3]. The influx of these air masses leads a notable transformation in the ambient conditions over the islands, including a pronounced rise in temperature and a marked decrease in the relative humidity especially during the hottest months of the year, a change in the direction of the dominant wind and the presence of a considerable amount of lithogenic particulate material. The analysis of the last item is the principal objective of the work referred to below. However, crustal material is not transported with every incursion of African air; its presence being extremely variable and irregular [4, 5]. On some days there are only small quantities, while on others there are massive intrusions that have a serious impact on various social, economic and environmental aspects of the Canaries.

The Canary Islands are situated on the northern fringe of the Saharan dust plume over the North Atlantic Ocean (Figure 1). Dust export from the Sahara towards the Atlantic has existed since the Lower Cretaceous, increasing in a series of steps during the Neogene and the Quaternary, notably during the desiccation of North Africa in the Upper Miocene (6-5 Ma), with the onset of mid-latitude glaciation (about 2.5 Ma) and between 1.6 and 1.2 Ma [6, 7]. In view of such continuous Saharan influence on the North Atlantic region, it is to be expected that Saharan dust is found in the soils and sedimentary deposits of the Canary Islands that date from the Quaternary.

Airborne dust derived from the Sahara Desert generates a dust plume that seasonally crosses the Canarian Archipelago and thus contributes to the sedimentation cycle on the islands and in the surrounding ocean. Complex processes of gravitational, kinetic, thermal, and aggregation forces are involved in delivering this particulate matter to water and land surfaces.

Figure 1. The geographical location of Canary Islands. Modified by Google Earth.

The potential health impact on the population of the Canary Islands of influxes of natural mineral aerosols in Saharan dust storms (local Spanish: calimas) has been recognized for some years [8]. Typically, the densest clouds of dust occur in the winter half-year [1, 9], sometimes considerably reducing daytime visibility (Figure 2) and contributing to hospital admissions in the colder part of the year.

Figure 2. Playa de Amadores, Mogán County, Gran Canaria. before and during a *calima* event in March 2005. Left image – visibility greater than 100km; right image – visibility 1.5 km.

2. Origin and Transport of Saharan Dust (Towards Canary Islands) and the Arrivals of Such Plume Dust to the Archipelago (*Pedro Dorta*)

2.1. Background

In this sense, these dust plumes are one of the most significant characteristics of the archipelago's climate, as evidenced by its mention in all kinds of texts that have been written throughout history [10]. In addition, academics and scientists from different European countries have documented a few very intense episodes of dust held in suspension above the archipelago observed when they were present on the islands. This is the case of Scott who speaks of one event in February, 1898 [11], and Smith who refers to another such event in February, 1920 [12].

However, sound scientific work focusing specifically on these atmospheric conditions in the Canarian region did not start appearing until the middle of the 20th Century. The first was published in 1950 by the Canarian meteorologist Inocencio Font Tulltot [13], entitled "Invasions of hot African air in the Canarian Archipelago". However, it was not until the end of the 1980s that further work on this topic began emerging [14]. Although incomplete, these early studies inspired researchers at the end of the 1990s [1] to begin examining this situation more thoroughly, culminating in a substantial number of precise and scientifically rigorous works published since 2000 and covering a broad range of aspects related to Saharan aerosols [3, 10, 15, 16, 17, 18,19, 20], to name but a few.). Over the past few years this work has included numerous doctoral theses and around twenty articles published in specialised magazines. This is consistent with investigations on general scale; the number of publications on this subject has increased at a higher percentage rate than work published on climate change; a growth commensurate with the importance of the subject [21].

This ongoing examination of the atmosphere over the Canary Islands region has provided a great deal of insight into a multitude of aspects of which we had almost no information just a decade ago. The meteorological causes of the incursions of Saharan air masses have been analysed along with their approximate frequency; the quantity of material that reaches the islands has been estimated; studies have been carried out on the trajectory of the air masses; correlations with the indices of different climate teleconnections have been examined; the multiple effects of Saharan aerosols on socio-economic and environmental aspects of the Canary Islands have begun to be examined; and, finally, all of this research is making it easier to predict the meteorological conditions that favour the arrival of dust clouds.

2.2. Meteorological Causes

The synoptic situations that cause advection of dust-transporting Saharan air have been analysed in various articles. At first, authors sought to classify different types of weather conditions using somewhat subjective techniques [1, 22], but recently they have started to incorporate more advanced methods and objectives [5, 23]. An analysis of this scientific bibliography demonstrates that the lithogenic material reaches the islands in different ways.

In most cases it is removed from the atmosphere and deposited on the surface through dry deposition, although occasionally wet deposition occurs when raindrops collide with the crustal particles and drag them to the surface [24]. In other words, the majority of the Saharan dust is deposited on the islands when the atmospheric conditions are stable. This also makes clear why even the most recent research on the atmospheric circulation systems that transport dust to the Canaries confirms the prevalence of anticyclones during Saharan intrusions.

The most common synoptic scenario is one in which high pressure located over south-western Europe or north-western Africa originates flows from the second quadrant over the islands along the south-western flank of the anticyclonic centre. This situation is especially frequent during the winter months when the dynamic conditions associated with high Atlantic pressure combine with a certain thermal influence from the European continent to promote anticyclonic conditions. This makes it possible for these conditions to persist over time, giving rise to episodes that last longer than those that occur during the warm months [4]. One example of these conditions took place in February of 2004 (Figure 3).

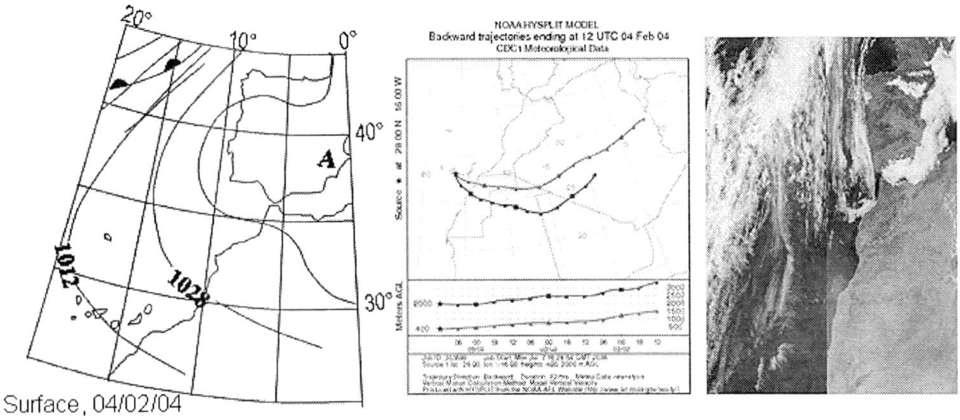

Figure 3. The intrusion of Saharan dust on February of 2004 (synoptic situation, backward trajectories and satellite image). Source: AEMET modified; NOAA: HYSPLIT model; http://oceancolor.gsfc.nasa.gov/

The most typical summer circulation is produced by a depression over the Sahara with a clear thermal influence, as a consequence of the intense desert heat during these months. At the same time, the Atlantic anticyclone extends out as a wedge toward the Mediterranean. When an easterly circulation heading toward the west coast of Africa is created and the Canary Islands are caught between the two pressure centres, intense heat waves are generated that sometimes lead to the transport of crustal material toward the archipelago. As in the earlier case, these are dry deposition episodes. Such an important advection of very hot desert air was recorded on 20 July 2005 with a considerable volume of Saharan dust (Figure 4).

A further synoptic situation is related to the presence of a depression near the archipelago, a situation referred to as "return episodes" [5] by some researchers. It is important to underscore the fact that African dust can remain over the islands for some time after the ocean circulation has been restored. Although the dominant flow has changed, the dust that remains in suspension continues to reach the island for a short period of time because of certain inertia [10, 23].

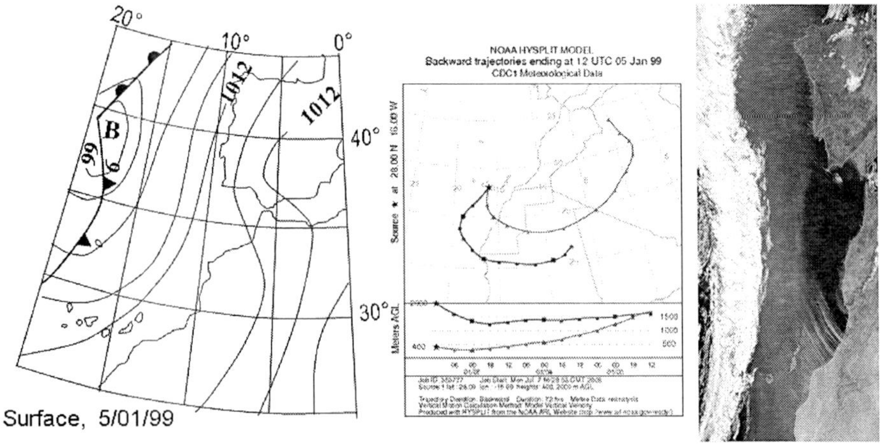

Figure 4. The intrusion of Saharan dust on July of 2005 (synoptic situation, backward trajectories and satellite image). Source: AEMET modified; NOAA: HYSPLIT model; http://oceancolor.gsfc.nasa.gov/

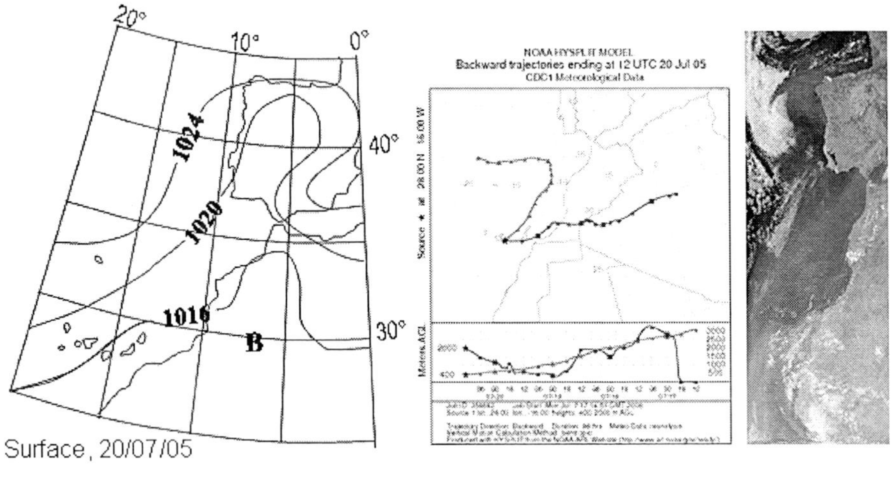

Figure 5. The intrusion of Saharan dust on January of 1999 (synoptic situation, backward trajectories and satellite image). Source: AEMET modified; NOAA: HYSPLIT model; http://oceancolor.gsfc.nasa.gov/

Occasionally, these situations provoke intrusions of Saharan air under conditions of atmospheric instability, which at times can be very intense [25], leading to wet depositional episodes when precipitation drags suspended material down to the surface. Good examples occurred in January 1998 and again in January 1999 (Figure 5). In these atmospheric situations, the low pressure draws in the Saharan air, incorporating it into the cyclonic circulation which creates a situation in which dust reaches the Canaries carried on winds coming from the Atlantic. However, far more material is accumulated on the surface through dry rather than wet deposition to the extent of between 10 and 30 times more [24].

Finally, Saharan aerosols can be dragged to the surface from the middle and upper layers of the troposphere through gravitational deposition. That is to say, material is transported at considerable altitudes during some episodes (which, as we shall see later, frequently occurs during warm months) while a north-easterly oceanic circulation is usually installed in the lower layers. The transfer of material above this first layer sometimes gives rise to a portion of the dust falling due to gravity [23]. This is another case of dry deposition.

2.3. Main Sources of Lithogenic Material

The transport of lithogenic material depends on various factors, the most significant of which are climatic conditions and geomorphological characteristics such as the nature of the wind, the type of substrate and the presence of topographic obstacles [26]. Certain anthropogenic activities such as improper pasturing or changes in land use, among others, can also influence the contribution made by different dust source regions [27]. Similarly, many different terrains across the planet are affected by the presence and transport of crustal material.

Today, the principal sources of lithogenic material contributing to atmospheric transport of aerosols are well known on a planetary scale. The extensive bibliography on this aspect allows us not only to identify different source regions [6, 26, 28, 29, etc.], but also to recognize a global dust belt. This dust source belt includes, in addition to the Sahara, a great proportion of the arid and semi-arid sectors of Asia in which are included the main deserts, such as the Taklamakan, Kyzyl Kum, Karakum, Thar, etc.

It has been estimated that the Sahara is the largest single source of crustal material on the planet, accounting for more than 50% of all the natural aerosols that are deposited on the oceans [27]. The most important sources of lithogenic material in North Africa have also been studied in detail. Today it is known that they are mostly located in the Bodélé Depression, the Sahel Belt and the Western Sahara; especially in an area between Eastern Mauritania, Western Mali and Southern Algeria [27].

Study of the sources of the lithogenic material transported to the Canary Archipelago has only recently begun, it being one of the vanguard lines of research in various research centres in the Canaries. The abundant information on teleconnection (there being various remote sensors that provide information either directly or indirectly about aerosols), the use of different indices to measure the quantities of particulate material in the atmosphere, the vast computational power of statistical tools and, above all, modelling systems have made it possible to analyse precisely the most relevant source regions.

One research technique that has greatly facilitated such studies is the use of backward trajectories created using the HYSPLIT (Hybrid Single-Particle Lagrangian Integrated Trajectory) model developed by NOAA (National Oceanic and Atmospheric Administration). This modelling system is being implemented because it is efficient, can be applied to many different areas and provides easy access to different sources of information. The first results in the Canaries using this technique are just starting to emerge [3, 18, 19, 23].

Initial findings indicate that the main sources of the lithogenic material that reaches the archipelago are located in parts of the Western Sahara closest to the islands [5]. This region accounts for more of the dust reaching the Canaries than others where global dust production is higher but transport distance is much greater, such as the Sahel Belt or the Bodélé Depression in Chad, the latter being, without a doubt, the region that emits most dust into the atmosphere [29].

Another important conclusion reached by the creation and analysis of backward trajectories is that the air masses which convey the Saharan dust clouds show a marked difference between the boundary layer and the free troposphere. The Canary Archipelago is frequently affected by a strong thermal inversion characteristic of the trade winds which cause two substantially different layers of air to emerge. The first is located beneath the inversion and its height is variable. The second is located above the inversion and has totally different thermohygrometric properties; its first few meters are warmer and, above all, much drier with notably lower relative humidity; its winds also blow in a different direction. These are strikingly different layers of air. Furthermore, this thermal inversion is markedly seasonal, in that its altitude is higher in winter than in summer and stronger during the warm months compared to the colder part of the year [9].

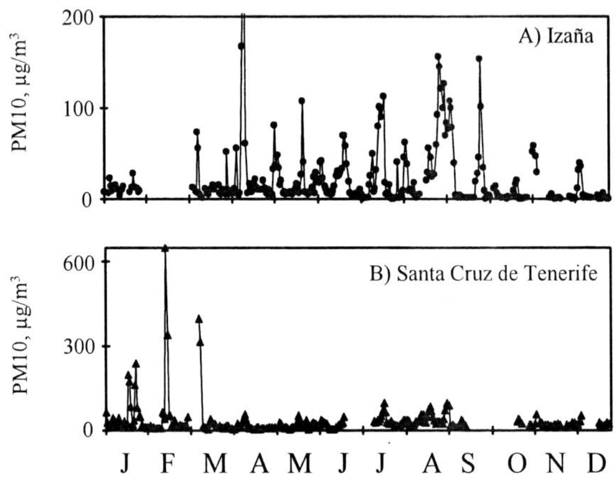

Figure 6. Daily mean concentrations of PM10 during 2005 at two sites in Tenerife Island: A) Izaña Observatory (2367 m a.s.l.; free troposphere) and B) Santa Cruz de Tenerife city Observatory (50 m a.s.l.; boundary layer). Data provided by the Izaña Atmospheric Research Centre. Courtesy of Sergio Rodríguez and E. Cuevas.

In this sense, the transport of crustal material is also clearly seasonal. Along with more frequent intrusions of Saharan air masses, the density of the Saharan aerosol in winter is noticeably higher at low and middle altitudes (mostly below 500 meters above the sea level), located in the boundary layer, where almost the entire population of the island is concentrated; nevertheless, because of its thickness, which may be considerable, the layer of dust sometimes reaches high altitudes, especially during very intense episodes [30]. Meanwhile, during warm months, dust is transported at high altitudes (mostly above 1000 meters above the sea level), above the thermal inversion, as indicated by numerous authors [5, 18]. In other words, during this period it is possible for Atlantic air to predominate in the populated regions, while a flow from the desert, laden with aerosols, is established above. In fact, the Izaña Atmospheric Research Centre, an eminent observatory located on the island at almost 2400 meters above sea level, has been able to confirm this circulation at different altitudes depending on the time of year. In light of this, the island of Tenerife is in a perfect situation to observe these phenomena (Figure 6). This Figure illustrates the different seasonal evolution of Saharan dust transport in the marine boundary layer and in the free troposphere. Observe how high PM10 occur in winter at Santa Cruz de Tenerife and in summer at Izaña

Observatory At Izaña, PM10 concentrations higher than 10μg/m^3 are due to Saharan dust transport. At Santa Cruz de Tenerife, PM10 concentrations higher than 30μg/m^3 are due to Saharan dust transport (S. Rodriguez, personal communication).

2.4. Evolution and Tendencies

As mentioned earlier, intrusions of lithogenic material are caused by the establishment of flows of Saharan air above the islands. Various studies have examined different series of data on these air masses that were recorded by stations on the ground [4] and analysed the tendencies and possible teleconnections. However, the research centred specifically on the load of crustal material is very recent [5, 19] and a large part of its results have not yet been published. Furthermore, the series are not excessively long and it is only possible to provide some estimates.

Findings to date lead to two important conclusions: first, there are significant correlations between the transport of lithogenic material and the ENSO (El niño South Oscillation) and NAO (North Atlantic Oscillation) indices; and second, there is an upward trend in the number of intrusions and the total quantity of particulate material.

Regarding the first conclusion, it can be observed that negative ENSO indices generally imply an increase in the advections of Saharan air masses [2]. It can be assumed that the incursions of crustal material occur similarly. An example of this probable correlation is the observation that there was an intense ENSO in the winter of 1997-1998, while 1998 was one of the years with the greatest amount of Saharan atmospheric dust incursions over the Canaries and a large number of days when Saharan air was prevalent over the islands [2, 15, 24].

There are also studies indicating that there is a link between the NAO and the African intrusions, in that a high positive index implies a greater concentration of dust over the Canaries [24]. Research that employs relatively long series of data obtains moderate but significant correlations, arriving at the conclusion that the pattern of the NAO teleconnection significantly influences the intensity of the winter incursions of Saharan dust [5]. This fact is consistent with what other authors indicate about an increase in the amount of dust being driven out toward the Atlantic during winters with positive NAO indices [31].

Finally, it must be noted that the studies have not found a statistically significant correlation between this phenomenon and another possible teleconnection, the Sahel Rainfall Index (SRI), despite the fact that this region is relatively close to the islands. The author of the main research on this subject points out that this finding demonstrates that the southern edge of the Sahara contributes a negligible proportion of the crustal material that reaches the Canaries, at least in the boundary layer [5].

Regarding the second conclusion about the upward tendency of the quantity of Saharan aerosols over the Canary Islands, it must be noted that this subject has been studied by researchers with diverse series and data. The most complete statistical study carried out to date shows an upward tendency, although analyses of tendencies is a complex undertaking given that different data sources or different indicators can be employed. Different statistical tests undeniably indicate that there has been a general increase in both the frequency and intensity of the Saharan dust episodes, particularly after the 1970s, and also beneath the thermal inversion, in the boundary layer [5].

Finally, it should be mentioned that the latest research is starting to use indirect methods to detect intrusions of Saharan dust in the Canaries [32, 33]. For instance, the high correlation that seems to exist between the concentration of some radioactive isotopes, like 137Cs and 40K, and the arrival of lithogenic material in the islands allows researchers to trace these dust intrusions by measuring the isotopes, a technique that will make the detection of Saharan dust in the atmosphere even more precise.

3. Past Records of Saharan Dust on the Canary Islands *(Hans Von Suchodoletz and Ludwig Zöller)*

3.1. Occurrence of Saharan Dust Deposits on the Canary Islands

Saharan dust is readily recognized in the Canaries by the presence of quartz, which is not authigenically formed by the local basic volcanism [34]. Consequently, the widespread existence of quartz in Canarian soils, or at least in their upper horizons, demonstrates that Saharan dust constitutes a major component of the parent material involved in recent pedogenesis [e.g. 34, 35, 36, 37]. The marginal infiltration of yellowish material containing quartz into lapilli and vesicular basaltic rock fragments that were ejected during volcanic eruptions in the middle of the 18th century on Lanzarote [38] (Figure 7) shows that dust trapping has continued until the present. Furthermore, the occurrence of quartz in palaeosols of different Quaternary ages found in various relief locations in Lanzarote and Fuerteventura shows that all these palaeosols evolved from material containing at least a certain proportion of Saharan dust [36, 39, 40] (Figure 8).

Figure 7. Saharan dust (white arrows) infiltrating a vesicular lava fragment from Lanzarote that was ejected during volcanic eruptions in the middle of the 18th century.

Figure 8. Soils of different ages in the eastern coastal plain of Lanzarote near Costa Teguise, both exhibiting quartz contents during XRD analyses: a) polygenetic soil profile < 170 ka (unpublished data) overlain by a desert pavement, b) palaeosoil fritted by a lava flow ca. 170 ka ago (unpublished data).

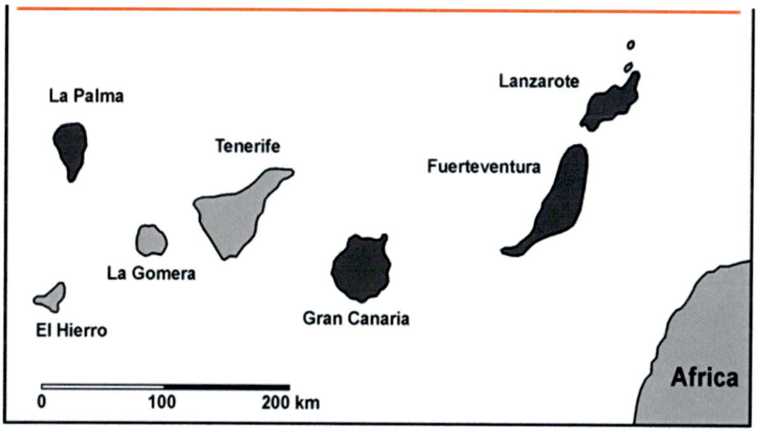

Figure 9. The Canary Islands. Those islands with described or recently discovered stacked Quaternary Saharan dust deposits are dark coloured.

In certain places, Saharan dust deposits form stacked sequences with thicknesses of some meters deposited during some long periods in the Quaternary. Until very recently, such stacked Quaternary sequences were reported only from the oldest and thus most eroded islands of Lanzarote and Fuerteventura on the eastern side of the archipelago [41, 42, 43]; they have not been detected on the younger, western islands, which exhibit a much steeper topography. However, this situation recently changed with the publication by Menéndez et al. [44] of an account of such deposits on Gran Canaria, as well as with the discovery of stacked sediment sequences composed of quartz-bearing material by L. Zöller on La Palma in 2008. Ideally, such deposits develop where old valleys or flat plains were dammed by younger volcanic material of Early Pleistocene to Holocene age, subsequently leading to sediment deposition up to a thickness of a few tens of meters (Figures 9, 10). Similar, but somewhat thinner deposits can be found on flat valley terraces, on the more gentle slopes or on plateau surfaces (e.g., sediments 10 m thick occur on a plateau in northern Lanzarote close to the Mirador del Río; 4 m of sediments are found on a gentle slope in eastern La Palma; Figure 11). However, owing to the high erodibility of the latter deposits hiatuses must be expected here. As is evident from the widespread presence of quartz, the majority of these stacked sediments consist of Saharan dust with only a minority of local volcanic material [44, 45, 46].

Figure 10. Outcrops of sediments containing Saharan dust mixed with local volcanic material trapped in dammed valleys (*vegas*). Quartz contents were determined by semi quantitative XRD analyses. *a)* Outcrop of 4 m of trapped sediments overlying a Late Quaternary basaltic flow [47] in the *vega* of La Oliva (northern Fuerteventura). The sediments have quartz contents between 7 and 30%. *b)* Outcrop of 6.3 m of trapped sediments in the *vega* of Femés (southern Lanzarote). The sediments contain between 6 and 39% of quartz

Figure 11. Sediments derived from Saharan dust on plateau positions and slope flattering. The presence of quartz is demonstrated by qualitative XRD analyses. a) 10 m of sediments deposited on a plateau close to Mirador del Río (northern Lanzarote). b) Almost 4 m of sediments deposited on a slope flattening above Santa Cruz (eastern La Palma).

These sediments show marked alternations between reddish-clayey and yellowish-silty layers, and are believed to constitute palaeoenvironmental archives [47] (Figure 10). Below we present results from some investigations of wind-lain dust archives in three lava-dammed valleys in Lanzarote.

3.2. Saharan Dust Deposits from Lanzarote as Palaeoenvironmental Archives

Lanzarote, the northeasternmost of the Canary Islands, has several valleys or flat plains that were dammed by volcanic material during the Quaternary and today are filled with sediment accumulations up to 50 m thick. Locally, these sediment traps are called vegas [48]. The location of vegas in Lanzarote known to the authors is shown in Figure 12.

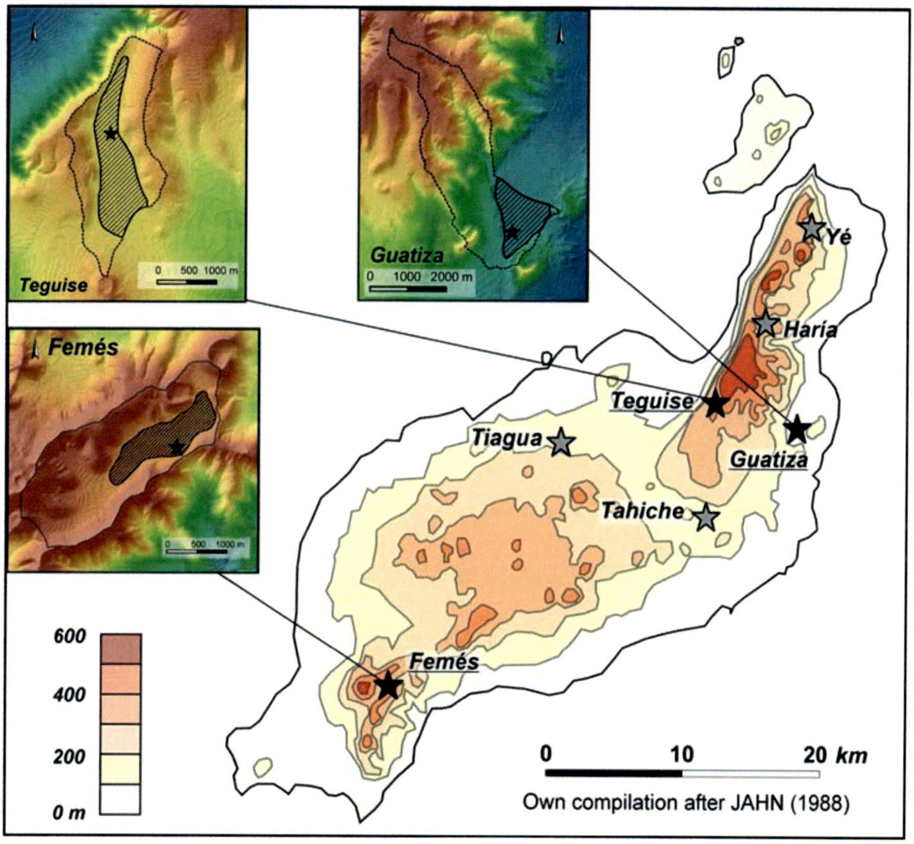

Figure 12. Location of outcropped *vegas* on Lanzarote known to the authors. Further investigated sites are shown in black and their name is underlined, whereas not investigated sites are indicated in grey and their name is not underlined. The insets show GIS-models of further investigated *vegas*, where dashed lines indicate the catchment areas and hatched surfaces highlight the valley bottoms. The location of investigated outcrops in the *vegas* is shown with black stars.

3.2.1. Dust Deposits as Archives of Palaeo-Soil Moisture

Loamy *vega* sediments were outcropped by the local population in order to gain material for agricultural purposes, and clearly show various distinct layers of reddish and yellowish material. This intercalation of different layers is believed to reflect archives of past palaeoenvironmental conditions. Thus, during a research project carried out by a research group from the University of Bayreuth (Germany), three *vegas* in Lanzarote with a well developed stratification (Guatiza, Teguise, Femés) were investigated with a view to revealing the palaeoenvironmental history of Lanzarote (Figure 12). The investigations were based on a multi-method approach that comprised geomorphological mapping, GIS-analyses, environmental magnetic measurements and sedimentologic-pedologic analyses. Chronostratigraphies of the *vegas* sediment accumulations were built up using luminescence dating, further supported by interprofile correlations and correlations with proxies from nearby marine cores [46].

The results show that outcropped sediments in the valley bottoms were almost continuously deposited between the Middle Pleistocene and the Holocene (Figure 13). Owing to the fact that the sediment archives are situated on the local erosional base level in the valleys or *vegas*, however, trapped sediments contain *in situ* eolian material (Saharan dust and volcanic fallout) as well as material that was originally deposited as eolian fallout on the slopes, but was later reworked and transported to the floors of the *vegas* by colluvial processes. Sedimentation rate and the proportion of *in situ* and reworked material in the archives directly depend on the valley bottom: catchment area ratio, which was determined by GIS-analyses (Figure 12, insets). Consequently, the outcrop in the Guatiza *vega* (Guatiza III) with a valley bottom: catchment area ratio of 1:6.2, has the highest proportion of reworked material and thus the highest sedimentation rate, followed by the Femés *vega* with a ratio of 1:5.4 and the Teguise *vega* with a ratio of 1:2.9 [49].

The crucial question raised by these archives is whether the colluvial transport from the slopes occurred in "low frequency/high magnitude" fashion or if it was a "high frequency/low magnitude" process. In the former, rare events would have deposited large quantities of sediment, strongly disturbing the chronological sequence of the sediments. In the latter case, however, frequent erosional events would have mobilised small amounts of material, causing only small disturbances of the chronological sequence. In the latter case, therefore, a palaeoenvironmental interpretation would be possible, albeit with a somewhat reduced temporal resolution compared to sediments exclusively deposited as eolian fallout. In order to address this problem, it is useful to recall a classic study of landscape stability related to annual precipitation. Langbein et al. [50] demonstrated that the highest erosion rates generally occur in a precipitation regime with annual precipitation rates of between 100 and 550 mm, conditions that prevail on the island of Lanzarote today (100 - 250 mm/a, [36]) and that are assumed to apply also to former moister palaeoclimatic (ca. 560 mm/a, see below; [51]). Furthermore, marine studies off NW-Africa show that the climate of the region during the last glacial cycle was strongly influenced by high-frequency millennial climate fluctuations, such as Heinrich-events and Dansgaard-Oeschger cycles [52, 53] that probably impeded continuous stabilisation of slopes by vegetation growth. Thus, the colluvial sedimentation most probably occurred as a "high frequency/low magnitude" process, with frequencies at a scale of hundreds to a maximum of some thousands of years.

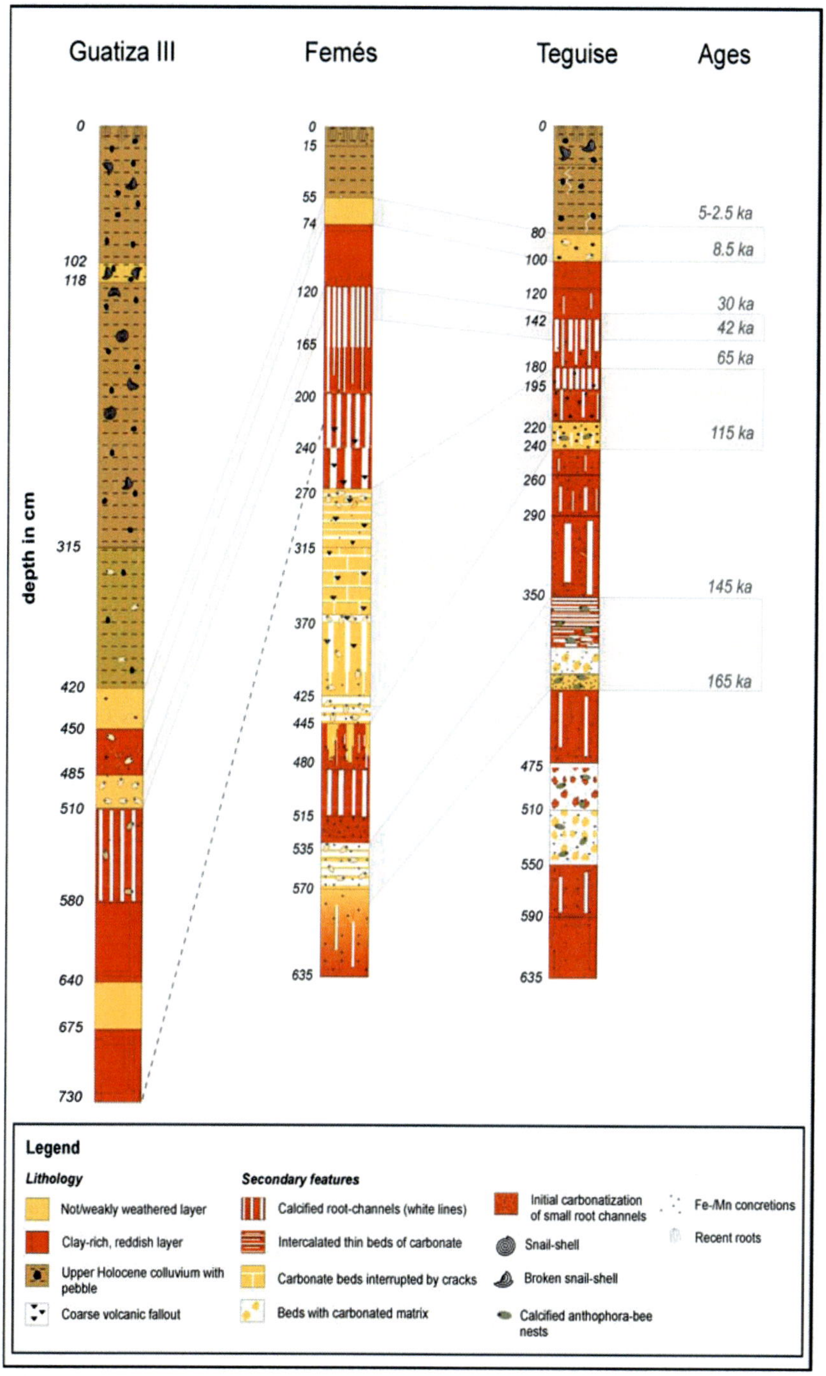

Figure 13. Investigated profiles in Guatiza III, Femés and Teguise with chronological frame. The upper horizons contain anthropogenically triggered colluvia of various thicknesses (from 55 to 420 cm) that can not be used for palaeoenvironmental inferences.

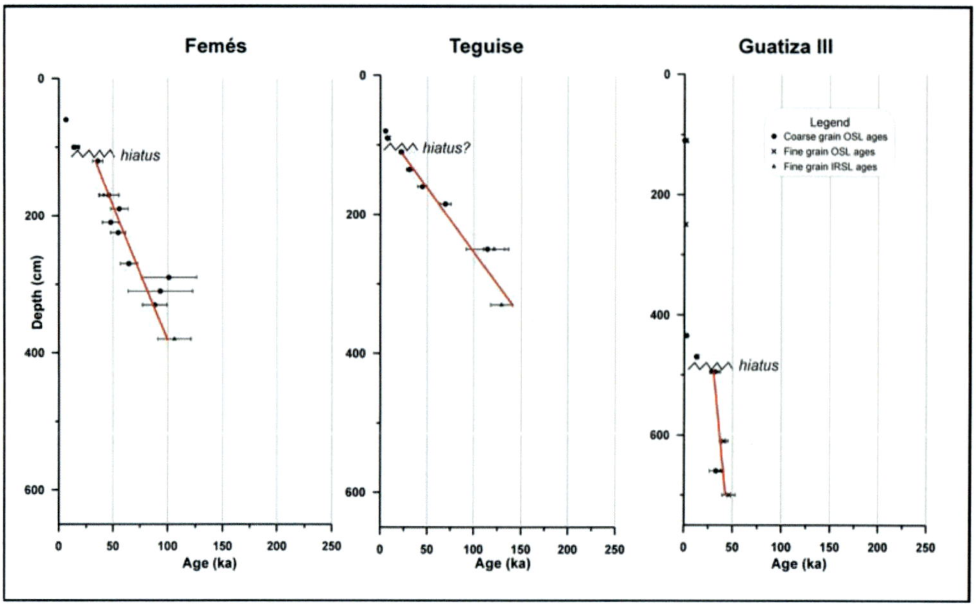

Figure 14. Luminescence dating results from the *vegas* of Lanzarote. Mean sedimentation rates are drawn as a red line in every *vega*. No large disruptions of this rate within error bars are detected arguing for a more or less continuous sedimentation into the valley bottoms.

This hypothesis is supported by our luminescence dating results, which indicate a largely continuous sedimentation rate in the valleys without notable disruptions (one exception being a conspicuous hiatus between ca. 15 and 25-30 ka in all *vegas*; Figure 14). This demonstrates that the *vega* deposits can be used as palaeoclimatic archives, even given that temporal resolution is unlikely ever to be better than some hundreds or thousands of years.

In order to investigate past palaeoenvironmental conditions, proxies sensitive to changes in palaeoenvironmental parameters have to be identified. In the case of the sediments from Lanzarote, the change from reddish-clayey to yellowish-silty layers is believed to have been caused by different intensities of pedogenesis, a process directly dependent on palaeo-soil moisture. However, because of the high proportion of colluvial material in the valley floor sediments, it is clear that reddish/clayey layers are mostly soil sediments derived from pedogenesis that occurred on the slopes, rather than as *in situ* palaeosols. Furthermore, a crucial question is whether the observed alternation of sediment properties was acquired *in situ* during pedogenesis on the island, or if these properties were inherited from a change in Saharan dust source(s). To investigate this problem, clay contents were measured in silty and clayey horizons, exhibiting values between 23 and 80% [51]. However, dust recently collected close to the Canary Islands shows clay contents of <15% [25, 54, 55], and a marine study 300 km north of the Canary islands revealed maximal clay contents of only 25% derived from Saharan dust during the last 250 ka [56]. This indicates that clay contents of up to 80% in the *vega* sediments are not directly inherited from Saharan dust, but were instead formed by pedogenesis in Lanzarote. It follows that the clay content can serve as a proxy of pedogenesis on Lanzarote during the Late Quaternary. The proportion of ultra-fine superparamagnetic particles in sediments can be identified by measuring frequency dependent magnetic susceptibility (κ_{fd}). These small particles are produced during pedogenesis and are often used as a proxy of pedogenesis in palaeopedological studies [57, 58]. This parameter is found to be strongly enhanced in reddish/clayey layers in the *vegas* of Lanzarote, with its depth dependent function running almost parallel to that of the clay content (Figure 15). This confirms the pedogenetic origin of superparamagnetic particles on the island, adding strength to the view that frequency dependent magnetic susceptibility can also be taken as a proxy of pedogenesis and thus of palaeo-soil moisture. When looking at the depth dependent distribution of these proxies, it emerges that parts of Marine Isotope Stages (MIS) 2, 3, 4 and 6 experienced a higher soil moisture than today, whereas most parts of MIS 1 and 5 were apparently as dry or even drier than the present (Figure 15). This pattern shows that global cold periods were generally characterized by moister, and globally warm periods by drier conditions on Lanzarote. Furthermore, we compared pedogenetic features from the *vegas* of Lanzarote with those of a catena developed on Holocene volcanic material on Tenerife [59], supported by a comparison of the occurrence pattern of ground-nesting anthophora-bees in the *vega* profiles with their ecological spectrum [39, 60] (Figure 13). These comparisons lead us to the conclusion that maximal soil humidity on Lanzarote during periods of strong pedogenesis was equivalent to that in a region of Tenerife with a recent precipitation amount of about 560 mm/a [51]. Due to the dependence of soil moisture on precipitation, temperature and insolation (cloudiness) this value is not directly transformable into palaeoprecipitation values. However, it does provide a broad indication of past humidity conditions on Lanzarote during generally cold phases of the Late Quaternary.

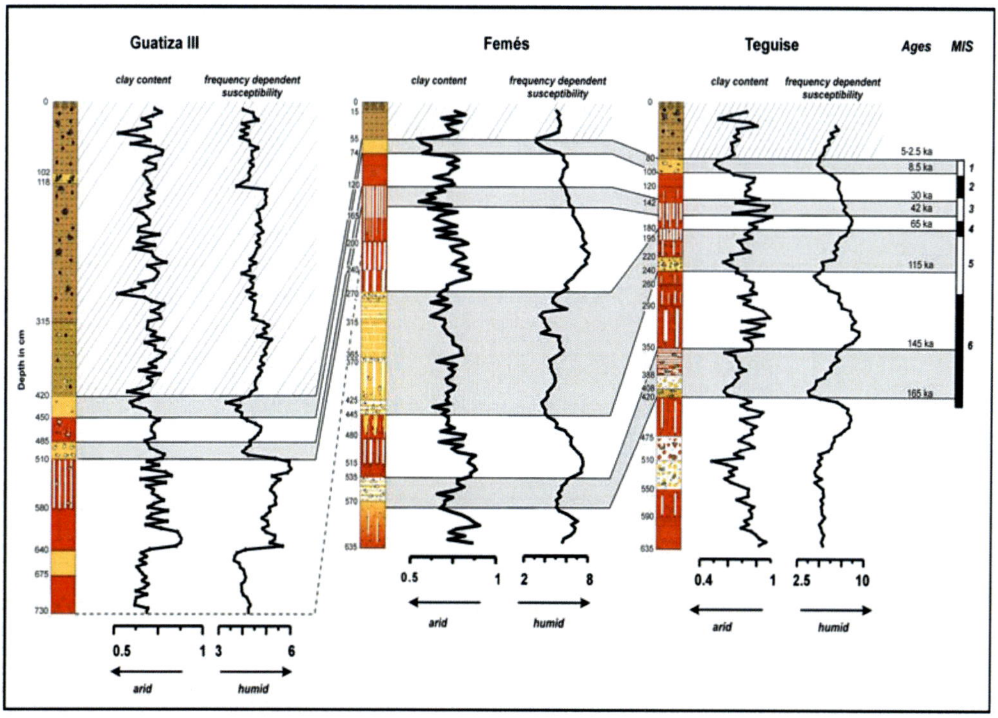

Figure 15. Compilation of profiles, ages and the pedogenetic (palaeoenvironmental) proxies relative clay content (arbitrary scale) and frequency dependent magnetic susceptibility (%). Shaded sections indicate more arid periods, hatched areas anthropogenically triggered colluvial. Ages and Marine Isotope Stages (MIS) are indicated on the right side.

3.2.2. Saharan Dust Deposits as Archives Of Palaeo-Dust Input

Using geomorphological mapping and GIS-analyses in combination with our chronostratigraphy, as well as with K/Ar-datings of volcanites from the same volcanic group that makes up the lava flows that dam the vegas [61, 62], it was possible to calculate a chronological sediment mass balance for the completely dammed vega of Femés [49] so as to provide a base for conclusions about changes of Saharan dust input on the island. In order to construct this mass balance, the floor of the Femés vega was divided into 10 different segments for which sediment volumes were calculated separately. Sediment thicknesses for every valley segment needed for this calculation were determined by extrapolating the slope forms into the valley bottom using GIS software (Figure 16). Results of geomorphological mapping were used to estimate surface and thickness of alluvial fan deposits as well as the remaining sediment thickness on eroded slopes.

The results of this investigation demonstrate that dust input steadily increased from about 1.0 Ma until the present, as already stated in former studies of the eastern Atlantic [6, 7] (Figure 17). Despite some smaller uncertainties in luminescence dating and in measurements of recent dust falls over the Canary Islands, it is apparent that the highest dust input of the period of the Quaternary studied here, occurred during the Holocene, whereas extremely high values exceeding all past magnitudes are found today [37, 55]. This would appear to indicate that the elevated values of dust input recorded during the Holocene and especially during the recent period have not only a climatic cause but must, in addition, have been triggered by factors other than aridity, e.g. human impact in the Saharan dust source areas.

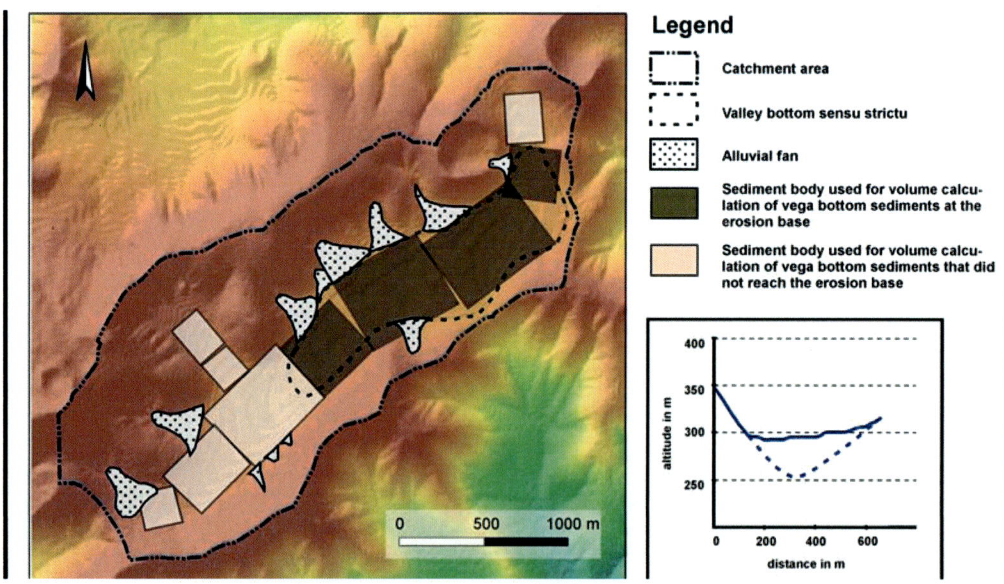

Figure 16. Calculation of the sediment budget of the *vega* of Femés: The *vega* bottom was divided into different sediment bodies whereof some are located at the erosion base (in the valley bottom *sensu strictu*) and others at somewhat higher elevations where the material must be regarded as temporarily stored. In order to calculate the depth of every sediment body, cuts were made through the valley and the sediment depth in the sediment bodies was finally determined by extrapolation of the slopes (see dashed line in the example of the inset). The limits of the valley bottom *sensu strictu* and of alluvial fans were determined by geomorphologic mapping.

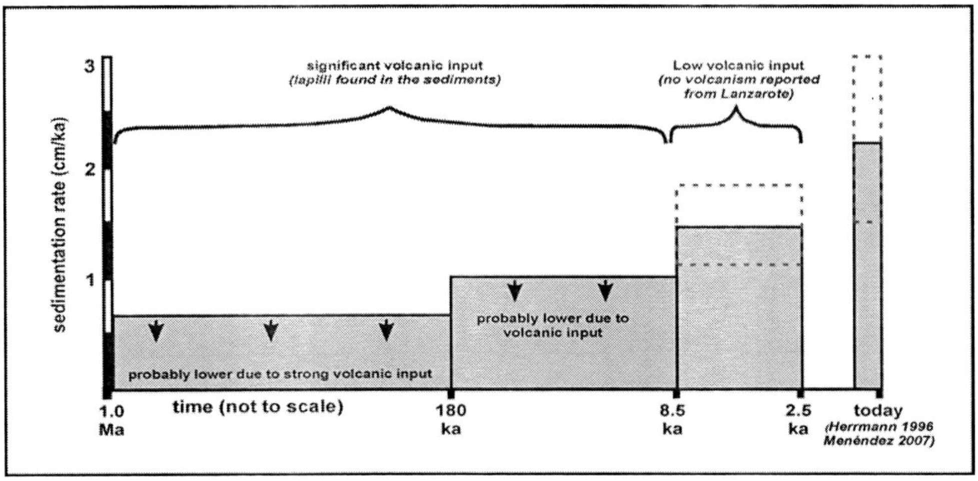

Figure 17. Evolution of dust input on Lanzarote based on the sediment mass balance of the *vega* of Femés. Dashed lines indicate the uncertainty of calculations during the Holocene (own data) and for the recent period [37, 55]. Between 180 and 8.5 ka as well as between 1.0 Ma and 180 ka dust input was probably lower than derived from the sediment mass balance due to volcanic input recognizable in the profiles. In the latter case this reduction was probably more important, since a layer of 30 cm almost exclusively composed of lapilli and tephra was found in a drilling below the recent surface of the outcrop [43].

4. Deposition Rates, Grain-Size Distribution and Mineralogy of Saharan Dust on the Canary Islands *(Inmaculada Menendez and Johann Engelbercht)*

4.1. Deposition Rates of Saharan Dust on the Canary Islands

4.1.1. Dust in the Air

Dust concentrations in air over the Canary Islands and their surrounding oceanic areas were measured over a decade ago in Fuerteventura [14] and in Tenerife (Izaña Observatory, 16°29'58'' W, 28°18'32'' N [63, 64, 65]), revealing equivalent mineral dust concentrations ranging from 0.14 to 285.7 μg m^{-3} (deduced from the average concentration of Al in crustal materials: 6-8%; [65, 66]). Direct measurements were made also during a specific Saharan event (July 2002: 396-345 μg m^{-3}; [67]), with a Saharan dust episode threshold in Tenerife later deduced, at a concentration up to 28.5 μg m^{-3} [23]. Since the 1990s, airborne dust measurements have been extended to Gran Canaria [15, 24, 68, 69] with a mean value of 47 μg m^{-3} and Saharan dust pulses exceeding 1000 μg m^{-3} (Figure 18).

One of the most interesting characteristics of this airborne particulate matter is its impact on the air quality. The Saharan dust influence is now analyzed in the two most populated islands of the Archipelago (Gran Canaria and Tenerife; [23, 70], with atmospheric particulate matter (PM10 and PM2.5) being monitored by the air quality network of the Canarian Government, following EU Air Quality Directive EC/30/1999.

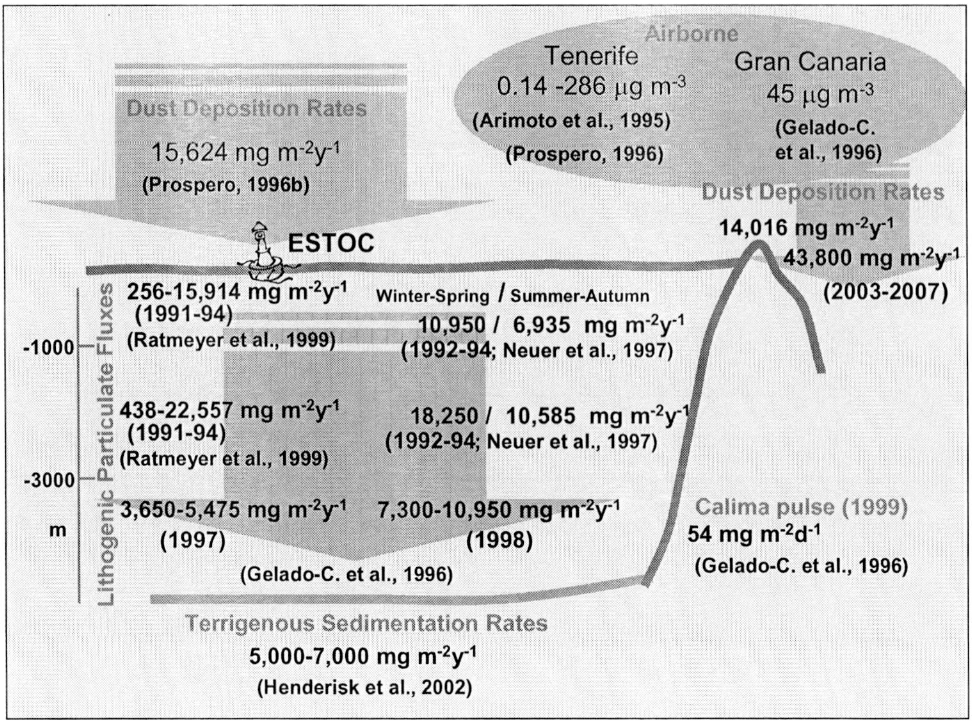

Figure 18. Process of Saharan dust distribution in the region of Canary Islands, from the air to the land surface and to the ocean surface, towards the water column and last sedimentation in the basin.

4.1.2. Dust Deposition Rates on The Islands

Deposition rates for the Canary Islands were estimated by Prospero ([64], in Table 2.a pp. 142, in a grid box of 20°-30°N/20°-10°W), following the GESMAP empirical model, with a value of 15624 mg m^{-2} y^{-1}. In contrast, current manually measured dust inputs on Gran Canaria (from 2003 to 2007) at different altitudes and geographical locations indicates deposition rates of between 14016 and 43800 mg m^{-2} y^{-1} . The highest rates were registered at the lowest altitudes [55] and in the areas with greater exposure to the trade winds (N-NE-E; Figure 19). The recycled product, occurring as local dust, of this deposited Saharan particulate matter in the more arid and populated coastal areas may account for the higher values obtained by manual measurements compared to previous estimates.

4.1.3. Lithogenic Particulate Fluxes in the Ocean

When airborne dust is deposited in the ocean it is called lithogenic particulate. Such Saharan-derived particulate fluxes add to the phytoplankton development [24, 68, 71]. The input of such airborne dust plumes to the oceanic fluxes probably favours physical aggregation and rate of settlement of oceanic particles [72, 73].

A number of lithogenic particle fluxes have been monitored in the proximity of the Canary Islands (European Station for Time-Series in the Ocean, Canary Islands, ESTOC Station 29°15'N-27°11'W; [69, 71. 74] (Figure 18). Deposition rates of airborne dust at the ocean surface defined for the Canary Islands region (15624 mg m^{-2} y^{-1}; [64]) falls in the maximal range obtained for lithogenic fluxes at 0-1000 m depth by Ratmeyer et al. ([74];

256-15914 mg m^{-2} y^{-1}). A marked seasonality was found in the study period carry out by Neuer et al. ([71]; sampling period from 1992 to 1994), clearly showing higher lithogenic fluxes in late winter-spring than in the summer-autumn period. This result matches the periodicity of the highest calima pulses in the Canarian region; here, also, the dust input peaks more frequently in winter-spring than in summer-autumn [1, 9].

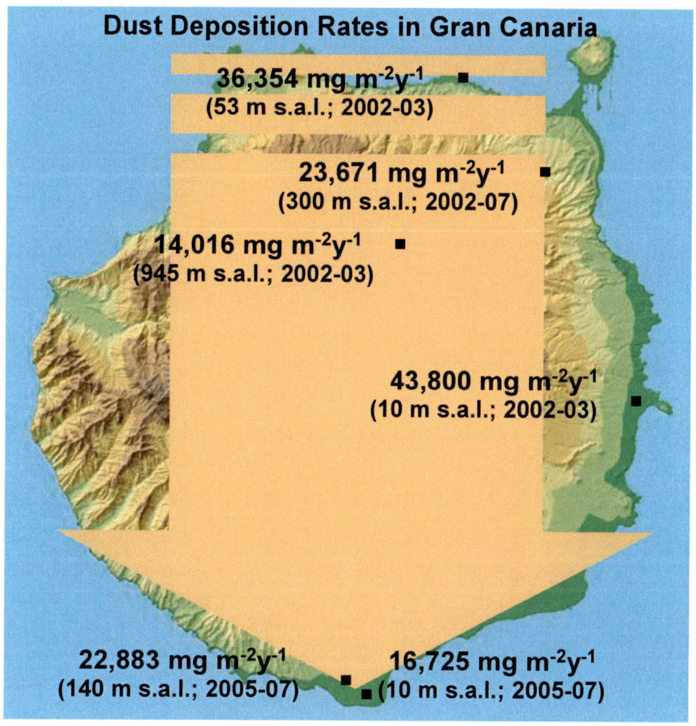

Figure 19. Dust deposition rates, in mg m^{-2} y^{-1}, on the Gran Canaria Island, at different altitudes and exposures.

4.1.4. Deposition Rates on the Seafloor

Airborne input to the ocean water column is integrated in the gross particulate production of the ocean (including mainly organic matter, carbonates and siliceous skeletons). Henderiks et al. [74] have defined a sedimentation rate for the Holocene in the Canary Basin of 1-2 g $cm^{-2}ka^{-1}$ (10000-20000 mg m^{-2} y^{-1}). Lithogenic particles (the terrigenous fraction, most of which is assumed to be aeolian in provenance) total 0.5-0.7 g $cm^{-2}ka^{-1}$ (5000-7000 mg m^{-2} y^{-1}). These rates are lower than measured at land sites, a result arising from the different densities and viscosities of the media, being higher in the ocean water column than in the air.

4.2. Grain-Size Distribution of Saharan Dust on Canary Islands

4.2.1. Particle Size of Airborne Dust

Measurements of airborne dust made in Gran Canaria (1980 m s.a.l.) in the period 1997-1998 reveal unimodal curves centred on the 4.9 or 2.7 μm range with a fine fraction 'tail'

(Figure 20; [24]). Apart from this characteristic, there does not appear to be a great difference in Saharan dust fall from one day to another. Grain size distribution in Tenerife, measured in July and August 2005, also produced a unimodal curve centred on 1.5 μm [76] (Figure 20). This same mode was also detected by Maring et al. [77] and a secondary mode, for aerosols, was found at around 0.2 μm, which increases with distance from the source (comparing Tenerife with, for example, Puerto Rico in Gran Canaria) because of the preferential removal of the coarser grain sizes during atmospheric transport.

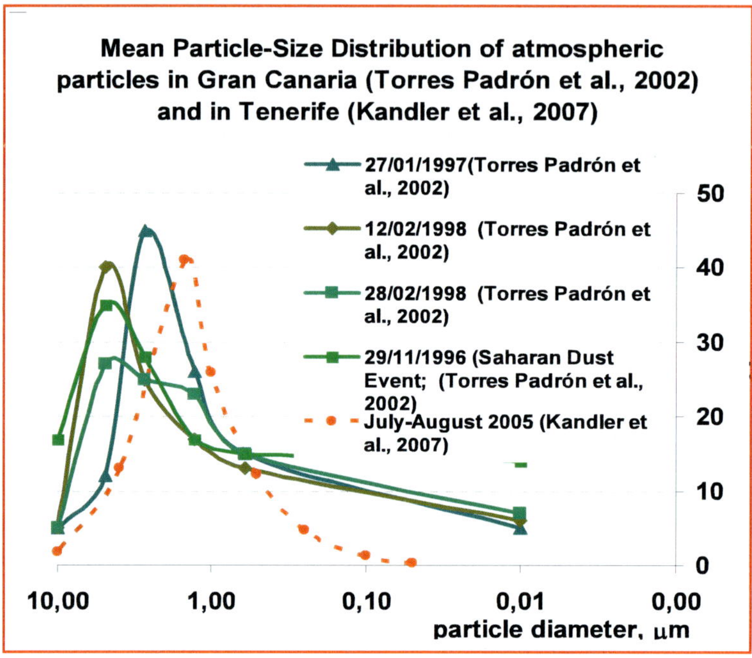

Figure 20. The particle-size distribution of the Saharan airborne dust at Gran Canaria and Tenerife Islands. It is remarkable the unimodal tendency.

Airborne plumes from Saharan sources are not composed of individual particles only, but include aggregates of particles [67]. The occurrence of these aggregates in airborne dust may well be a factor in the overestimation of the Stokes gravitational settling rates of large particles during atmospheric transport from North Africa across the tropical North Atlantic and the Caribbean, detected by Maring et al. [77].

The presence of Saharan dust in urban/industrial areas may enhance the secondary particles formation due to heterogeneous reactions. For example, the formation of sulphate and nitrate onto Saharan dust particles has been documented on Santa Cruz de Tenerife [67]. This implies that Saharan dust impairs the air quality by increasing PM10 and PM2.5 concentrations by two ways: 1) by increasing the concentrations of mineral compounds, and 2) by enhancing the SO_2 (gas) → sulphate (PM) and NOx/HNO_3 → NO_3^- (PM) conversion. The latter process alters the surface of the dust particles and consequently its impact on the environment (S. Rodríguez, *personal communication*).

4.2.2. Dust Grain-Size on the Islands

The granulometry of samples of deposited dust collected from sites on Gran Canaria Island during 2003-2004 also display a bimodal curve with a principal mode around the medium silt size (0.016 mm), with another, secondary peak in the very fine silt size (0.004 mm), and a coarser fraction 'tail' (Figure 21). Comparison of the granulometry of airborne dust and that deposited on the land surface poses no problem because each has a distinctive size range (airborne >10 μm - <0.01μm compared to >250 μm - <2 μm for land-deposited dust). Both types share a common size range (>10 μm - <2 μm) as shown by the two measured points in deposited dust (8 μm and 2 μm). Comparison of this size range in both sample types clearly shows a tendency for airborne dust to peak towards the finer fraction and the converse to be true of dust deposited on the land surface. This is consistent with the gravitational settling effect in which the coarser particles tend to settle before the finer fractions, as well as the greater abundance of coarse fractions in deposited dust.

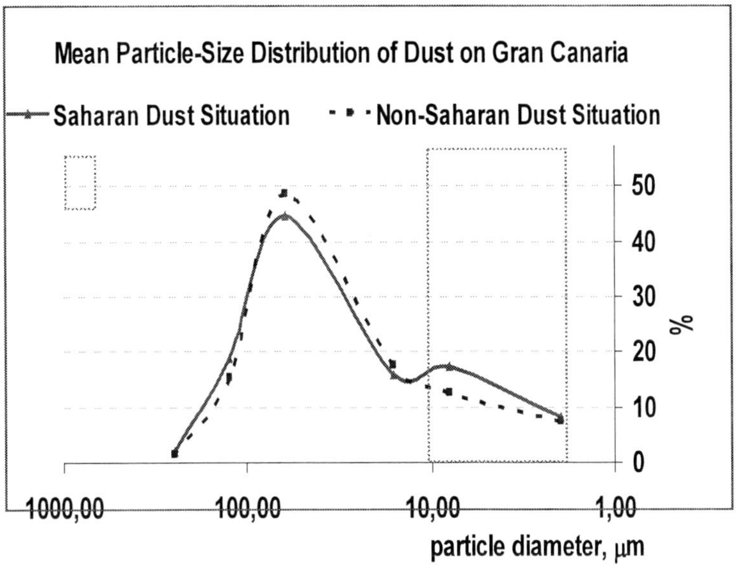

Figure 21. The particle -size distribution of the deposited dust on Gran Canaria Island, differentiating the Saharan dust presence and absence in the atmosphere of the island. The dotted area is the particle range comparable with the particle distribution of the airborne dust of the islands.

As with airborne dust samples, numerous aggregates (Figures 22-24) made up of mixed size particles are present in land-lain samples that form aggregations in which the medium-coarse silt fraction is dominant. The aggregates quantified using SEM-EDX images are frequently more numerous than individual particles, estimated at around 60%. Such a high percentage is important because the particular physical characteristics of the aggregates gives them a lower bulk density and slower fall velocity when compared with single mineral particles of the same dimensions.

Figure 22. SEM-EDX image of the deposited dust on the Gran Canaria Island, showing mineral particles and aggregates.

Figure 23. Other example of the EDX image from the deposited dust on the Gran Canaria Island, showing mineral particles and aggregates.

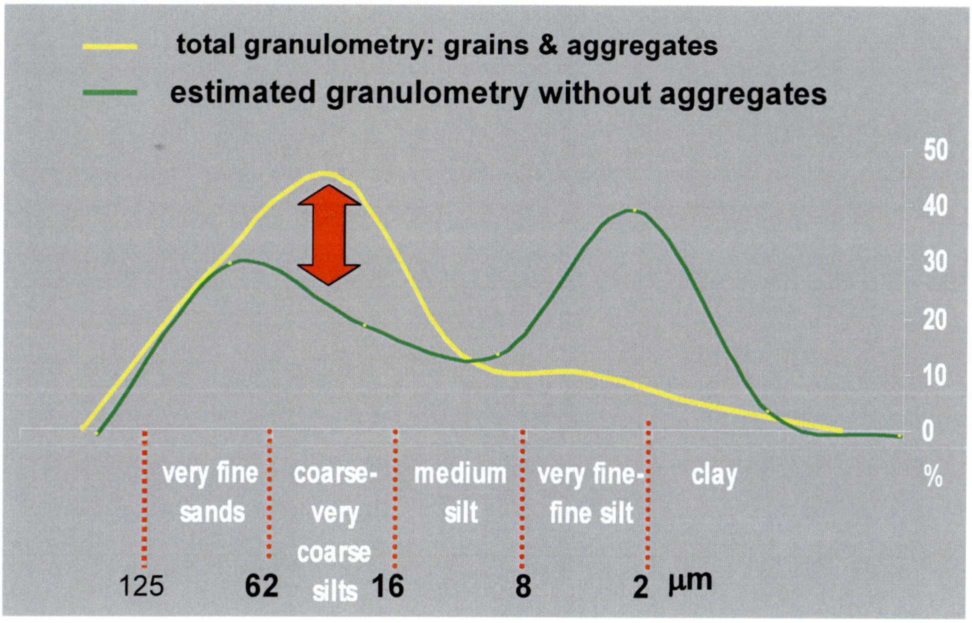

Figure 24. The polymodal curve of particle size distribution in deposited dust samples on Gran Canaria, considering the total particles (grains and aggregates) and the estimated curve of the disaggregated sample, with a consequent increment of the finer fractions.

4.2.3. Lithogenic Grain-Size Within the Ocean and on the Seafloor

Using a variety of instrumental methods including Sedigraph, FRITSCH laser and CIS-GALAI particle analyzer, Ratmeyer et al. [74] found that particle size of dust samples varied greatly according to the instrument used. For example, the < 6 µm particle fraction in a Gran Canaria site was 68, 43, and 7% according to the Sedigraph, FRITSCH laser and CIS-GALAI particle analyzer, respectively. The < 2 µm showed even greater variability (49, 18 and 0% for Sedigraph, FRITSCH laser, CIS-GALAI particle analyzer, respectively). Not surprisingly, these authors rejected the < 6 µm particles size for the purpose of comparison with airborne dust, but they considered that the > 10 µm grades were sufficiently reliable for valid comparisons. Our work inclines us to agree with this decision because grains > 10 µm in diameter represent 80 % of the sizes in a sample [55].

Other authors [78] (in their study of the Namibian Slope) also give more weight to the coarse-grained component of their two cores as reflecting the airborne dust input to the ocean.

4.3. Mineralogy of Saharan Dust on the Canary Islands

4.3.1. Mineralogy of Airborne Dust

African dust collected on Tenerife and analyzed using XRD (X-ray diffraction) and SEM-EDAX (scanning electron microscope-electron dispersion and X-ray; [67]) was found to be made up of clay minerals and quartz (47% and 23%, respectively). The majority of the particles appear to be coated with sulphates, with an average thickness of 60 nm; aggregation of these particles was observed clearly by SEM-EDX (Figure 5 of [67]). Examination of the mineralogy and grain size of airborne samples finer than 0.5 μm from Izaña (Tenerife) showed that sulphates and carbonaceous material dominate aerosol composition. Above this grain size, the dominant composition is dominated by the silicates [76].

Saharan dust may act as cloud condensation nuclei (CCN). It is not well understood at the moment the hydroscopic properties of dust and if this is able for absorb significant amounts of water for acting as CCN. This will depend on the solubility of dust, which may be enhanced if this is coated by soluble compounds such as sulphate (S. Rodríguez, *personal communication*) [76].

4.3.2. Dust Mineralogy on the Islands

Using semiquantitative methods, the main components of the mineralogy of dust deposited on Gran Canaria during 2003-04, analyzed by XRD and SEM-EDX, includes quartz (42% ± 22%), carbonates (40% ± 13%), feldspars (11% ± 8%), magnetite (7% ± 8%) and small quantities of halite, sulphates and clay minerals (illite, kaolinite-chlorite, palygorskite; [55]). Comparison of this mineral composition with the composition in air collected on Tenerife Island (Figure 25) is only partially possible, because the grain size range is so different, so that the proportions differ. However, the minerals found in both fractions are similar (excepting magnetite) while, with respect to the mineral proportions, there is an increase in quartz and carbonates and a decrease in clay minerals in deposited dust compared to airborne particulates. The reason for this difference may arise from the smaller bulk particle size of the airborne dust samples in which the clays are concentrated. Magnetite is a heavy mineral (5200 kg m-3), its high density explaining why it is not detected in the airborne dust collected in Tenerife.

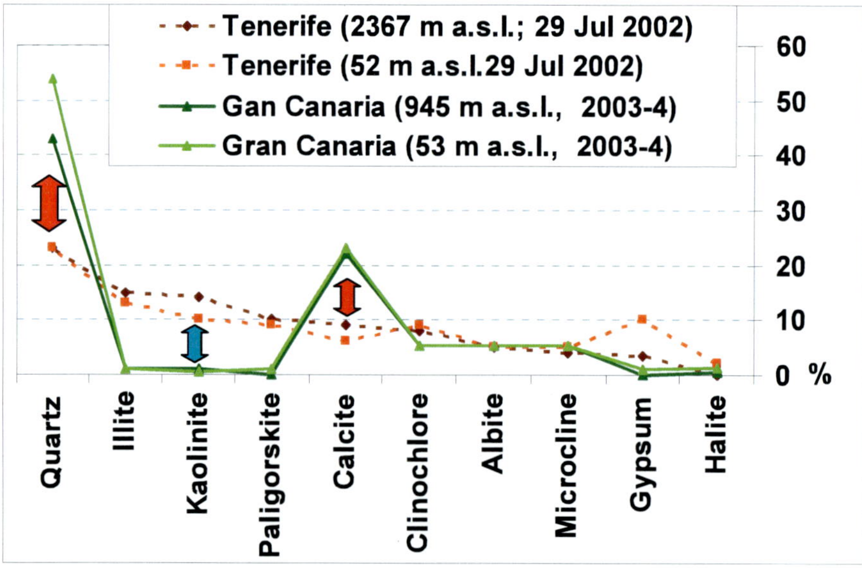

Figure 25. The mineral composition and abundance on the airborne dust samples collected in Tenerife and deposited dust on Gran Canaria Islands, at different altitudes.

The silica-subsaturated nature of potential protoliths in the Canary Islands [79], characteristic of an oceanic island volcanic setting, precludes source of the widespread quartz presence by paragenesis of the underlying rocks. The origin of the quartz is clearly allochthonous since the volcanic rocks in this archipelago rarely contain this primary mineral. Yet the overwhelmingly external origin of the airborne dust minerals does not prevent some recycling circulation in the islands. This can be affirmed from the fact that the concentration of quartz in airborne dust sampled on days with a Saharan dust pall is very similar to dust collected in clear days with no such dust pall, quartz concentration being only 10% higher under Saharan dust pall conditions, thus confirming both an external and a recycled input of Saharan dust [55].

There are eight major minerals in most mineral dusts and soils of arid areas: i.e. quartz, feldspar, calcite, gypsum, illite, kaolinite, smectite and hematite, for both the clay and silt fractions. On a global scale, there seems to be a reasonable relationship between the dust and soil mineralogies [80]. However, this is not always the case for soils from the Canary Islands where there is no local source for quartz.

4.3.3. Lithogenic Mineralogy Within the Ocean and on the Seafloor

Lithogenic accounts of particle fluxes in the ocean usually detail its terrigenous fraction [56, 81]. It mainly consists of quartz and clay particles reflecting the mineral composition and climatic conditions of the source. The highest frequency of airborne input in Canary Islands seems to correspond to glacial-interglacial transitions [56].

5. Potential Impact on Human Health of Saharan Dust Palls Over the Canary Islands *(Edward Derbyshire, Teresa Carrillo and Felipe Rodríguez-De Castro)*

5.1. Potential Pathology

Pathological effects associated with prolonged or chronic inhalation of mineral dust are subject to many factors. Particle size and composition are of particular importance. Inhaled mineral particles larger than 10µm (PM10) tend to lodge in the upper respiratory tract and are rejected by expectoration. However, the potential impact of coarser particulates may have been underestimated in the face of increasing legislative attention being concentrated upon the PM2.5 fraction; we need to know to what extent different mineral species are toxic wherever they may lodge in the respiratory system. The component of airborne dust finer than 10µm (the *respirable fraction*) may remain suspended in the atmosphere for weeks. These finer fractions and especially those finer than 4 µm (including the conventional PM2.5) are able to penetrate deeply into the narrower lung passages. In many parts of the world, case studies strongly suggest that rates of chronic respiratory disease increase with denser ambient dust loads and longer (decadal) periods of exposure [82].

Prolonged inhalation of the <4 µm mineral dust fraction eventually leads to deposition and accumulation in the pulmonary alveoli (where oxygen exchange takes place) in which chronic lung disease is initiated. Although the precise nature of the pulmonary response to mineral dust particles is complex, a number of factors are known to influence potential toxicity. These include the presence of specific minerals within the respirable fraction, the shape of particles (which influences clearance from the lungs), the presence or absence of mineral coatings, and particle surface characteristics including surface area and surface chemistry.

The numerous agents found within natural mineral dust palls have the potential to cause cell damage if absorbed by animal and human tissue. Important naturally occurring groups present as particles, encased within larger particles or as particle coatings, include the silicate minerals (notably quartz), asbestiform minerals, and metals and metalloids such as arsenic, cadmium, iron, led and mercury. The toxicity of minerals varies with their valences and species [83]. Examples include divalent iron (Fe^{2+}) as particle surface coatings and the quartz species cristobalite in volcanic situations (*e.g.* [84]). Some metals are known to cause lung cell damage because of a potential for oxidative stress (imbalance involving increased oxidants compared to antioxidants) and the generation of free radicals (groups of atoms with one unpaired electron - hence unstable and highly reactive), which interfere with cell function and cause DNA mutation [85], leading to pathology.

Taking a regional view, quartz (SiO_2: 'free silica') is the dominant mineral in natural dust derived from the extensive North Africa-Middle East drylands. The measured quartz content in major dust storms emanating from parts of North Africa (60.95%) is closely similar to the average quartz content of Earth's crustal rocks (58.98%), although the percentage varies with distance from source and dust pathway, as shown by the reported range of quartz content in dust palls over the Canary islands (>60% to <25%).

Free silica is a highly fibrogenic agent in the lungs; prolonged exposure to it can lead to the progressive and seriously disabling disease known as silicosis. Most of what is known about silicosis (including the sequence inflammation-low level fibrosis-massive fibrosis) has been drawn from *occupational* situations, however; non-occupational silicosis has received relatively little attention even in locations where large communities are exposed to high concentrations of ambient dust [82]. At the same time, some contaminants in quartz may reduce its toxicity, notably aluminium salts with which it commonly occurs. Association of silicosis with tuberculosis has been recognized for many years, notably in inhabitants of some dryland regions.

Some forms of biogenic silica (mainly occurring as diatoms, phytoliths and testate amoeba with diameters <10μm), many associated with palaeo-lake basins (Figure 26, modified of a world map of potential lake areas by Tegen [86]), are quite commonly seen in deposited mineral aerosols including Saharan dust over the Canaries (Figure 27). Typically, they contain a number of impurities, including alkali metals, alkaline earth oxides and Fe ions. Little research has been undertaken on this group but, because they lack surface radicals, their impact on lung tissue is considered to be benign [87], although a view exists that biogenic silica forms found in soils and aerosols may be a potential health hazard, albeit an unproven one (*e.g.* [88]).

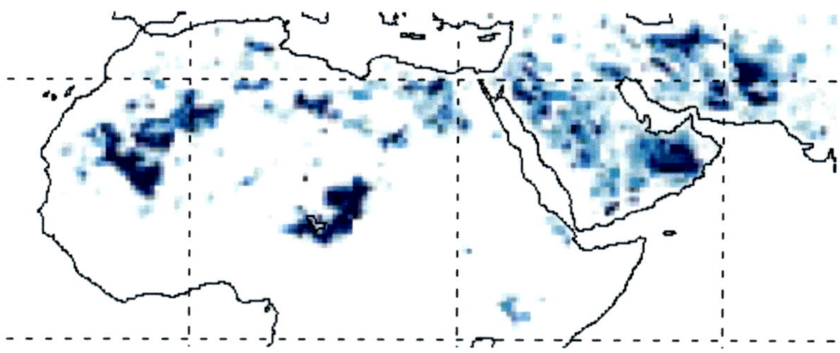

Figure 26. Areal coverage of preferential dust sources in the Greater Saharan Desert calculated from the extent of potential lake areas, excluding actual lakes but including palaeo-lake basins. (Fragment of a world map of potential lake areas by Tegen [86]).

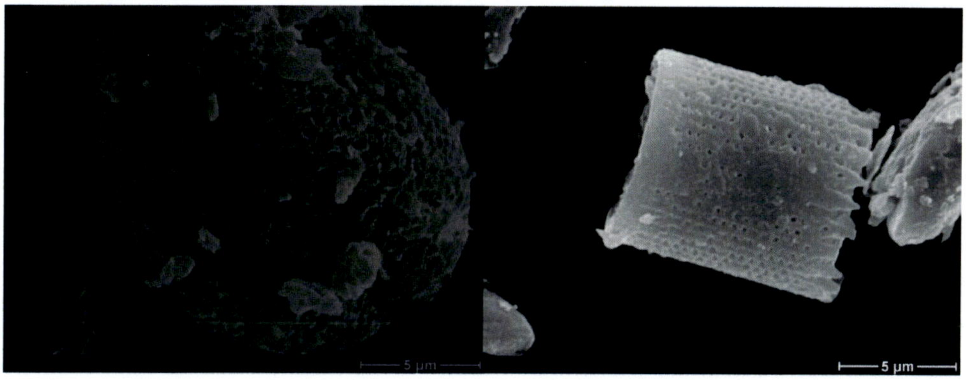

Figure 27. Diatoms typical of those deflated from desiccated Saharan-Sahelian lakes, *e.g. Aulacoseira granulata*, deposited on Gran Canaria in dense *calima* conditions on 4 March 2004.

The impact of airborne mineral dust on lung tissue may be further exacerbated by the presence within fine dust particles of bacteria, fungi, viruses and other micro-organisms (*e.g.* [89]). The long-distance transport of such organisms has the potential to damage plants and animals, including human subjects (*e.g.* [90, 91, 92]). While most micro-organisms are destroyed by exposure to ultra-violet radiation, a proportion will survive if lodged in cavities and cracks within suspended mineral dust particles. A comprehensive assessment of micro-organisms in dust storms and their health implications has recently been published by Griffin [93].

5.2. Saharan Dust over the Canaries: Recent Progress and Future Potential

The perceived importance of natural dust influxes such as those experienced in the Canary Islands and the Mediterranean Basin has led the European Commission to pass air quality legislation (Directive 1999/30/EC) that sets the standard for controlling emissions of particulates finer than 10 µm (PM10). This directive has been recently replaced by a new directive which establishes limit values for PM10 and PM2.5.(Directive 2008/50/EC). For this reason, PM10 and PM2.5 concentrations are already been monitored by the government of the Canary Islands (there as 3 stations at the moment, and about 4 new stations will be installed in the next year; S. Rodriguez, personal communication).

Al these actions has stimulated scientific work designed to measure the transportation and deposition of all types of atmospheric particles [89, 91, 93, 94, 95, 96, 97, 98, 99]. As already indicated, the vast majority of PM10 particles found in the Canaries are of Saharan origin. Their quantitative study and analysis is thus a priority in Canarian sampling and measurement stations such as the Center for Atmospheric Research Izaña from which initial results are now available [23]. An important outcome of the Directive concerns the measurement of the impact upon populations in many of the regions surrounding the Sahara, such as the Mediterranean Basin, where frequent incursions of mineral aerosols are of Saharan origin. Ways and means of measuring and quantifying the crustal material that reaches the archipelago are now emerging, in order to distinguish natural input from those arising from anthropogenic activities [100]. High quality meteorological data for the detection of imminent dust incursions is also a priority, there being several general prediction models such as SKIRON (http://forecast.uoa.gr/dustindx.html) of the University of Athens, MEIDEX (http://luna.tau.ac.il/~peter/MEIDEX/english.htm) of the University of Tel Aviv and DREAM (http://www.bsc.es/plantillaH.php?cat_id=519; Dust Regional Atmospheric Model) based in the Division of Earth Sciences in the supercomputing center in Barcelona. All provide real-time information and are available online.

When the respirable dust percentage is high during dust storm events in the Canaries, some 35% - 40% by volume may be finer than 10μm with about 20% being finer than 5μm ([101]; Figure 28). This may give rise to increased hospital admissions for cases of breathing disorders, including asthma, from early winter to early summer. Dust samples collected regularly during Saharan dust falls at several sites on the island of Gran Canaria, including a summit site (1930 m), contain measurable amounts of several elements implicated in lung function disorders and disease. Silicon is dominant in all cases (quartz > 60% [102]; Gelado et al. unpublished; Table 1).

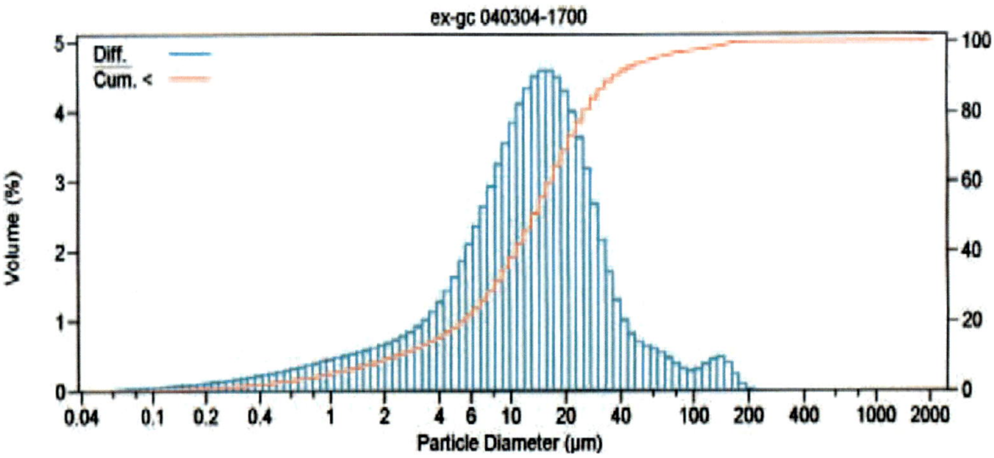

Figure 28. Modal and cumulative percentage volume curves for natural Saharan dust deposited on the island of Gran Canaria in the first week of March 2004. The PM10 fraction makes up almost 40%, and the PM2.5 almost 20% of the total volume. [101] (original analysis courtesy of Kenneth Pye).

Table 1. Gran Canarian Dust - Trace Aerosol Composition (at Taliarte site – near sea level). Potential links between selected elements in Saharan dust over Gran Canaria and some known lung function conditions and diseases. Key to colours. Red = cancer. Purple = cancer suspected. Brown = cancer and asthma. Turquoise = emphysema. Yellow = asthma. Other agents such as As (cancer) and V (asthma): no data for Gran Canaria. Geochemical data kindly provided by Dr. Maria Dolores Gelado, Universidad de Las Palmas de Gran Canaria.

	N	Minimum	Max	Median	Dep. typical
Mn (ng/m3)	63	2.12	1096.02	50.14	141.69
Fe (μg/m3)	101	0.9	57.00	1.98	5.85
Co (ng/m3)	101	0.16	23.00	1.13	2.46
Pb (ng/m3)	101	1.51	44.78	9.90	7.83
Cu (ng/m3)	101	1.76	63.93	11.64	12.90
Cd (ng/m3)	101	0.01	2.75	0.24	0.40
Mg (μ/m3)	100	0.03	429.39	17.56	59.94
Al (μ/m3)	100	0.30	148.60	4.40	15.06
Ca (μ/m3)	78	0.23	786.10	23.82	119.56
Na (μ/m3)	100	0.27	802.48	70.84	103.81
Cr (ng/m3) [but species critical]	41	1.20	154.38	18.72	29.62
Zn (μ/m3)	79	0.01	11.92	0.50	1.36
Ni (ng/m3)	38	0.32	31.72	3.38	5.71
Ti (μ/m3)	38	0.05	0.46	0.15	0.09

In general, the impact of natural wind-borne dust on human health in the form of chronic pathology reflects long-term exposure of subjects, commonly measured in decades. However, certain short-term high-density dust palls may give rise to a number of responses, especially in vulnerable subjects such as the young, the old and the infirm. Some results from the Canary Islands illustrate such short-term responses. For example, a recent unpublished analysis of emergency cases admitted to the Dr. Negrín Hospital in Gran Canaria provides a comparison between ailments recorded both outside of and within Saharan dust (calima) situations for the period 2005 to 2008. It appears that the admissions for tonsillitis, asthma, breathlessness, nasopharyngitis and pneumonia do not appear to correlate with the incidence of calima conditions, but that cases of bronchitis, sore throat and other non-specific respiratory problems do show a correlation (Table 2). In a study of admissions on the island of Tenerife, the control group was ten times larger and the total number of patient cases much greater than in the Gran Canaria case, such that the Tenerife results may eventually prove to be more indicative. However, further research will be necessary to explain these preliminary results.

Table 2. Incidence of hospital cases of selected clinical conditions in Gran Canaria in relation to *calima* events from January 2005 to March 2008 inclusive.

	Number of clinical cases in hospital emergency			
	Mean	sd	sum	n
	1	0	5	5
AMIGDALITIS ACUTE	1	0	6	5
	2	1	115	62
ASTHMA	2	1	77	42
	6	3	448	76
ACUTE BRONCHITIS	4	3	268	60
DISNEA ALTERATIONS IN	5	3	425	80
RESPIRATORY SYSTEM	5	4	316	61
	2	1	21	13
FARINGITIS ACUTE	1	0	8	7
NASOFARINGITIS ACUTE	1	1	15	11
(COMMON COLD)	1	0	8	7
PNEUMONIA, AGENT NOT	3	2	183	71
SPECIFY	3	2	164	55
OTHER RESPIRATORY and	4	2	281	74
Chronic obstructions	3	2	199	61

calima situation

The potential for major advances in the study of this branch of the multi-disciplinary science of medical geology is substantial. Given their location in relation to the greatest single source of geogenic dust on Earth, the Canary Islands are well placed to contribute to this rapidly emerging and societally relevant field.

There is much to be discovered using the aggregated skills and expertise of Earth, atmospheric, medical and social scientists. Notwithstanding the half century record and limited number of field and clinical case studies of silica-related diseases, for example, together with more recent laboratory experiments, understanding of the nature of the reactive mechanisms of environmental particulate silica at the molecular level remains incomplete. So far, none of the physicochemical features of silica used to explain its evident pathogenicity appear to be compatible with the available toxicity data [87]. Lung cell response to silica toxicity varies with dust sources, mineral species and even different surfaces of the same mineral specimen [103], suggesting an important degree of variability in 'the quartz hazard' [104].

6. Conclusion

The Canary Islands constitute a privileged observatory for investigation of aerosols. Its geographic situation on the western edge of the Sahara, its topography, with measuring stations sited at different altitudes and their scientific infrastructure, with three important research centers including two universities and the Izaña Atmospheric Research Centre, facilitates contributions leading to substantial advances in this scientific topic. This is so, especially in studies about the transport of lithogenic material, the effects on the population, and the properties and role of these aerosols in the energy balance of the planet and, therefore, in climatic change.

The work carried out to date gives rise to the following important conclusions: The Canary Islands lie along the northern edge of the habitual Saharan dust plume, so that the archipelago is only occasionally invaded by intrusions of lithogenic material, although massive arrivals are not unknown. Irregularity is one of the main characteristics of these intrusions. Most of these events take place in winter at low levels with summer events generally being higher. In both cases the most common synoptic situation is essentially anticyclonic. The main source of lithogenic material that reaches the islands is one sector of the Western Sahara nearest to the archipelago. Finally, it must be noted that a relation exists between a greater dust concentration on the Canary Islands and an intense ENSO in the Pacific, as well as high positive NAO indices.

Deposits derived from Saharan dust are ubiquitously found on the Canary archipelago, being more widespread on the eastern than on the western islands due to the old, smooth topography of the eastern islands. In some places, this dust forms stacked sequences which, in cases where the valley floor sediments have been securely dammed by volcanic action, the sediments can be used as valuable palaeoenvironmental archives indicative of local humidity conditions as well as for tracing the input of Saharan dust to the archipelago. Although numerous studies exist based on marine cores along the African coast that reveal different aspects of the palaeoclimatic (e.g. [105, 106, 107, 108]), long continuous terrestrial archives indicating changing humidity conditions are generally lacking in the NW African region. Thus, Saharan dust deposits on the Canary Islands are the only quasi-continuous palaeoenvironmental archives in a broad region stretching from southern Spain (pollen sequence of Padul; [109]) to tropical West Africa (lake Bosumtwi; [110]), partially filling a conspicuous gap in palaeoenvironmental knowledge. In contrast to monsoonal West Africa where warm periods were characterized by higher humidity and cold periods by arid conditions [111, 112], humidity on Lanzarote was generally enhanced during cold and lowered during warm periods. Our soil moisture record also contrasts with the palaeoclimatic record in the Mediterranean region. Whereas the Early and Mid-Holocene are described as humid in the Mediterranean [113], vega sediments from Lanzarote show continuous aridity during that time. However, the observed soil moisture pattern of Lanzarote clearly shows that it is dominantly controlled by northern high latitude processes controlled by glacial/interglacial cycles, rather than by a low latitude monsoonal influence triggered by the precessional cycle [114].

The buildup of a sediment mass balance in the completely closed Femés vega provides insights into the evolution of Saharan dust input to the Canary Islands region located, as it is, on the northern margin of the Saharan dust plume. Unlike regions crossed by the central and southern dust plume [6, 7], long-term quantitative changes of dust input have not previously been investigated in this northern area. Our investigations in the Canarian region confirm the general trend of increasing dust input as found in the central East Atlantic during the last 1.0 Ma [6, 7]. However, it appears that strong dust mobilization, probably triggered by human activity in the Sahara, had already begun by the Early Holocene, whereas an increase in dust mobilisation as shown by marine cores taken from off the Southern Sahara and the Sahel is recorded only during the Middle Holocene [112].

These results demonstrate that Saharan dust deposits on the Canary Islands constitute a valuable palaeoenvironmental archive that cast light upon several palaeoenvironmental problems yet to be thoroughly investigated in the NW African region.

Under current arid conditions prevailing for this interglacial epoch, with the contributions from human activity in the last centuries, over the recent decades, deposition rate of Saharan dust on the islands increased substantially to more than a dozen grams per square meter per year (14016 to 43800 mg m-2 y-1 [55]. For social and environmental reasons airborne dust on the Canary Islands is being investigated by several groups [15, 23, 24, 67, 68, 69], The mean value of 47 □ m-3, with 1000 micrograms per cubic meter during Saharan dust events was measured by the EU Air Quality Directive EC/30/1999. This obligated the Canary Island government agencies to set up a well regulated air quality network for the Canary Islands.

Other subsequent studies are the influence of Saharan dust on biogeochemical fluxes in the ocean surrounding the Canary Islands, and the increased marine fertilization with lithogenic fluxes during Saharan dust events [69, 71, 74].

Particle-size distribution of airborne dust in Canary Islands remained consistent, showing a unimodal distribution about the limit of silt-clay size (2 □m; [24, 76, 77]). However, the particle-size distribution of dust deposits on the Canary Islands display a bimodal distribution about the mean particle size of silt. Gravitational settling processes can explain such an increase of particle size and distribution of deposited Saharan dust. Alternatively, this can be explained by a process of aggregation, commonly seen in the dust deposits and aerosols [67].

Quartz is a diagnostic tracer mineral for Saharan dust, due to its absence in the mineral paragenesis of these volcanic islands. Quartz in local dust predominates due to it being recycled.

There must also be taken into account that Saharan dust may acts as a "carrier" providing surface where microorganims (pathogens: bacteria, fungi, etc.) may be transported from North Africa to distant regions. Some examples the studies performed by Griffin et al. [91, 94, 95] and Kellogg et al. [89, 98] in the Caribbean As it is shown in these articles, there is an open debate on if the impact of Saharan dust on human health is caused by "minerals" or by the "micro organism" transported by dust (S. Rodríguez personal communication).

Respecting to potential impact on human health of Saharan dust palls over the Canary Islands, the limited number of examples mentioned in this chapter, neither epidemiological nor clinical studies are widely available on many aspects of this important societally-centred subject. Understandably, such gaps in knowledge still give rise to some scepticism about whether or not chronic exposure to natural dusts generally leads to disease. To overcome this situation, an unprecedented level of collaboration among a range of conventional scientific disciplines will be necessary. The Canary Islands are well placed to share in this process.

Acknowledgments

We wish to express our gratitude to Silvia Alonso-Pérez who contributed to the section "Origin and Transport of Saharan Dust (Towards Canary Islands) and the Arrivals of Such Plume Dust to the Archipelago" with her doctoral thesis.

We thank to the Dr. Sergio Rodriguez (srodriguez@inm.es; Izaña Atmospheric Observatory, and University of Huelva, Spain), and Dr. Jan-Berend Stuut (jbstuut@ marum.de; University of Bremen) for their scientific review and contributions, that improved the quality of the manuscript.

References

[1] Dorta, P. Las invasiones de aire sahariano en Canarias. Consejería de Agricultura, Pesca y Alimentación del Gobierno de Canarias y Caja Rural de Tenerife Publ.: *Santa Cruz de Tenerife*, 1999, pp. 1-287.

[2] Dorta, P.; Gelado, M.D.; Hernández, J.; Cardona, P.; Collado, C.; Mendoza, S.; Rodríguez, M.J. Siruela, V.; Torres, ME. Frecuencia, estacionalidad y tendencias de las advecciones de aire sahariano en Canarias M. *Investigaciones Geográficas* 2005a, 38, 23-45.

[3] Viana, M.M. Niveles, composición y origen del material particulado atmosférico en los sectores Norte y Este de la Península Ibérica y Canarias. Ph.D. *Thesis. Universitat de Barcelona*, 2003.

[4] Dorta, P.; Gelado, M.D.; Criado, C.; Cardona, P.; Collado, C.; Hernández-Brito, J.J.; Mendoza, S.; Torres, M.E.; Rodríguez, M.J.; Siruela, V.F. Sahara air incursions and dust deposition over the Canary Islands and its impact on the terrestrial and oceanic environment. In Proceedings of the VIII Seminario *Internacional Forum Unesco*. El Patrimonio Mundial Natural: Los Espacios Naturales, Desarrollo, Sostenibilidad y Ética, 2005b, pp. 276-282.

[5] Alonso-Pérez, S. Caracterización de las intrusiones de polvo africano en Canarias. Ph.D. *Thesis. Universidad de La Laguna*, 2007.

[6] Goudie, A.S.; Middleton, N.J. Saharan dust storms: nature and consequences. *Earth-Sci Rev* 2001, 56, 179-204.

[7] Pflaumann, U.; Sarnthein, N.; Ficken, K.; Grothmann, A.; Winkler, A. Variations in eolian and carbonate sedimentation, sea-surface temperature, and productivity over the last 3 Ma at site 958 off Northwest Africa. In Proceedings of the Ocean Drilling Program, *Scientific Results; Firth*, J.V. Ed.: 1998, 159T, pp. 3-16.

[8] García, J.; Hernández, A.; Blasco, A.; Rodríguez, C.; Rancaño, E.; Núñez, S. Invasión de viento sahariano y su impacto en la asistencia sanitaria urgente. *Emergencias* 2001, 13, 372-376.

[9] Dorta, P. Las inversiones térmicas en Canarias. *Investigaciones Geográficas* 1996, 15, 109-124.

[10] Dorta, P.; Cardona, P.; Collado, C.; Criado, C.; Gelado, Mª.D.; Hernández, J.J.; Mendoza, S.; Siruela, V.; Torres, M.E. Algunas consideraciones sobre la importancia del polvo de origen sahariano en el clima del archipiélago canario y su aporte a las

aguas superficiales oceánicas: el episodio de abril de 2002. *In Proceedings of the III Congreso de la Asociación Española de Climatología: El agua y el clima; Asociación Española de Climatología: Palma de Mallorca, Spain,* 2002, pp.13-24.

[11] Scott, R.H. Note of a remarkable dust haze experienced at Tenerife, Canary Islands, february 1898. *Royal Meteorological Office* 1900, 33-36.

[12] Smith, W.C. Note on a fall of dust, "Blood rain" at Gran Canaria, 8th to 11th February 1920. In The Canary Island, their History, *Natural History and Scenery Appendix A at the Bannerman;* David, A. Ed.; London, 1922.

[13] [13] Font Tullot, I. Las invasiones de aire caliente africano en el Archipiélago Canario. *Revista de Geofísica* 1980, IX (36), 334-349.

[14] Bergametti, G.; Gomes, L.; Coude-Gaussen, G.; Rognon, P.; Le Coustumer, M.-N. African dust observed over Canary Islands: Source-regions identification and transport pattern for some summer situations. *J Geophys Res* 1989, 94 (D12), 14855-14864.

[15] Pérez-Marrero J.; Llinas O.; Maroto L.; Rueda M.J.; Cianca A. Saharan dust storms over the Canary Islands during winter 1998 as depicted from the advanced very high-resolution radiometer. *Deep-Sea Res Pt II* 2002, 49 (17), 3465-3479.

[16] Gelado, M.D.; Brito De Acevedo, E.M.; Hernández, J.J.; Dorta, P.; Díaz, J.P.; Somoza, MJ.; Collado, C.; Cardona, P.; Siruela, V.; Bergasa, O. Caracterización de los eventos de polvo africano en Gran Canaria (Islas Canarias). In Proceedings of the 4° Simpósio de Meteorologia e Geofísica da APMG/6° *Encontro luso-espanhol de Meteorologia, Lisboa*, 2005a, pp. 195-199.

[17] Gelado, M.D.; Rodríguez, M.J.; Hernández, J.J.; Brito De Acevedo, E.M.; Llinás, O.; Collado, C.; Cardona, P.; López, P. Dust deposition fluxes to north-eastern subtropical Atlantic (Canary Islands). *Geophys Res Abstr* 2005b, 7, 08399.

[18] Díaz, A.M.; García, O.E.; González, J.; Guirado, C.; Expósito, F.J.; Díaz, J.P.; Guerra, J.C. Estructura vertical de las invasiones de polvo mineral sobre las Islas Canarias durante 2004. *Revista de Teledetección* 2005, 24, 89-94.

[19] Díaz, A.M. Caracterización de los aerosoles atmosféricos y su influencia en los niveles de radiación UV en la región de Canarias. Ph.D. Thesis. Universidad de La Laguna, 2006.

[20] Díaz, A.M.; Díaz, J.P.; Expósito, F.J.; Hernández-Leal, P.A.; Savoie, D.; Querol, X. Air masses and aerosols chemical components in the free troposphere at the subtropical Northeast Atlantic region". *J Atmos Chem* 2006, 53 (1), 63-90.

[21] Kaufman, Y.J.; Koren, I.; Remer, L.A.; Tanre, D., Ginoux, P.; Fan, S. Dust transport and deposition observed from the Terra-Moderate Resolution Imaging Spectroradiometer (MODIS) spacecraft over the Atlantic Ocean. *J Geophys Res* 2005, 110, D10S12.

[22] Dorta, P. Et Al. Selección de tipos de tiempo en Canarias. Un ejemplo: las invasiones de aire sahariano. In Proceedings of the 1er Encuentro sobre Meteorología y Atmósfera de Canarias; Ministerio de Medio Ambiente: *Santa Cruz de Tenerife*, 2003, pp. 77-79.

[23] Alonso-Perez, S.; Cuevas, E.; Querol, X.; Viana, M.; Guerra, J.C. Impact of the Saharan dust outbreaks on the ambient levels of total suspended particles (TSP) in the marine boundary layer (MBL) of the Subtropical Eastern North Atlantic Ocean, *Atmos Environ* 2007, 41, 9468-9480.

[24] Torres-Padrón, M.E.; Gelado-Caballero, M.D., Collado-Sánchez, C.; Siruela-Matos, V. F.; Cardona-Castellano, P. J.; Hernández-Brito, J. J.; Variability of dust inputs to the CANIGO zone. *Deep-Sea Res Pt II* 2002, 49 (17), 3455-3464.

[25] Criado, C.; Dorta, P. An unusual "blood rain" over the Canary Islands (Spain). The storm of January 1999. *J Arid Environ* 2003, 55, 765-783.

[26] Middleton, N. Desert dust. In: Arid Zone Geomorphology Process. Form and Change in Drylands. David S.G. Thomas Eds.; *Whiley: Chichester*, 1997, pp. 413-436.

[27] Middleton, N.J.; Goudie, A.S. Saharan dust: sources and trajectories. *T I Brit Geogr* 2001, 26, 165-181.

[28] Arimoto, R. Eolian dust and climate: relationships to sources, tropospheric chemistry, transport and deposition. *Earth-Sci Rev* 2001, 54, 29-42.

[29] Prospero, J.M.; Ginoux, P.; Torres, O.; Nicholson, S.E.; Gill, T.E. Environmental Characterization of Global Sources of Atmospheric Soil Dust Identified with the Nimbus 7 Total Ozone Mapping Spectrometer (TOMS) Absorbing Aerosol Product. *Rev Geophys* 2002, 40,1.

[30] Gelado, M.D.; Dorta, P.; Torres, M.E.; Hernández, J.J.; Collado, C.; Siruela, V.F.; Cardona, P.; Rodríguez, M.J. Caracterización del aerosol sahariano en Gran Canaria". In Proceedings of the 1er Encuentro sobre Meteorología y Atmósfera de Canarias; Ministerio de Medio Ambiente: *Santa Cruz de Tenerife*, 2003, pp. 155-157.

[31] Amato, T.E.; Heidinger, A.K.; Knippertz, P. Analysis of winter dust activity off the coast of West Africa using a new 24-year over-water advanced very high resolution radiometer satellite dust climatology. *J Geophys Res* 2006, 111, D12210.

[32] Hernández, F.; Alonso-Pérez, S. Hernándes-Armas, J.; Cuevas, E. Influence of major African dust intrusions on the 137Cs and 40K activities in the lower atmosphere at the Island of Tenerife. *Atmos Environ* 2005, 39 (22), 4111-4118.

[33] Hernandez, F.; Rodríguez, S.; Karlsson, C.L.; Alonso-Pérez, S. López-Pérez, M.; Hernandez-Armas, J.; Cuevas, E. Origin of observed high 7Be and mineral dust concentrations in ambient air on the Island of Tenerife. *Atmos Environ* 2008, 42 (18), 4247–4256

[34] Mizota, C.; Matsuhisa, Y. Isotopic evidence for the eolian origin of quartz and mica in soils developed on volcanic materials in the Canary Archipelago. *Geoderma* 1995, 66, 313-320.

[35] Schwaighofer, B. Zur Verwitterung vulkanischer Gesteine - ein Beitrag zur Halloysit-Genese. Mitt. *Geol. Gesellsch. Wien*, 1973/74, 66/67, 221-236.

[36] Jahn, R. Böden Lanzarotes. PhD-thesis, *Stuttgart-Hohenheim*, 1988.

[37] Herrmann, L.; Jahn, R.; Stahr, K. Identification and quantification of dust additions in perisaharan soils. In The impact of desert dust across the Mediterranean; Guerzoni, S.; R. Chester Eds.; Kluwer Academic Publishers: Netherlands, 1996, pp. 133-151.

[38] Fúster, J.M.; Cendrero, A.; Gastesi, P., Ibarrola, E.; López-Ruiz, J. Geología y volcanología de las Islas Canarias, Lanzarote. Instituto "Lucas Mallada", *Consejo Superior de Investigaciones Científicas:* Madrid, 1968, pp. 1-239.

[39] Edwards, N.; Meco, J. Morphology and palaeoenvironment of brood cells of Quaternary ground-nesting solitary bees (Hymenoptera, Apidae) from Fuerteventura, Canary Islands, Spain. *P Geologist Assoc* 2000, 111, 173-183.

[40] Criado, C.; Guillou, H.; Hansen, A.; Hansen, C.; Lillo, P.; Torres, J.M.; Naranjo, A. 2004 Geomorphological evolution of parque natural de Las Dunas de Corralejo

(Fuerteventura, Canary Islands). In Contribuciones recientes sobre geomorfología. Actas de la VIII Reunión de Geomorfología, Toledo; Benito, G. & A. Diez-Herrero Eds.; *SEG & CSIC: Madrid.*, 2004, pp. 291-297.

[41] Coudé-Gaussen, G. Les poussières sahariennes. Edition *John Libbey Eurotext, Paris*, 1991; 485 pp.

[42] Grousset, F.E.; Rognon, P.; Coudé-Gaussen, G.; Pédernay, P. Origins of peri-Saharan dust deposits traced by their Nd and Sr isotopic composition. *Palaeogeogr Palaeocl* 1992, 93, 203-212.

[43] Zöller, L.; Suchodoletz, H. von; Küster, N. Geoarchaeological and chronometrical evidence of early human occupation on Lanzarote (Canary Islands). *Quaternary Sci Rev* 2003, 22, 1299-1307.

[44] Menéndez, I.; Cabrera, L.; Sánchez-Pérez, I.; Mangas, J.; Alonso, I. Characterization of two fluvio-lacustrine loessoids deposits in Gran Canaria Island, Canary Islands. *Quatern Int* 2008, in press, doi:10.1016/j.quaint.2008.05.011.

[45] Zöller, L.; Suchodoletz, H. von; Blanchard, H.; Faust, D.; Hambach, U. Reply to the comment by J.C. Carracedo et al., on 'Geoarchaeological and chronometrical evidence'. *Quaternary Sci Rev* 2004, 23, 2049-2052.

[46] Suchodoletz, H. von; Fuchs, M.; Zöller, L. Dating Saharan dust deposits at Lanzarote (Canary Islands) by luminescence dating techniques and their implication for paleoclimate reconstruction of NW Africa. *Geochem Geophy Geosy* 2008a, 9, doi:10.1029/2007GC001658.

[47] Rothe, P. Kanarische Inseln. Sammlung geologischer Führer. *Gebrüder Bornträger* Berlin, Stuttgart, 1996; 307 pp.

[48] Romero, C. El relieve de Lanzarote. Rubicón. *Cabildo de Lanzarote*, 2003; 242 pp.

[49] Suchodoletz, H. von; Faust, D.; Zöller, L. Geomorphological investigations of sediment traps on Lanzarote (Canary Islands) as a key for the interpretation of a palaeoclimate archive off NW Africa. *Quatern Int* 2008b (accepted). doi:10.1016/j.quaint. 2008.03.014.

[50] Langbein, W.B.; Schumm, S.A. Yield of sediment in relation to mean annual precipitation. *Transactions of the American Geophysical Union* 1958, 39 (6), 1076-1084.

[51] Suchodoletz, H. von; Kühn, P.; Hambach, U.; Dietze, M.; Zöller, L.; Faust, D. Loess-like and palaeosol sediments from Lanzarote (Canary Islands/Spain) – indicators of palaeoenvironmental change during the Late Quaternary. *Palaeogeogr Palaeocl* 2008c (resubmitted).

[52] Moreno, A.; Cacho, I.; Canals, M.; Prins, M.A.; Sanchez-Goni, M.-F.; Grimalt, J.O.; Weltje, G.J. Saharan dust transport and high latitude glacial climatic variability: the Alboran Sea record. *Quaternary Res* 2002, 58, 318-328.

[53] Moreno, A.; Canals, M. The role of dust in abrupt climate change: insights from offshore Northwest Africa and Alborean Sea sediment records. *Contributions to Science* 2004, 2 (4), 485-498.

[54] Holz, C.; Stuut, J.B.; Henrich, R. Variability in terrigenous sedimentation processes off NW Africa and its relation to climate changes: inferences from grain-size distributions of a Holocene marine sediment record. *Sedimentology* 2004, 51 (5), 1145-1154.

[55] Menéndez, I.; Díaz-Hernández, J.L.; Mangas, J.; Alonso, I.; Sánchez-Soto, P.J. Airborne dust accumulation and soil development in the North-East sector of Gran Canaria (Canary Islands, Spain). *J Arid Environ* 2007, 71, 57-81.
[56] Moreno, A.; Taragona, J.; Henderiks, J.; Canals, M.; Freudenthal, T.; Meggers, H. Orbital forcing of dust supply to the North Canary Basin over the last 250 kyr. *Quaternary Sci Rev* 2001, 20, 1327-1339.
[57] Maher, B.A. Magnetic properties of modern soils and Quaternary loessic paleosols: paleoclimatic implications. *Palaeogeogr Palaeocl* 1998, 137, 25-54.
[58] Matasova, G.; Petrovsky, E.; Jordanova, N.; Zykina, V.; Kapicka, A. Magnetic study of Late Pleistocene loess/palaeosol sections from Siberia: Palaeoenvironmental implications. *Geophys J Int* 2001, 147, 367-380.
[59] Fernandez-Caldas, E.; Quantin, P.; Tejedor-Salguero, M.L. Séquences climatiques des sol dérivés de roches volcaniques aux Iles Canaries. *Geoderma* 1981, 26, 47-62.
[60] Genise, J.F.; Edwards, N. Ichnotaxonomy, origin, and paleoenvironment of Quaternary insect cells from Fuerteventura, Canary Islands, Spain. *J Kansas Entomol Soc* 2003, 76(2), 320-327.
[61] Coello, J.; Cantagrel, J.-M.; Hernán, F.; Fúster, J.-M.; Ibarrola, E.; Ancochea, E.; Casquet, C.; Jamond, C.; de Téran, J.-R.D.; Cendrero, A. Evolution of the eastern volcanic ridge of the Canary Islands based on new K-Ar data. *J Volcanol Geoth Res* 1992, 53, 251-274.
[62] Instituto Tecnológico y Geominero de España. Mapa geológico de España, Escala 1:100.000, Memoria de la hoja geológica de la Isla de Lanzarote; 2005, *Madrid, ISBN*: 84-7840-606-9.
[63] Arimoto, R.; Duce, R. A.; Ray, B. J.; Ellis Jr., W. G.; Cullen, J. D.; Merrill, J. T. Trace elements in the atmosphere over the North Atlantic. *J Geophys Res* 1995, 100 (D1), 1199–1213.
[64] Prospero J.M.; Saharan dust transport over the North Atlantic Ocean and Mediterranean: An Overview. In The impact of desert dust from Northern Africa across the Mediterranean; S. Guerzoni and R. Chester Eds.; Kluwer *Academic Publishers: The Netherlands,* 1996, pp. 133-151.
[65] Prospero, J.M.; Barrett, K.; Church, T.; Dentener, F.; Duce, R.A.; Galloway, J.N.; Levy H.; Moody, J.; Quinn, P. Atmospheric deposition of nutrients to the North Atlantic Basin. *Biogeochemistry* 1996, 35, 27-73.
[66] Taylor, S.R.; McLennan, S.M. The continental crust: Its composition and evolution. *Blackwell Scientific Pub*., Palo Alto, CA, 1985; 328 pp.
[67] Alastuey, A.; Querol, X.; Castillo, S.; Escudero, M.; Avila, A.; Cuevas, E.; Torres, C.; Romero, P.M.; Expósito, F.; Garcia, O.; Diaz, J.P.; Dingenene, R.V.; Putaude, J.P. Characterisation of TSP and PM2.5 at Izaña and Sta. Cruz de Tenerife (Canary Islands, Spain) during a Saharan dust episode (July 2002). *Atmos Environ* 2005, 39, 4715-4728.
[68] Gelado, M. D.; Torres-Padrón, M. E.; Hernández-Brito, J. J.; Herrera-Melián J. A.; Pérez-Peña J. Aluminium distributions in Central East Atlantic waters (Canary Islands). *Mar Chem* 1996, 51 (4), 359-372
[69] Neuer, S.; Torres-Padrón, M. E.; Gelado-Caballero, M. D.; Rueda, M. J.; Hernández-Brito, J.; Davenport, R.; Wefer G. Dust deposition pulses to the eastern subtropical North Atlantic gyre: Does ocean's biogeochemistry respond? *Global Biogeochem Cy* 2004, 18, GB4020.

[70] Viana, M.M.; Querol, X.; Alastuey, A.; Cuevas, E.; Rodriguez, S. Influence of African dust on the levels of atmospheric particulates in the Canary Islands air quality network. *Atmos Environ* 2002, 36, 5861-5875.

[71] Neuer, S.; Ratmeyer, D.; Davenport, R.; Fischer, G.; Wefer, G. Deep water particle flux in the Canary Island region: seasonal trends in relation to long-term satellite derived pigment data and lateral sources. *Deep-Sea Res Pt I* 1997, 44, 1451-1466.

[72] Buat-Menard P.; Davies, J.; Remoudaki, E.; Miquel, J.; Bergametti, C. G.; Lambert, C. E.; Etzat, U.; Quetel, C.; Rosa, J. L.; Fowler, S. W. Non-steady-state biological removal of atmospheric particles from Mediterranean surface waters. *Nature* 1989, 340, 131-134.

[73] Hamm, C.E. Interactive Aggregation and Sedimentation of Diatoms and Clay-Sized Lithogenic Material. *Limnol Oceanogr* 2002, 47 (6), 1790-1795

[74] Ratmeyer V.; Fischer, G.; Wefer, G. Lithogenic particle fluxes and grain size distributions in the deep ocean off northwest Africa: Implications for seasonal changes of aeolian dust input and downward transport. *Deep-Sea Res Pt II* 1999, 46, 1289-1337.

[75] Henderiks, J.; Freudenthal, T.; Meggers, H.; Nave, S.; Abrantes, F.; Bollmann, J.; Thierstein, H. Glacial-interglacial variability of particle accumulation in the Canary Basin: a time-slice approach. *Deep-Sea Res Pt II,* 2002, 49, 3675-3705.

[76] Kandler K.; Benker, N.; Bundke, U.; Cuevas, E.; Ebert, M.; Knippertz, P.; Rodriguez, S.; Schütz, L.; Weinbruch, S. Chemical composition and complex refractive index of Saharan Mineral Dust at Izaña, Tenerife (Spain) derived by electron microscopy. *Atmos Environ* 2007, 41 (37), 8058-8074.

[77] Maring, H.; Savoie, D. L.; Izaguirre, M. A.; Custals, L.; Reid, J. S. Mineral dust aerosol size distribution change during atmospheric transport, *J Geophys Res* 2003, 108 (D19), 8592.

[78] Pichevin, L.; Cremer, M.; Giraudeau J.; Bertrand, P. A 190ky record of lithogenic grain-size on the Namibian slope: forging a tight link between past wind-strength and coastal upwelling dynamics. *Mar Geol* 2005, 218, 81-96

[79] [79] Carracedo, J.C.; Pérez Torrado, F.J.; Ancochea, E.; Meco, J.; Hernán, F.; Cubas, C.R.; Casillas, R.; Rodríguez Bardiola, E.; Ahijado, A. Cenozoic volcanism II: The The Canary Islands. In The geology of Spain; W. Gibbons and T. Moreno Eds.; The Geological Society: London, 2002; pp. 439-472.

[80] Claquin T., Schulz, M., Balkanski Y.J. Modelling the mineralogy of atmospheric dust sources. *J Geophys Res* 1999, 104 (D18), 22243-22256.

[81] Holz, C.; Stuut, J.B.; Henrich, R.;.Meggers, H. Variability in terrigenous sedimentation processes off northwest Africa and its relation to climate changes: Inferences from grain-size distributions of a Holocene marine sediment record. *Sediment Geol* 2007, 202, 499-508.

[82] Derbyshire, E. Natural aerosolic mineral dusts and human health: potential effects. In Essentials of Medical Geology; Selinus, O., Alloway, B., Centeno, J.A., Finkelman, R.B., Fuge, R., Lindh, U., Smedley, P. Eds. *Elsevier Academic Press: Amsterdam*; 2005, Chapter 18, 459-480

[83] Fubini B.; Fenoglio, I. Toxic potential of mineral dusts. *Elements* 2007, 3:407-414.

[84] Horwell, C.J.; Fenoglio, I.; Ragnarsdottir. K.V.; Sparks, R.S.J., Fubini. B. Surface reactivity of volcanic ash from the eruption of Soufrière Hills volcano, Montserrat, West indies with implications for health hazards. *Environ Res* 2003, 93, 202-215.

[85] Fubini, B.; Hubbard, A. Reactive oxygen species (ROS) and reactive nitrogen species (RNS) generation by silica in inflammation and fibrosis. *Free Radical Bio Med* 2003, 34, 1507-1516.

[86] Tegen, I. Modelling the mineral dust aerosol cycle in the climate system. *Quaterary Sci Rev* 2003, 22, 1821-1834.

[87] Fubini, B. Surface chemistry and quartz hazard. *Ann. Occup. Hyg.* 1998, 42, 521-530.

[88] Clarke, J. The occurrence and significance of biogenic opal in the regolith. *Earth-Sci Rev* 2003, 60, 175-194

[89] Kellogg, C.A., Griffin, D.W., Garrison, V.H., Peak, K.K., Royall, N., Smith, R.R., Shinn, E.A. Characterization of aerosolized bacteria and fungi from desert dust events, in Mali, West Africa. *Aerobiologia* 2004, 20, 99-110.

[90] Howitt, M.E.; Naibu, R.; Roach, T.C. The prevalence of childhood asthma and allergy in Barbados. *Am J Respir Crit* 1998, 157, A624.

[91] Griffin, D.W.; Kellogg, C.A.; Garrison, V.H.; Shinn, E.A. The global transport of dust. *Am Sci* 2001, 90: 228–235.

[92] Gyan, K.; Henry, W.; Lacaille, S.; Laloo, A.; Lamsee-Ebanks, C.; McKay, S.; Antoine, R.M.; Monteil, M.A. African dust clouds are associated with increased paediatric asthma accident and emergency admissions on the Caribbean island of Trinidad. *Earth Environ Sci* 2005, 49, 371-376.

[93] Griffin, D.W. Atmospheric movement of microorganisms in clouds of desert dust and implications for human health. *Clin Microbiol Rev* 2007, 20, 459-477.

[94] Griffin, D.W.; Kellogg, C.A.; Garrison, V.H.; Lisle, J.T., Borden, T. C.; Shinn, E.A. Atmospheric microbiology in the northern Caribbean during African dust events. *Aerobiologia* 2003, 19: 143-157.

[95] Griffin, D.W.; Kellogg, C.A. Dust Storms and Their Impact on Ocean and Human Health: Dust in Earth's Atmosphere. *EcoHealth* 2004, 1, 284-295.

[96] Querol X.; Alastuey, A.; Rodríguez, S.; Viana, M.M.; Artiñano, B.; Salvador, P.; Mantilla, E.; García do Santos, S.; Fernández Patier, R.; de La Rosa, J.; Sánchez de la Campa, A.; Menéndez, M. J.; Gil., J.I. Levels of particulate matter in rural, urban and industrial sites in Spain. *Sci Total Environ* 2004, 334-335, 359-376.

[97] Querol X.; Alastuey, A.; Viana, M.M.; Rodríguez, S.; Artiñano, B.; Salvador, P.; García do Santos, S.; Fernández Patier, R.; Ruiz, C.R.; de la Rosa, J.; Sánchez de la Campa, A.; Menéndez, M.; Gil, J.I. Speciation and origin of PM10 and PM2.5 in Spain. *J Aerosol Sci* 2004, 35, 1151-1172.

[98] Kellogg, C.A., Griffin, D.W. Aerobiology and the global transport of desert dust. *Trends Ecol Evol* 2006, 21 (11), 638-644.

[99] Rodríguez, S.; Querol, X.; Alastuey, A.; de la Rosa, J. Atmospheric particulate matter and air quality in the Mediterranean: a review. *Environ Chem Lett* 2007, 5, 1-7.

[100] Escudero, M.; Querol, X.; Pey, J.; Alastuey, A.; Ferreira, F.; Cuevas, E.; Alonso-Pérez, S. A methodology for the quantification of the net African dust load in air quality monitoring networks. *Atmos Environ* 2007, 41, 5516-5524.

[101] Derbyshire, E. Natural minerogenic dust and human health. *Ambio* 2007, 36, 74-78.

[102] Gelado. M.D.; Dorta. P.; Hernández. J.J.; Collado. C.; Rodríguez, M.J.; Cardona P.; Siruela, V. Characterisation of African dust outbreaks in Gran Canaria (Canary Islands). *Geophys Res Abstr* 2004, 6, 05227.

[103] Daniel, N.L.; Mao, Y.; Wang, C.-T.; Markey, C.J.; Markey, S.P.; Shi, X.; Saffiotti, U. DNA strand breakage, thymine glycol production, and hydroxyl radical generation induced by different samples of crystalline silica in vitro. *Environ Res* 1995, 71: 60-73.

[104] Donaldson, K.; Borm, P.J.A. The quartz hazard: a variable entity. *Ann occup Hyg* 1998, 42, 287-294.

[105] Matthewson, A.P.; Shimmield, G.B.; Kroon, D. A 300 kyr high-resolution aridity record of the North African continent. *Paleoceanography* 1995, 10/3, 677-692.

[106] Bozzano, G.; Kuhlmann, H.; Alonso, B. Storminess control over African dust input to the Moroccan Atlantic margin (NW Africa) at the time of maxima boreal summer insolation: a record of the last 220 kyr. *Palaeogeogr Palaeocl* 2002, 183, 155-168.

[107] Zhao, M.; Dupont, L.; Eglinton, G.; Teece, M. n-Alkane and pollen reconstruction of terrestrial climate and vegetation for N.W. Africa over the last 160 kyr. *Org Geochem* 2003, 34, 131-143.

[108] Kuhlmann, H.; Meggers, H.; Freudenthal, T.; Wefer, G. The transition of the monsoonal and the N Atlantic climate system off NW Africa during the Holocene. *Geophys Res Lett* 2004, 31, L22204.

[109] Pons, A.; Reille, M. The Holocene- and upper Pleistocene pollen record from Padul (Granada, Spain): *A new study. Palaeogeogr Palaeocl* 1988, 66, 243-263.

[110] Shanahan, T.M.; Overpeck, J.T.; Winston Wheeler, C.; Warren Beck, J.; Pigatie, J.S.; Talbot, M.R., Scholz, C.A.; Peck, J.; King, J.W. Paleoclimatic variations in West Africa from a record of late Pleistocene and Holocene lake level stands of Lake Bosumtwi, Ghana. *Palaeogeogr Palaeocl* 2006, 242, 287-302.

[111] Yan, Z.; Petit-Maire, N. The last 140 ka in the Afro-Asian arid/semiarid transitional zone. *Palaeogeogr Palaeocl* 1994, 110, 217-233.

[112] deMenocal, P.; Ortiz, J.; Guilderson, T.; Sarnthein, M. Coherent high- and low-latitude climate variability during the Holocene warm period. *Science* 2000, 288, 2198-2202.

[113] Verschuren, D.; Briffa, K.R.; Hoelzmann, P.; Barber, K.; Barker, P.; Scott, L.; Snowball, I.; Roberts, N.; Battarbee, R.W. Climate variability in Europe and Africa: A PAGES-PEP III time stream I synthesis. *In Past climate variability through Europe and Africa;* Battarbee, R.W.; Gasse, F.; C.E. Stickley Eds.; Springer: Dordrecht, 2004; chapter 19, pp. 567-582.

[114] deMenocal, P.B.; Ruddiman, W.F.; Pokras, E.M. Influences of high- and low-latitude processes on African terrestrial climate: Pleistocene eolian records from Equatorial Atlantic Ocean Drilling Program site 663. *Paleoceanography* 1993, 8(2), 209-242.

In: Airborne Particulates
Editor: Ming Cheng and Wen Liu

ISBN: 978-1-60692-907-0

Chapter 3

ANTIMONY IN AIRBORNE PARTICULATES: A REVIEW ON ENVIRONMENTAL MONITORING AND POTENTIAL SOURCES

Akihiro Iijima[1a,c], ***Keiichi Sato***[b] ***and Naoki Furuta***[c]

[a] Gunma Prefectural Institute of Public Health and Environmental Sciences, 378 Kamioki, Maebashi, Gunma 371-0052, Japan.

[b] Data Management Department, Acid Deposition and Oxidant Research Center, 1182 Sowa, Nishi-ku, Niigata 950-2144, Japan

[c] Faculty of Science and Engineering, Department of Applied Chemistry, Chuo University, 1-13-27 Kasuga, Bunkyo-ku, Tokyo 112-8551, Japan

ABSTRACT

Antimony (Sb) and its compounds have distinguished physicochemical properties. Hence, they have been widely used as functional materials; moreover, awareness regarding its negative impact on the environment and ecosystem has increased considerably. This chapter aims at providing an overview of the available studies on the recent enrichment of Sb in airborne particulates and its potential sources. The analyses of samples of environmental archives such as peat and ice cores have revealed marked changes in the concentrations of various atmospheric pollutants, probably due to the changes in human activities after the Second Industrial Revolution. In fact, historical studies have revealed that the temporal trend of the anthropogenic impact of Sb on the environment between the mid- and late 20th century was strongly dominated by two emission sources: coal combustion and metallurgical processes. Interestingly, another remarkable enrichment of Sb has been observed during the last several decades, implying the emergence of some new anthropogenic Sb sources. This emergence is most likely to be related with the recent expansion of Sb applications as functional materials, such as a flame retardant for plastics and chemical fablics and as a lubricant for automotive brake friction materials. In this chapter, first, the latest knowledge on the fates of Sb through

[1] Corresponding author: Akihiro Iijima, Gunma Prefectural Institute of Public Health and Environmental Sciences, 378 Kamioki, Maebashi, Gunma 371-0052, Japan.
Tel: +81-27-232-4881, Fax: +81-27-234-8438, E-mail: iijima-akihiro@pref.gunma.jp

mining, fabricating, usage, and emission to the atmosphere and the toxicities of various Sb compounds are briefly summarized with the aim of providing an in-depth understanding of this topic. Then, the temporal trends of Sb in the environmental samples are reviewed for the purpose of discussing the relationship between atmospheric Sb behaviors and human activities. Subsequently, current studies on airborne Sb, which are carried out in various regions of the world, are overviewed, and the potential sources of Sb are characterized. Moreover, abrasion dust from automotive brakes is highlighted to be an important Sb source that potentially affects highly-trafficked areas. Finally, we focus on the outlook on the demand for Sb compounds and subsequent changes in the Sb concentration in airborne particulates.

1. Introduction

Modern society has been supported by the consumption of various kinds of earth resources. Since trace elements are ubiquitously distributed in various raw materials such as fossil fuels and metal ores (Pacyna & Pacyna, 2001), a considerable amount of these elements have been built into the ecological cycle because of human activities. Several useful functions of these trace elements have been unearthed, and hence, their utility has consequently increased. Concurrently, however, the increased exposure to potentially toxic trace elements and their by-products has lead to a number of pollution-related diseases. Thus, the expansion of the applications of trace elements and the increased concern regarding their negative impacts on the ecosystem such as health disorder and environmental contamination are mutually correlated issues.

Antimony (Sb) is generally found in various sulfide minerals (Boyle & Jonasson, 1984), and its compounds have been used for medical and/or cosmetic purposes since ancient times (Léonard & Gerber, 1996; Emsley, 2001; Smichowski, 2008). Recently, the multifunctional utilities of Sb compounds have been discovered. Namely, its compounds are being widely used as flame retardants for plastics or chemical fablics, lead alloys for batteries, lubricants for friction materials; in addition, Sb compounds find wide ranging applications such as semiconductor devices, paints, pigments, opacifying and glazing agents, catalysts, and therapeutic drugs for tropical diseases or AIDS (Léonard & Gerber, 1996; IARC, 2001; Qi *et al.*, 2008). The possibilities for exposure to Sb compounds in our daily life have, thus, been gradually increased; hence, the awareness on the negative impact of Sb on the environment and ecosystem has also increased.

Currently, the toxicity of Sb compounds is emerging. Although, studies on humans have been inconclusive in deciding whether Sb exposure leads to cancer, studies on animals have shown evidence for Sb_2O_3 inhalation causing lung tumors in rats. On the basis of various research conducted on the toxicity of Sb compounds, USEPA (1979) and the Council of the European Communities (1976) have not classified Sb as a carcinogen; however, it is currently considered to be a priority pollutant. Hence, from the environmental viewpoint, Sb compounds are treated as potentially toxic substances.

However, till date, the environmental impacts, cycles, and fates of Sb have not sufficiently been highlighted as those of As, Cr, Hg, Pb, and Sn have been (Gómez, 2005). Moreover, various scientific disciplines have revealed that the environmental significance of Sb has been considerably underestimated (Shotyk *et al.*, 2004b). From these backgrounds, "First International Workshop on Antimony in the Environment," was held in Heidelberg,

Germany in 2005. This workshop reported that the diversity of the commercial and industrial uses of Sb, the numerous sources of Sb emission, its dynamic behavior in the environment, and the potential toxicity of different Sb species call for in-depth studies on the biogeochemical cycle of this remarkable element (Shotyk *et al.*, 2005a). Since developed countries have set the maximum permissible limit of Sb concentration in drinking water from the concerns for chronic toxicities, particular attention is being paid to the behavior of Sb in aquatic environment. Filella *et al.* (2002a, 2002b, 2007) reviewed more than 400 research papers on Sb in natural water. In the case of research on atmospheric environment, we can find lots of papers reporting the data of Sb as a part of results including the data of other trace elements. However, there are very few papers particularly focusing on the behavior of atmospheric Sb. Nonetheless, it is noteworthy that those few papers have commonly indicated substantial Sb enrichment in airborne particulates, most probably as a result of human activities (Dietl *et al.*, 1997; Shotyk *et al.*, 2004a; Shotyk *et al.*, 2005b; Cloy *et al.*, 2005; Furuta *et al.*, 2005; Gómez *et al.*, 2005; Krachler *et al.*, 2005). Therefore, Sb in airborne particulates is one of the most important topics in recent issues concerning anthropogenic air pollutants. This chapter is intended to provide an overview of the available studies on the recent enrichment of Sb in airborne particulates and its potential sources.

2. Antimony in the Environment: General Consideration

• Production and Usage of Antimony Compounds

Antimony and its compounds have been closely related to human activities since ancient civilizations. In ancient Egypt, antimony trisulfide (Sb_2S_3) was used in cosmetics, casting, and household articles. In the Middle Age, Sb was used for producing cast metal printing plates, light-reflecting materials, pigments, *etc.* (Li *et al.*, 1992). Until the 20th century, the supply and demand for Sb had remained small. At the beginning of the 20th century, the development of the lead storage battery, in which Sb was used as a hardening agent, resulted in a marked increase in Sb consumption. Figure 1 illustrates the historical trend of the Sb mining output in the 20th century (Butterman & Carlin, 2004). The mining output of Sb suddenly increased during World Wars I and II because of the increased demand for Sb-hardened lead alloy bullets and flame retardants used for tents and clothes (Schrader, 1922; Mote, 1948). Moreover, the rapid growth of the automobile industry at the beginning of the 20th century called for an enormous demand for high-performance lead storage batteries in which Sb-containing lead alloy plates were installed. Consequently, the world mining output of Sb continued to increase with the industrial growth in the 20th century, whereas the amounts of Sb produced from the mines increased markedly in the late 20th century. The rapid growth of the plastic industry has also resulted in a very high demand for Sb, which is used as a flame retardant in packaging materials, clothes, pigments, adhesive materials, *etc.* (Green, 2004). The historical trend of the mining output during the last couple of decades implies a substantial increase in the emission of Sb into the environment. Figure 1 also shows that China has been the largest supplier of Sb since the beginning of the 20th century. In the 1980s and the early 1990s, the mine output in China increased very rapidly, until by 1999, it accounted for more than 80% of the world's mine output of Sb.

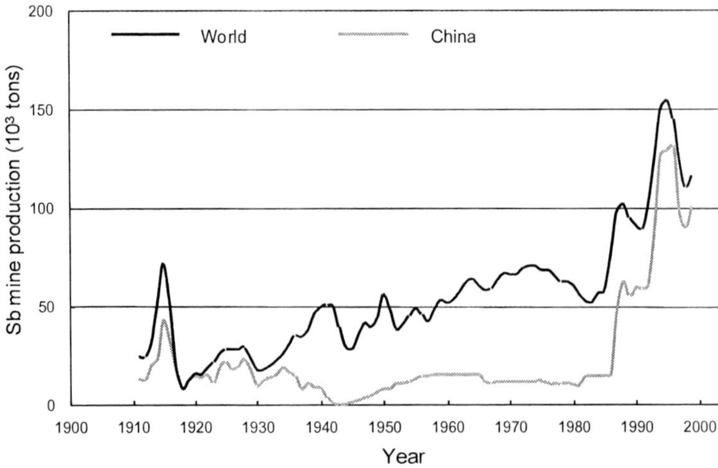

Figure 1.Antimony mine production from 1911 to 2000 (reproduced from Butterman & Carlin (2004)).

We next focus on the current status of the world mine production of Sb and the reserves and reserve base for Sb. The data are summarized in Table 1 (USGS, 2008). The world mine production of Sb is dominated by very few countries, and this trend has not been changed since the beginning of the 20th century. As mentioned above, China produces more than 80% of the world's output of Sb; however, Sb production has suffered a setback following a fatal accident in the mines in July 2001 (Carlin, 2002). Therefore, China's mine output of Sb remained constant between 2006 and 2007. However, at a global level, several new Sb mining projects are being developed. Bolivia has been a major supplier of Sb since the beginning of the 20th century; in the 1950s, South Africa and U.S.S.R. (including Russia and Tajikistan) emerged as important Sb suppliers. By 2007, the world Sb reserve and reserved base were estimated to be 2.1 and 4.3 megatons, respectively. The reserve-production ratio of Sb is approximately 16 years. Since the available mining output of Sb is expected to be limited, recycling of lead battery plates has also been attempted.

Table 1. World mine production, reserves, and reserve base of antimony (tons)

Country	Mine production		Reserves	Reserve base
	2006	2007		
United States	–	–	–	90,000
Bolivia	6,600	7,000	310,000	320,000
China	110,000	110,000	790,000	2,400,000
Guatemala	1,000	1,000	NA[b)]	NA[b)]
Russia[a)]	3,500	4,000	350,000	370,000
South Africa	6,000	6,000	44,000	200,000
Tajikistan	2,000	2,000	50,000	150,000
Thailand	940	1,500	420,000	450,000
Other countries	4,000	4,000	150,000	330,000
World total	134,040	135,500	2,114,000	4,310,000

Data were cited from USGS (2008). [a)] On the basis of recoverable amounts. [b)] Not assessed.

Table 2 shows the time trends of the consumption and supply of Sb in Japan from 2001 to 2005 (JOGMEC, 2007). A vast majority of Sb was imported from China and was produced as primary Sb in smelters, whereas some of the secondary Sb was recycled from lead battery plates. Antimony is industrially used in its elemental form (Sb) or in the form of compounds such as antimony trioxide (Sb_2O_3) and Sb_2S_3. A mixture of Sb_2O_3 and organic halogen compounds exhibits excellent flame retardancy (Touval, 1988). Principally, domestic use of Sb in Japan is dominated by flame retardant for plastics in electric appliances and chemical fablics. We also mention here the applications of Sb bundled in the others in Table 2(a). Antimony compounds find wide ranging applications such as semiconductor devices, a catalyst in the production of PET (polyethylene terephthalate) containers (Shotyk *et al.*, 2006); further, Sb_2S_3 is characteristically used in the manufacture of brake pads, where it functions as a lubricant (Jang & Kim, 2000).

Table 2. Domestic use and supply of antimony in Japan (tons)

(a) Use of antimony

Use / Year	2001	2002	2003	2004	2005
Buttery	203	316	348	184	114
Antimonial lead casting	99	87	88	80	47
Flame retardant	6573	7153	6168	6788	5720
Paint and pigment	141	154	138	230	210
Glass	73	48	26	35	26
Others	272	138	253	283	284
Total	7361	7896	7021	7600	6401

(b) Supply of antimony

Category / Year	2001	2002	2003	2004	2005
Import	7270	6799	7194	8092	7327
Recycling *	330	330	330	330	330
Total	7600	7129	7524	8422	7657

Data were cited from JOGMEC (2007). * Recycling data are estimated values.

• Worldwide Natural/Anthropogenic Emission Inventory of Sb to the Atmosphere

In general, the principal natural sources of trace metals in the atmosphere are presumed to be wind-borne soil particles, volcanoes, sea-salt sprays, and wild forest fires. Nriagu (1989) estimated that a total of 2,400 tons/yr of Sb could be emitted from natural emission sources. In this estimation, wind-borne soil particles, volcanoes, and sea-salt sprays accounted for 33% (780 tons/yr), 30% (710 tons/yr), and 23% (550 tons/yr) of the total natural emission inventory, respectively. However, the results reported in different papers are completely different from the results of Nriagu's estimation. Hinkley *et al.* (1999) reported that volcanoes release approximately 5 tons/yr of Sb to the atmosphere. Likewise, Shotyk *et al.* (2004b) estimated that volcanic emissions account for only 3–5% of the global Sb emission. Moreover, Shotyk *et al.* (2004a) obtained the background depositional flux of Sb from ancient core sample (pre-anthropogenic period (8020–5320 BP)) collected from Swiss peat bog and determined the global flux of natural Sb to be 154 tons/yr. In another approach, Patterson & Settle (1987) estimated the global flux of natural Sb to be 90 tons/yr from the background Pb/Sb ratio in pre-anthropogenic Swiss peat bog and natural depositional flux of Pb. These values are considerably lower than those reported by Nriagu (1989) in his current estimation. On the basis of the ancient evidence obtained from peat bog samples, Shotyk *et al.* (2004b, 2005b) alleged that the natural flux of Sb to the global atmosphere might have been overestimated by a factor of ten.

On the other hand, Pacyna & Pacyna (2001) estimated the anthropogenic emission of Sb, Cr, Hg, Mn, Se, Sn, and Tl to the atmosphere from individual anthropogenic emission sources (see Figure 2). The total Sb emission was estimated to be 1,561 tons/yr. It is well known that Sb and Pb have common emission sources such as coal combustion and mining and/or smelting of Pb ores (Nriagu, 1983; Qi *et al.*, 2008). In fact, stationary fossil fuel combustion (mainly coal combustion) and non-ferrous metal production are the dominant sources of Pb and Sb emission. Krachler *et al.* (2008) estimated the depositional fluxes of Sb from Arctic ice core and snow pit samples. From the marked divergence between the natural background flux observed in ancient (pre-anthropogenic period) ice core and the current depositional flux observed in snow pit sample, Krachler *et al.* concluded that approximately 99.8% of the Sb deposited in the Arctic today originates from anthropogenic activities.

In a nutshell, it is more proper to state that currently, Sb emission from anthropogenic activities greatly exceeds that from natural sources.

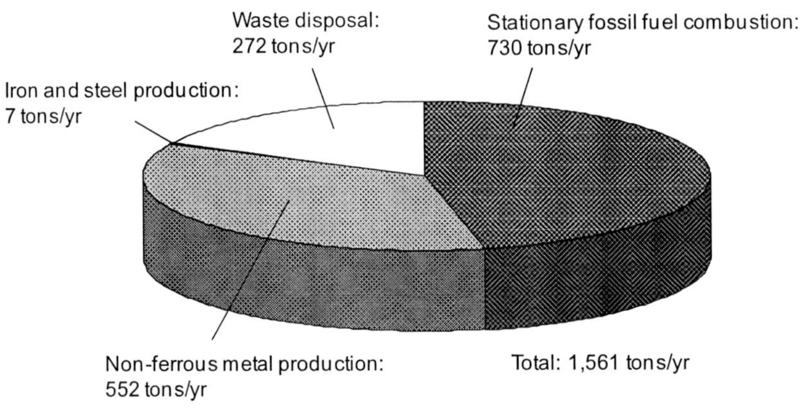

Figure 2.Worldwide emissions of Sb from major anthropogenic source categories to atmosphere in the mid-1990s. Data were cited from Pacyna & Pacyna (2001).

• Toxicity of Antimony Compounds

Antimony and arsenic are congeneric elements and thus have common physicochemical properties. Both inorganic Sb and As exhibit generally trivalent (Sb(III) and As(III)), as well as pentavalent (Sb(V) and As(V)) oxidation states, and their toxicities are basically dependent on the oxidation state. Gebel (1997) reviewed the relationship between the oxidation states of these two elements and their biological effects. After application of inorganic Sb(III) such as $SbCl_3$ or Sb_2O_3, the species appearing which is biologically active is most likely to be $Sb(OH)_3$. Thus, inorganic Sb(III) in biological media was found to be present as an uncharged species. On the other hand, inorganic Sb(V) is present in the form of $Sb(OH)_6^-$ complex in biological media. On the basis of inhalation experiments on hamsters, Felicetti *et al.* (1974) showed that the uncharged Sb(III) has a greater affinity for erythrocytes than the negatively charged Sb(V). Moreover, Gebel (1998) found that both Sb(III) and As(III) are accumulated in mammalian V79 cells. Prior to this report, Gebel (1997) had mentioned that ionic charge has an important effect on the biological activity of Sb compounds.

The genotoxicity of Sb compounds has been extensively studied, and reviewed by Boeck *et al.* (2003). Both Sb(III) and Sb(V) compounds tested negative results in non-mammalian genotoxicity tests (Kuroda *et al.*, 1991; Lantzsch & Gebel, 1997; Elliott *et al.*, 1998; Knasmuller *et al.*, 1998), while mammalian tests usually gave positive results for Sb(III) and negative results for Sb(V) (Kuroda *et al.*, 1991; Gebel *et al.*, 1997). However, the mechanisms underlying the genotoxicity caused by Sb(III) compounds have not been clearly elucidated. The generation of active oxygen species caused by external stress *via* the exposure to the Sb compounds and the subsequent interference of DNA repair systems were identified as the possible causes for the expression of genotoxicity (Gebel *et al.*, 1997; Cavallo *et al.*, 2002; Boeck *et al.*, 2003). Based on the evidence that confirms the genotoxicity of Sb(III) compounds, therapeutic drugs containing Sb(III) compounds for leishmaniasis have been replaced with those containing the less toxic Sb(V) compounds (Boeck *et al.*, 2003).

The currently available data are regarded as inconclusive regarding the carcinogenicity of Sb compounds (McCallum, 2005). Watt (1983) reported that Sb_2O_3 exhibits fibrogenic and oncogenic activity in the lungs of the female Fisher 344 rats (only females were tested) when administered by inhalation for 1 year. Groth *et al.* (1986) also reported the occurrence of neoplasms in female Wister-derived rats exposed to Sb_2O_3 or Sb_2S_3. However, Boeck *et al.* (2003) criticized the lack of representativeness in these two studies. Since Watt studied only one sex group, and Groth *et al.* examined only one mode of administration, it is difficult to apply these two studies for general risk assessment (Boeck *et al.* 2003). On the other hand, Newton *et al.* (1994) studied the subchronic and chronic inhalation toxicity of Sb_2O_3 in rats and reported that microscopic changes in the lungs were limited to subacute-chronic interstitial inflammation, increased numbers of macrophages, granulomatous inflammation, and fibrosis. However, because no neoplasms were found, they concluded that Sb_2O_3 was not a carcinogen. To analyze the carcinogenicity of Sb to humans, Boeck *et al.* (2003) and McCallum (2005) reviewed various epidemiological studies on the occupational exposure to Sb compounds. Epidemiological studies on individuals who were occupationally exposed to Sb compounds have demonstrated increased risks of death caused by various types of cancer, cardiovascular diseases, and cerebrovascular diseases compared to the control groups (Wingren & Axelson, 1993; Gerhardsson & Nordberg, 1993; Jones, 1994; Schnorr *et al.*, 1995). However, the individuals under study would be concurrently exposed to other carcinogenic substances such as As compounds or polycyclic aromatic hydrocarbons in their occupational environment. Therefore, the increased risk of cancer is not directly correlated with the exposure to Sb compounds alone. On the basis of inadequate evidence for the carcinogenicity of Sb_2O_3 and Sb_2S_3 in humans, sufficient evidence for the carcinogenicity of Sb_2O_3 in experimental animals, and the limited evidence available for the carcinogenicity of Sb_2S_3 in experimental animals, Sb_2O_3 was considered to be potentially carcinogenic to humans (Group 2B); however, Sb_2S_3 could not be concluded to be carcinogenic to humans (Group 3) (IARC, 2001).

3. Historical Records of Antimony in the Environment

• Temporal Trend of Anthropogenic Impact over the Past Centuries

As described in the previous section, the historical records of Sb concentration in environmental samples would be associated with the changes in human activities. A better understanding of the historical trend, therefore, will help understand the impact of human activities on the environment.

Peat core from an ombrotrophic bog is one of the excellent archives that can be used to record changes in air quality. Shotyk *et al.* (2004a) showed the vertical profiles of Sb and Pb concentrations over the past 14,500 years in the ombrotrophic peat bog core samples collected from Switzerland. As mentioned previously, the common emission sources of Sb and Pb are mining and/or smelting of Pb ores and coal combustion. In fact, the vertical profiles of Sb concentrations were found to be parallel to those of the Pb concentrations in peat core samples. Interestingly, the maximum increase in the concentrations of both Sb and Pb was observed since the late 19th century. Moreover, Sc-normalized enrichment factors for Sb and Pb also concurrently increased up to approximately 80 and 300, respectively, indicating that these elements were clearly emitted by anthropogenic sources. These distinctive chronologies were also found in the other peat bog core samples collected in northwest Scotland, Shetland Islands (Shotyk *et al.*, 2004a), and Faroe Islands (Shotyk *et al.*, 2005b). From these observations, Shotyk *et al.* concluded that the Second Industrial Revolution (1850–1914) marked the onset of the most remarkable changes in Sb concentrations in airborne particulates in Europe (Shotyk *et al.*, 2005b).

Cloy *et al.* (2005) reported the vertical profiles of the atmospheric deposition of Sb and Pb over the past 2,500 years in the ombrotrophic peat bog core samples collected from Flanders Moss, Scotland. To distinguish between anthropogenic inputs of heavy metals and natural inputs *via* wind-borne soil dust, Cloy *et al.* traced the Sc-normalized trends for Sb and Pb. In ancient times, the trends for Sb/Sc and Pb/Sc were influenced by metallurgical activities during the Mediaeval and Roman/pre-Roman periods. Similar trends were also found in other peat core samples collected in Switzerland (Shotyk *et al.*, 2004a) and Faroe Islands (Shotyk *et al.*, 2005b). Recently, the trends of atmospheric depositional fluxes of anthropogenic Sb and Pb peaked during the period 1920–1960 owing to an increase in coal consumption and military procurement in World War II. Cloy *et al.* (2005) also measured the $^{206}Pb/^{207}Pb$ ratio in order to identify the contribution of leaded gasoline, whose lead isotope ratio is known to be lower than that in coal. During the 1920s to 1960s, the $^{206}Pb/^{207}Pb$ ratio showed an obvious decline, indicating that leaded gasoline also affects the anthropogenic Pb deposition. From the late 1960s to the early 1980s, Sb/Sc decreased owing to a decrease in coal combustion. During this period, Pb/Sc also appeared to be influenced both by the decrease in coal combustion and increase in exhaust emissions from vehicles running on leaded gasoline. It is interesting to note that after the 1980s, the Sb/Sc ratio apparently increased again, while the Pb/Sc ratio started to level off or decline. Cloy *et al.* inferred that the most recently obvious enrichment of Sb is probably caused by the emergence of some new anthropogenic Sb sources such as wear of automotive brake pads and incineration of wastes containing plastic products. In fact, world mine production of Sb showed drastic increase from mid-1980s (see Figure 1).

On the other hand, ice core samples are also treated as one of the excellent archives that can be used for recording changes in air quality. Krackler *et al.* (2008) reported the vertical profiles of Sb and Sc over the past 16,000 years in the ice core sample collected from Devon Island, Arctic Canada. Maximum concentrations of Sb and Sc were observed during the Younger Dryas cold climate event centered around 12,350 years BP. This was probably caused by the climatological characteristics in this period, such as sparse vegetation cover, strong winds that favored the transport and subsequent deposition of a greater number of larger dust particles into the Arctic region. In fact, the Sb/Sc ratio was almost stable during this period, indicating that Sb deposition during this period primarily originated from lithogenic sources. The Sb/Sc ratio remained remarkably constant until approximately 1,000 years ago, and subsequently, the Sb concentration started increasing continuously and significantly.

The latest 160-years chronologies of Sb and Pb deposition on the Arctic ice are discussed in detail in another study published by Krackler *et al.* (2005). The vertical profiles of Sb and Pb concentrations in the ice core sample continued to be flat till 1900. Drastic increases in the Sb and Pb concentrations were found after the 20th century, probably reflecting the growing importance of anthropogenic activities such as coal combustion, and mining and smelting of Pb and Cu ores. The Sb trend peaked in the late 1950s owing to the economic boom after the World War II, and then gradually declined, possibly because of the recent introduction of flue-gas filter technologies and emission reduction efforts. Meanwhile, since 1930, the temporal trend of Pb concentration has shown a drastic increase, peaking in the 1960s, owing to the increased use of leaded gasoline. Subsequently, because of the decreased use of lead in gasoline since the 1970s, the Pb concentrations reduced by 80%. To identify the anthropogenic impacts, Krachler *et al.* also showed the Sc-normalized trends for Sb and Pb in an ice core sample. Interestingly, the Sb/Sc trend showed a 50% increase during the last two to three decades, whereas the Pb/Sc trend continued to decline. This recent another enrichment of Sb was very similar to that shown in the peat core collected in Scotland (Cloy *et al.*, 2005), and also consistent with the increase of world mine production of Sb (see Figure 1). Krachler *et al.* inferred that these phenomena are likely to be associated with the increased global consumption of Sb compounds (as flame retardant for plastic products, lubricant for automotive brake pads, *etc.*) and the gradual elimination of gasoline lead additives. On the basis of the above-mentioned Arctic observations, Krachler *et al.* stated that Sb burdens to the environment more than Pb because the Sb is more enriched in Arctic atmosphere than Pb, which is assumed as the same toxicity as Pb.

• Temporal Trend of Antimony in Airborne Particulates during Recent Several Decades

Only a few research has focused on the long-term trend of Sb in airborne particulates among the current air monitoring research. In this section, two in-depth works which were carried out in Tokyo Metropolitan Area, Japan are shown. The National Air Surveillance Network (NASN) project established by the Japanese Government is one of the most prominent studies focused on the long-term trend of airborne elements in the late 20th century (Japan Environmental Sanitation Center, 1975–1998). In this project, during 1974 to 1997, PM_{10} samples were collected on a monthly basis at 16 Japanese monitoring sites, which were located in different climatic and geographical zones, and 31 elements (Na, Al, Cl, K, Ca, Sc, Ti, V, Cr, Mn, Fe, Co, Ni, Cu, Zn, As, Se, Br, Ag, Cd, Sb, Cs, Ba, La, Ce, Sm, Lu, Hf, W, Pb, and Th) were analyzed. From the NASN project, Var *et al.* (2000) reported the temporal and seasonal profiles of 6 crustal elements (Al, Ca, Sc, Ti, Mn, and Fe), 7 anthropogenic elements (V, Cr, Ni, Cu, Zn, As, and Pb), and marine element (Na). Since the temporal trend of airborne Sb concentration was not documented in the paper published by Var *et al.*, we have depicted the trend observed in Tokyo, Japan, in Figure 3(a), on the basis of the NASN data report (Japan Environmental Sanitation Center, 1975–1998). Additionally, the temporal trend of airborne Pb concentration in Tokyo is also shown in Figure 3(b) for comparison with the chronology of Sb, in the same idea as the above reviewed literatures (Shotyk *et al.*, 2004a; Shotyk *et al.*, 2005b; Cloy *et al.*, 2005; Krachler *et al.*, 2005; Krachler *et al.*, 2008). The concentrations of Sb and Pb generally decreased in summer and increased in winter, probably due to the meteorological conditions. Typically, the climate of Tokyo is characterized as high precipitation and low atmospheric stability in summer, and low precipitation and high atmospheric stability in winter. The drastic decrease in the airborne Pb concentration was observed from the late 1970s to the early 1980s (see Figure 3(b)). This trend is consistent with the Pb chronology provided by the peat or the ice core samples (Cloy *et al.*, 2005; Krachler *et al.*, 2005). This was probably due to the replacement of coal with petroleum or electricity and the discontinuation of the use of leaded gasoline. In Japan, the use of leaded regular gasoline was discontinued from 1975. However, no drastic decrease could be observed in the Pb concentration after the mid-1980s. On the other hand, the concentration of Sb in airborne particulates appeared to increase slightly from the mid-1980s (see Figure 3(a)). Figures 3(c) and (d) show the trends of the Al-normalized enrichment factors for Sb and Pb in PM_{10}. The enrichment factors were calculated by using the elemental composition of the Earth's crust as a reference sample (Wedepohl, 1995). It should be noticed that the absolute values of enrichment factors depend on the reference sample composition. In fact, Krachler *et al.* (2005) demonstrated that the enrichment factor varies an order of magnitude among different reference samples. We hereby focus on temporal trends of the enrichment factors. As shown in figures 3(c) and (d), the enrichment factor for Sb tended to slightly increase after the end of the 20th century, whereas that for Pb showed a drastic decrease in the late 1970s. These disparities in the late 20th century in the trends between the Sb and Pb behaviors in Tokyo Metropolitan Area were also consistent with those observed in the peat and ice core samples reported by Cloy *et al.* (2005) and Krachler *et al.* (2005). The trend of substance flow of Sb compounds in Japan estimated by Tsunemi & Wada (2008) indicates that the domestic supplies of Sb compounds significantly increased from the early 1980s to 2000 (5,000 tons/yr in 1980 to 14,000 tons/yr in 2000) owing to the increased use of Sb as a flame retardant for plastics in electric appliances, IT and OA devices, *etc.* Since the household wastes containing these plastic products were partly incinerated in Japan, Sb emission from incineration plants might have also increased accordingly. The trend of airborne Sb in the late 20th century is surely one of the most interesting topics in recent air pollution issues.

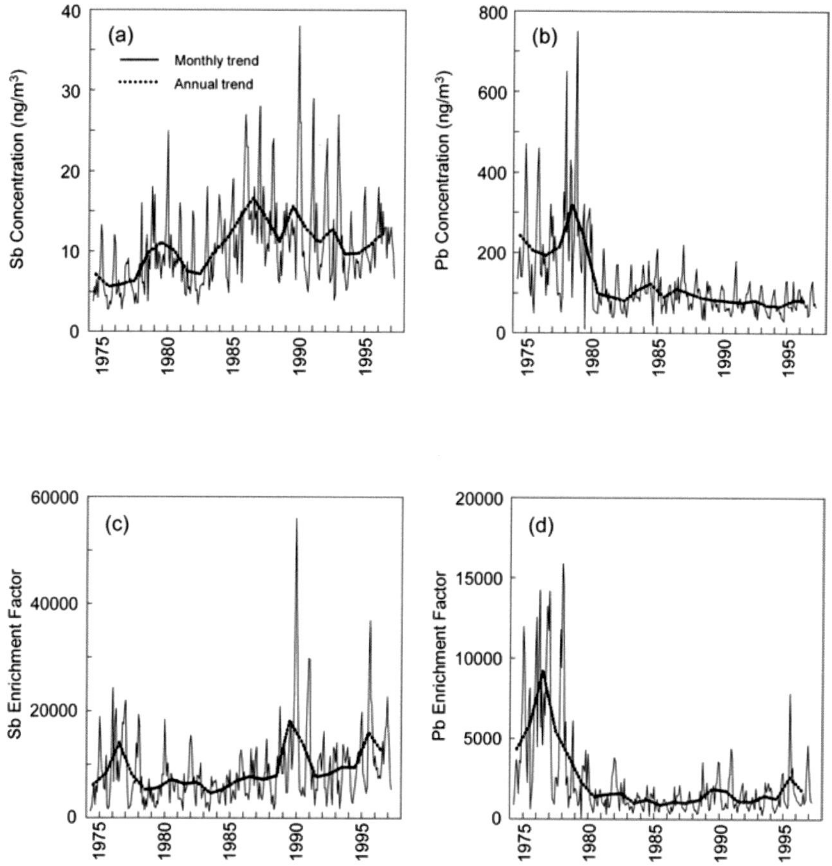

Figure 3.Long-term trends of elemental concentrations for (a) Sb and (b) Pb, and Al-normalized enrichment factors for (c) Sb and (d) Pb in PM_{10} collected in Tokyo, Japan from 1975 to 1997. Field observation was carried out under the project named National Air Surveillance Network of Japan (NASN). Analytical data were cited from NASN data reports published from Japan Environmental Sanitation Center (1976–1998).

Furuta and co-workers have continued the long-term monitoring of airborne particulates in central Tokyo, Japan. This work is one of the latest studies focused on the long-term trend of airborne elements. Airborne particulates of various sizes (d < 2 μm, 2–11 μm, > 11 μm) have been collected on a monthly basis since 1995, and 23 elements (Li, Be, Na, Mg, Al, K, Ca, Ti, V, Cr, Mn, Fe, Co, Ni, Cu, Zn, As, Se, Mo, Cd, Sb, Ba, and Pb) in the particulates have been analyzed (Furuta *et al.*, 2005). Figure 4(a) shows the trend of airborne Sb concentrations for different particle size fractions collected from May 1995 to December 2007. For comparison with the results reported in previously reviewed papers, the trends of airborne Pb concentrations for different particle size fractions are also depicted in Figure 4(b). As evident from Figures 4(a) and (b), intense seasonal variations (increase during winter and decrease during summer) were observed in fine (d < 2 μm) fractions. The temporal trends of airborne Sb and Pb concentrations were fairly similar to each other. Interestingly, both Sb and Pb concentrations showed a sharp decrease from 2000, and remained almost with slight seasonal variations from 2003. According to the estimation by Tsunemi & Wada (2008), the trend of the domestic supplies of Sb compounds in Japan peaked from 1995 to 2000 (14,000 tons/yr), and subsequently showed a gradual decrease. Additionally, since January 2000, the

Law Concerning Special Measures against Dioxins for controlling the emission of dioxins from incinerators was enforced in Japan. Following this regulation, the efficient filtration technologies such as the use of bag filters were promptly developed, and by the end of 2002, these had become popular. As a result, the emission of air pollutants from waste incinerators was significantly reduced, thus possibly causing a sharp decrease in airborne Sb concentrations in the fine fractions from 2000. To identify the contribution of anthropogenic emission sources, Furuta *et al.* (2005) obtained the Al-normalized enrichment factors by using the elemental composition of the Earth's crust as a reference sample (Wedepohl, 1995). Figures 4(c) and (d) show the trends of the enrichment factors for Sb and Pb in the different particle size fractions. The temporal trend of the Sb enrichment factor in the fine fraction peaked in 1999 (annual average: approximately 25,000), and then gradually declined to approximately 4,000. Despite the obvious decrease in the airborne Sb concentration, it is noteworthy that Sb is now still highly enriched in the atmosphere than in the Earth's crust. Moreover, from morphological observations using a scanning electron microscope equipped with an energy-dispersive X-ray analyzer, it was identified that a vast majority of edge-shaped particles contained high concentrations of Sb (Furuta *et al.*, 2005). This feature was consistent with that of automotive brake abrasion dusts. From this evidence, Furuta *et al.* (2005) concluded that automotive brake abrasion dusts would be one of the most important Sb sources affecting the recent atmospheric environment.

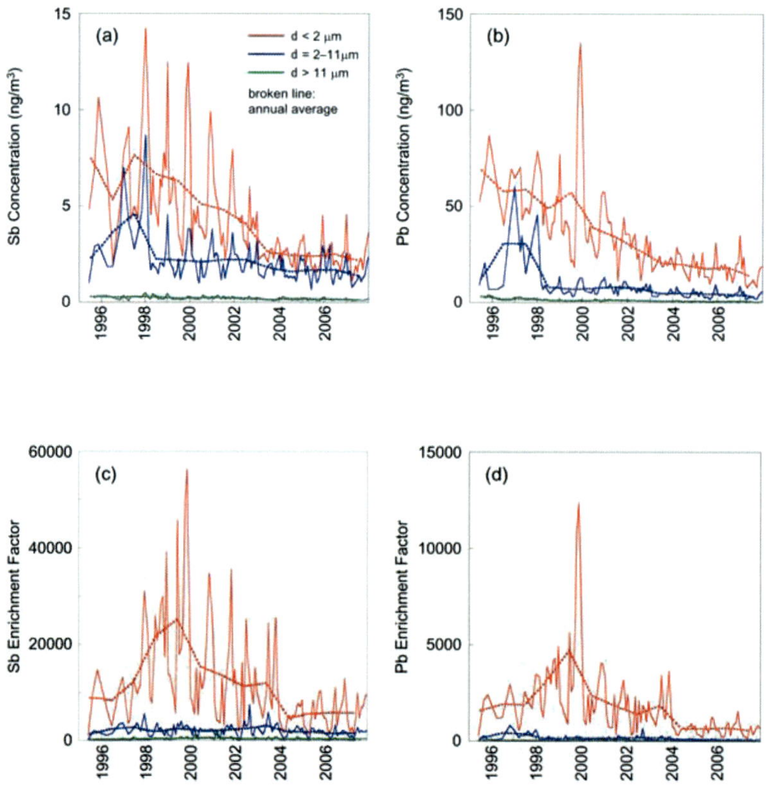

Figure 4.Long-term trends for elemental concentrations for (a) Sb and (b) Pb, and Al-normalized enrichment factors for (c) Sb and (d) Pb in size classified airborne particulates collected in Tokyo, Japan from 1995 to 2007.

• Summary

As stated in the above reviewed literatures, human activities have clearly affected the behaviors of Sb in the atmospheric environment. Several studies have reported the appearance of new anthropogenic emission sources of Sb and their impact on the recent atmospheric environment. General understandings on the relationship between chronology of Sb in the environment and human activities are summarized as follows;

1. In ancient times (Mediaeval and Roman/pre-Roman periods), metallurgical activities probably played a major role in Sb deposition. The onset of the greatest changes in airborne Sb deposition might have begun after the Second Industrial Revolution, namely, from the mid-19th century.
2. Drastic increases in Sb deposition were observed in the mid-20th century owing to the growing importance of anthropogenic activities such as coal combustion, mining and smelting of Pb ores, and military procurement in World War II .
3. From the 1960s to the 1980s, a rapid decline in Sb deposition was observed, probably owing to the decreasing influence of coal combustion and the recent introduction of flue-gas filter technologies and emission reduction efforts.
4. Another remarkable enrichment of Sb was found in the peat and ice core samples from the 1980s to the end of the 20th century. Similar trend was also found in airborne particulates collected in Tokyo, Japan. These observations implied the appearance of new anthropogenic Sb sources due to the increased consumption of Sb compounds (as flame retardants for plastic products, lubricant for automotive brake pads, *etc.*).
5. The remarkable increase in the enrichment factor for Sb in fine airborne particulates in Tokyo was likely to be associated with the increase in Sb emission from incineration plants treating household wastes containing plastic materials. After the enforcement of the Law Concerning Special Measures against Dioxins, the enrichment factor for Sb in fine particulates remarkably decreased and currently leveled off. However, Sb is still considerably enriched in airborne particulates with reference to the elemental composition in the Earth's crust
6. Automotive brake abrasion dusts are likely to be one of the most important Sb sources affecting the recent atmospheric environment in highly urbanized cities such as Tokyo, Japan.

4. Overview of Current Monitoring Studies on Antimony in Airborne Particulates

• Concentration of Antimony in Airborne Particulates and Exploration of its Sources

Since reactivity of metallic elements in the atmosphere is generally lower than that of other reactive atmospheric components such as nitrogen compounds, sulfuric compounds, and organic compounds, atmospheric concentration of metallic components will retain during transport processes such as dispersion and deposition. Therefore, the trace metal composition of airborne particulates can be used as a fingerprint of their emission sources. Multielement monitoring studies have been extensively carried out because some of fingerprint elements are effective to identify the emission sources of airborne particulates. Recently, various analytical methodologies for the Sb compounds and the monitoring researches of airborne elements were closely reviewed (Smichowski, 2008). The following part is intended to provide an overview of the Sb concentration in airborne particulates observed over the world. A total of 28 monitoring studies on airborne Sb concentrations were reviewed; these data are briefly listed in Table 3. It is difficult to directly compare the analytical data obtained from different sites because the respective studies were carried out at different time periods, at different locations, and for different particle size fractions. However, this overview will enable us to understand the current pollution level, the relationship between Sb concentration and its sampling location, and the potential sources of Sb emission.

North/South America

Gao *et al.* (2002) reported the characterization of 10 elements in $PM_{2.5}$ collected at 3 different locations in New York–New Jersey Harbor Bight Area, US, which is surrounded by industrial sectors and highways. The Sb concentration tended to be higher at the urban sites (0.88 and 2.1 ng/m^3) than at the coastal site (0.63 ng/m^3). Moreover, the Al-normalized enrichment factors for Sb in $PM_{2.5}$ were found to be 20,000 times higher than that expected from the Earth's crust. Factor analysis extracted three significant factors and Sb had a large loading to the first factor explaining the most of the variance (72%). Gao *et al.* interpreted that this factor was likely to represent waste incineration, oil burning, and nonferrous metal production.

Qureshi *et al.* (2006) reported the temporal trend of 15 elements in $PM_{2.5}$ collected at the urban site in New York, US (Sb conc.: 1.5 and 1.9 ng/m^3). The concentration of Sb was generally around 1 ng/m^3 during the observation. However, some events with noticeably higher Sb concentrations were also observed (18.7 ng/m^3 at the maximum), which is probably associated with anthropogenic Sb emissions. They also performed an extraction test for $PM_{2.5}$ with a dilute nitric acid solution. Since a relatively higher soluble (approx. 70%) fraction of Sb in a dilute nitric acid extract was observed, Sb components in this fraction probably exist as soluble forms such as chlorides or oxides originated from incinerators.

Kim *et al.* (2004) studied the emission sources of $PM_{2.5}$ collected at the site located in industrial and commercial area in Atlanta, US, by using positive matrix factorization (PMF) analysis, and showed a dataset containing the concentration of 16 elements (Sb conc.: 3.4 ng/m^3). Likewise, Qin *et al.* (2006) also published a dataset of 46 elements in $PM_{2.5}$ collected at 5 different locations in New York, US (Sb conc.: 3–6 ng/m^3), which was used for PMF analysis. Although quite a large number of daily samples were analyzed in these two studies, the trends of Sb concentrations were not discussed. The Sb concentrations shown in them are similar to those observed in the other urban sites in different countries.

Godoy *et al.* (2005) observed the concentrations of 42 elements in size-classified airborne particulates at 12 urban sites around a coal-fired power plant in Santa Catarina, Brazil. Volatile elements such as Sb, As, Se, Cd, and Pb are supposed to be concentrated in fine fly ash particles. Actually, relatively high concentrations of Sb were found in PM_2 (0.10 ng/m^3 for the data collected in 1999, 0.47 ng/m^3 for the data collected in 2000) as compared to those in $PM_{2\text{–}10}$ (0.05 ng/m^3 (1999), 0.21 ng/m^3 (2000)). The Fe-normalized enrichment factors for

Sb in PM_2 (111.5 (1999), 512.7 (2000)) demonstrated that anthropogenic emissions, probably originated from the power plant, had a considerable impact on the environment. Despite the vicinity of significant emission source, however, the observed Sb concentrations were not so high when compared to those reported in the other research, and this difference could no be accounted for. Godoy *et al.* stated that the observation data of many elements tended to be roughly equal to or less than that reported in the other studies carried out in different countries. However, the reason for this could not be explained.

Gómez *et al.* (2005) observed the concentrations of 13 elements in PM_{10} at 9 downtown sites in Buenos Aires, Argentina. In addition to the Sb concentrations (3.6–6.0 ng/m^3) shown in Table 3, they presented the range of Sb compositions (58.4–165 μg/g) at these 9 sites. Since the Sb compositions in PM_{10} exceeded those in the Earth's crust (0.2–0.3 μg/g, (Taylor, 1964; Wedepohl, 1995)), they inferred that anthropogenic origins such as household wastes and/or automotive brake pads had a strong impact on the environment. Indeed, they reported that the Sb composition tended to be higher at two sites which were actually located in heavily-trafficked streets. Moreover, significant correlations among the concentrations of Sb, Cu, and Mo were observed; these correlations indicated the effect of anthropogenic sources such as automotive brake pads (Weckwerth, 2001). From these results, Gómez *et al.* concluded that vehicular pollution primarily contributed to the Sb content in airborne particulates in Buenos Aires.

Asia/Oceania

As mentioned in the previous section, the long-term monitoring of 31 elements in PM_{10} was carried out at 16 locations in Japan in the NASN project (Japan Environmental Sanitation Center, 1975–1998). The Sb concentrations in the large cities (2.4–13.7 ng/m^3) and/or industrialized sites (3.9–12.2 ng/m^3) were roughly higher than those in the small cities (2.0–10.2 ng/m^3) or remote sites (1.2–3.3 ng/m^3).

Mizohata *et al.* (2000) compared the trends of elemental concentration in size-classified airborne particulates between the roadside and residential sites in Tokyo, Japan. At the residential site, Sb was abundantly found in PM_2 (5.2 ng/m^3) than $PM_{2\text{-}11}$ (1.4 ng/m^3). However, it is noteworthy that at the roadside sites, the Sb concentrations in both PM_2 (5.5–14.0 ng/m^3) and $PM_{2\text{-}11}$ (8.6–9.6 ng/m^3) were comparable. Mizohata *et al.* pointed out that road dust and brake abrasion dust were the major sources of Sb in the coarse fraction.

The latest long-term monitoring of 23 elements at the urban site in Tokyo, Japan, which was continued by Furuta *et al.* (2005), has already been reviewed in the previous section. Here we note the average concentrations of Sb in size-classified airborne particulates (4.93 ng/m^3 for PM_2, 2.29 ng/m^3 for $PM_{2\text{-}11}$, and 0.23 ng/m^3 for $>PM_{11}$). To identify the most important emission sources of airborne Sb, the elemental composition of Sb in several potential sources such as fly ash, chemical fibers, automotive brake pad, and diesel exhaust particles was determined. Moreover, from morphological observations using a scanning electron microscope equipped with an energy dispersive X-ray analyzer, it was identified that a vast majority of edge-shaped particles contained high concentrations of Sb. These features were consistent with those of automotive brake abrasion dusts. Besides the above observation, Iijima *et al.* (2008b) also performed the monitoring of airborne particulates at the suburban and mountainous sites nearby the Tokyo Metropolitan Area. They characterized the regional difference in the elemental concentration in PM_2 among the urban (data were cited from

Furuta *et al.* (2005) and Sato *et al.* (2008)), suburban, and mountainous sites. The Sb concentrations at the urban and suburban sites (2.47 and 2.17 ng/m^3, respectively) were considerably higher than that at the mountainous site (0.50 ng/m^3), which implies a prominent relationship between the airborne Sb concentration and human activities.

Hioki *et al.* (2006) reported the concentrations of 16 elements in size-classified airborne particulates collected at the mountainous site in Kyoto, Japan. Data were divided into two periods; one was the period when the Kosa (yellow sand) events occurred (Sb conc.: 0.607 ng/m^3 for $PM_{2.5}$) and the other was the Non-Kosa period (Sb conc.: 0.109 ng/m^3 for $PM_{2.5}$). Kosa is occasionally transported from arid area around Inner Mongolia to Japan from March to May. Hioki *et al.* pointed out that the concentrations of Sb, as well as the other various elements, tended to be more concentrated than usual during the Kosa events.

Okuda *et al.* (2004) reported the trend of the concentrations of 16 elements in PM_{10} collected on a daily basis over 2 years at the urban site in Beijing, China. In this period, the average PM_{10} concentration was 171 μg/m^3. This value was five times higher than that observed in Yokohama, Japan, during the same period. The average Sb concentration during the overall monitoring period was 33 ng/m^3. The daily variation of Sb concentration was quite large and a considerably high concentration (more than 200 ng/m^3) of airborne Sb was occasionally observed. However, the source of Sb was not discussed in this literature. Because the chemical mass balance analysis identified a significant contribution of coal combustion (25%) to PM_{10}, its effect to the enhancement of airborne Sb concentration should be carefully assessed.

Chen *et al.* (2008) reported the concentrations of 21 elements and lead isotope ratios in $PM_{2.5}$ collected on a daily basis over 1 year at 4 sites in Shanghai, China. The average Sb concentration in the industrial site (33 ng/m^3) was relatively higher than those in the residential (22 ng/m^3), commercial (21 ng/m^3), and control (22 ng/m^3) sites. Chen *et al.* also calculated the Ti-normalized enrichment factors and reported that the values for Sb in all the sites exceeded 1,000. This clearly indicates that airborne Sb is derived from anthropogenic sources. Zheng *et al.* (2004) also carried out similar investigation in Shanghai and reported that Sb concentrations in PM_{10} ranged from 15.8 to 34.6 ng/m^3 (data were not shown in Table 3). The results reported from China (Okuda *et al.* (2004); Chen *et al.* (2008); Zheng *et al.* (2004)) indicate that Sb in airborne particulates observed in urbanized areas in China has been concentrated by a factor of ten as compared to that observed in the other urbanized cities in the world. Chen *et al.* remarked that the characteristic enhancement of airborne Sb concentration in China is most likely associated with coal combustion.

Zheng *et al.* (2007) characterized the concentrations of 18 elements in total suspended particulates (TSP) collected during dry and wet seasons at the site located in industrial and commercial area in Hong Kong (Sb conc.: 2.53 ng/m^3). The northern winds blowing from the urbanized areas in China resulted in higher concentrations of Sb, As, Cd, and Pb than did the southern winds blowing from the coastal areas. From this observation, above-described 4 elements were treated as tracers to identify the effects from polluted areas in China.

Mouli *et al.* (2006) showed the concentrations of 15 elements in PM_{10} collected at the urban site in India. The concentrations of Sb (0.51 ng/m^3) appeared to be lesser than those collected at other urban sites in different countries. Factor analysis extracted three significant factors and Sb had a large loading to the second factor explaining 12% of variance. This factor was assigned to be the metallurgical industry.

Ghauri *et al.* (2001) reported the concentrations of 26 elements in TSP collected at the remote mountainous site in Pakistan (Sb conc.: 0.54 ng/m^3). They also studied the elemental solubility in cloud water samples. A high amount (94%) of Sb was found in the soluble fractions in the cloud water samples, indicating the existence of anthropogenic soluble substances such as chlorides and oxides of Sb.

Al-Momani *et al.* (2005) showed the concentrations of 19 elements in size-classified airborne particulates collected at the urban site in Al-Hashimya, Jordan. Interestingly, extremely high Sb concentrations were found in >$PM_{2.2}$ (179.3 ng/m^3). Since the observation sites were surrounded by arid regions, the concentrations of crustal elements such as Al and Ca generally tended to increase with the wind speed, particularly in the dry season. However, this tendency was not found in the case of Sb, indicating that this element is not derived from the re-suspension of soil dusts. The Al-normalized enrichment factors for Sb in both $PM_{2.2}$ and >$PM_{2.2}$ exceeded 10,000, confirming that Sb was derived from anthropogenic sources. From the factor analysis, Al-Momani *et al.* concluded that Sb might have originated from smelters.

Doğan *et al.* (2008) reported the concentrations of 31 elements in PM_{10} collected at the urban site in Antalya, Turkey (Sb conc.: 0.343 ng/m^3). PMF extracted seven significant factors and Sb was apportioned to the fifth factor as well as Cr, Ni, and Pb. This factor was interpreted as mixed anthropogenic sources, including motor vehicles and smelters.

Beavington *et al.* (2004) reported the elemental concentration in TSP collected at the suburban site in the vicinity of a Cu smelter in Port Kembla, Australia. The study was carried out before (in 1978) and after (during 2001 to 2002) the installation of a modern precipitator. The Sb concentration showed a drastic decrease due to the improvement of flue-gas treatment systems (15.8 ng/m^3 for the data before improvement, 0.82 ng/m^3 for the data after improvement), confirming that the Cu smelter dominantly contributed to the Sb concentration in this area.

Europe/Africa

Querol *et al.* (2004) compared the elemental concentrations in size-classified airborne particulates collected at 6 different locations in Spain, and found a remarkable regional difference. The Sb concentration at the curbside site (10 ng/m^3) was 10 times higher than that at the rural background site (1 ng/m^3), suggesting that Sb is a traffic-related element.

Moreno *et al.* (2007) reported the concentrations of 21 elements in size-classified airborne particulates collected at the industrial site in Puertollano, Spain (Sb conc.: 3.2 ng/m^3 for $PM_{2.5}$ and 4.2 ng/m^3 for PM_{10}). Industrial activities such as fossil fuel burning power plants and coal-fired power plants were considered to be the potential sources of air pollutants in this city. The local coal used in this region is known to contain relatively higher concentrations of Sb and Pb than the other samples of coal mined in different regions. In fact, Sb was the most enriched element in airborne particulates, and relatively high concentration of Sb was found in the fine fractions ($PM_{2.5}/PM_{10}$ = 0.76). From these results, Moreno *et al.* speculated that coal combustion in the primary source of airborne Sb.

Dietl *et al.* (1997) studied the characteristics of Sb in size-classified airborne particulates collected at the roadside and residential sites with different traffic intensities in Munich, Germany. The Sb concentrations showed large variations in the data, which was probably due to during sampling at locations with different traffic intensities. In both $PM_{2.5}$ and $PM_{2.5\text{–}10}$, the Sb concentrations were found to be significantly higher at the heavily-trafficked area

(2.40 ng/m^3 for $PM_{2.5}$ and 11.1 ng/m^3 for $PM_{2.5-10}$) than at the control location (1.26 ng/m^3 for $PM_{2.5}$ and 1.93 ng/m^3 for $PM_{2.5-10}$). At the roadside site, moreover, the Sb concentration in the coarse fraction was five times as much as that in the fine fraction. Dietl *et al.* detected a considerably high concentration of Sb (1–4%) in automotive brake friction materials, and they concluded that the mechanically abraded dust from these materials was the primary cause for the high concentration of Sb on the roads.

Voutsa *et al.* (2002) studied the temporal and spatial variability of 17 elements in PM_{10} collected at 3 different locations named S1, S2, and S3 in Thessaloniki, Greece. The monitoring sites of S1 and S3 were located in the vicinity of the road, and the concentrations of Sb, Br, Cu, and Pb were found to be significantly higher at both the sites S1 (Sb conc.: 8.3 ng/m^3) and S3 (Sb conc.: 5.9 ng/m^3) than at the site S2 (Sb conc.: 3.2 ng/m^3). Moreover, the Sc-normalized enrichment factor for Sb was found greater than 1,000. Voutsa *et al.* inferred that the enrichment for Sb in airborne particulates might be due to vehicular emissions such as automotive brake abrasion dusts.

Pakkanen *et al.* (2001) reported the concentrations of 25 elements in size-classified airborne particulates collected at the urban and rural sites in Helsinki, Finland. The concentrations of Sb, Ba, Fe, and V were found to be higher at the heavily-trafficked urban site (Sb conc.: 0.77 ng/m^3 for $PM_{2.3}$, 0.78 ng/m^3 for $PM_{2.3-15}$) than at the rural site (Sb conc.: 0.36 ng/m^3 for $PM_{2.3}$, 0.036 ng/m^3 for $PM_{2.3-15}$). Al-normalized enrichment factor for Sb at the urban site was approximately 1,400, indicating the existence of large anthropogenic Sb sources. Moreover, it is noteworthy that the comparable concentrations of Sb were found in both $PM_{2.3}$ and $PM_{2.3-15}$ at the urban site, whereas the Sb concentration was dominant in $PM_{2.3}$ at the rural site. Furthermore, a significant correlation was found between the concentrations of Sb and Cu, which is probably due to the effect of automotive brake abrasion dusts.

Salma *et al.* (2001) reported the concentrations of 40 elements in size-classified airborne particulates collected at the urban background and downtown sites in Budapest, Hungary. The Sb concentrations were markedly high in the heavily-trafficked downtown site (2.50 ng/m^3 for PM_2 and 6.60 ng/m^3 for PM_{2-10}) than those at the urban background site (0.69 ng/m^3 for PM_2 and 0.18 ng/m^3 for PM_{2-10}). The Al-normalized enrichment factors for Sb in PM_2 and PM_{2-10} were approximately 10,000 and 2,000, respectively, indicating that Sb was mainly derived from anthropogenic sources. It is interesting to note that the Sb concentration was dominant in the coarse fraction (PM_{2-10}) at downtown site, whereas the concentration of Sb in the fine fraction (PM_2) at the urban background site was more. This tendency is consistent with that shown by Dietl *et al.* (1997) and Pakkanen *et al.* (2001).

Hueglin *et al.* (2005) showed the concentration of 23 elements in size-classified airborne particulates collected at 6 different sites in Switzerland. The elemental concentrations of Sb, Ba, Cu, and Mo gradually decreased from the urban curbside (Sb conc.: 5.6–8.3 ng/m^3 for PM_{10}) to the urban (Sb conc.: 1.4 ng/m^3 for PM_{10}), suburban (Sb conc.: 0.66 ng/m^3 for PM_{10}), and rural (Sb conc.: 0.19–0.26 ng/m^3 for PM_{10}) sites, indicating that vehicular traffic is the main emission source of these elements. At the suburban and rural sites, relatively high concentrations of Sb were found in $PM_{2.5}$ than in $PM_{2.5-10}$. On the other hand, the concentration of Sb in $PM_{2.5-10}$ was higher that in $PM_{2.5}$ at the urban and curbside sites. This tendency was consistent with that shown in several of the aforementioned literatures, and was believed to be due to the contribution of re-suspended road dust and the re-suspension of mechanically generated particles such as tire wear and brake abrasion dusts to comprise coarse fractions.

Maenhaut & Cornille (1989) reported the concentrations of 42 elements in $PM_{2.5}$ collected at 2 remote sites in Arctic Norway. Considerably low concentrations of Sb, As, V, and Zn were observed in summer (Sb conc.: 0.0024–0.0126 ng/m^3) when the atmosphere is covered with clean air masses transported from the polar region. On the contrary, the elemental concentrations in winter (Sb conc.: 0.048–0.22 ng/m^3), which were affected by an aerosol transport from the polluted continental regions of southern Europe, were considerably higher than those in summer. Very similar results were obtained for the elemental concentrations in the Canadian Arctic regions (Barrie & Hoff, 1985; Krachler et al., 2005), indicating that Sb is strongly associated with human activities.

Nyanganyura *et al.* (2007) reported the concentrations of 26 elements in size-classified airborne particulates at the remote site in Zimbabwe. The Sb concentrations in both dry and wet seasons appeared to be very low (0.02–0.03 ng/m^3), similar to those at the other remote sites. However, insufficient data were available from the other regions in Africa.

• Summary

On the basis of various air monitoring studies performed all over the world, our current understandings of Sb in airborne particulates could be summarized as follows.

1 Depending on the location of the observation site, the concentration of Sb in airborne particulates is of the order of 10^{-3} to 10^{2} ng/m^3.
2 In urbanized areas, the concentrations of Sb in airborne particulates are typically of the order of 10^{0} to 10^{1} ng/m^3, and the enrichment factors are remarkably high (in some cases more than 10,000).
3 The enrichment of Sb in airborne particulates is considered to be dominated by several human activities such as vehicular traffic, coal combustion, waste incineration, and mining and smelting of metallic minerals. The predominant Sb sources are dependent on the location of the observation site.
4 Relatively high concentrations of airborne Sb are generally found at areas with high vehicular traffic; in particular, the Sb concentrations in coarse (typically $d > 2$ µm) particles are markedly higher than those at the other background sites.
5 Significant correlation can be found among the Sb, Cu, Mo, and Ba concentrations, which are treated as tracers for automotive brake pads in the highly-trafficked observation sites, indicating that automotive brake abrasion dust is the predominant Sb source in these areas.

Table 3. Overview of recent monitoring studies for antimony in airborne particulates

Region / Nation / City	Site	Location	PM fraction	Sb concentration (ng/m^3)	Number of samples	Sampling / Measurement unit	Monitoring period	Analytical method [a)]	Reference
North/South America									
US									
Monmouth County, NJ	Sandy Hook	Coastal	$PM_{2.5}$	0.63	59	24 h	Jan. 1998 – Jul. 1999	ICP-MS	Gao *et al.* (2002)
Jersey City, NJ	Liberty Science Center	Urban	$PM_{2.5}$	2.1	45	24 h	Oct. 1998 – Jan. 2000	ICP-MS	
Middlesex County, NJ	New Brunswick	Urban	$PM_{2.5}$	0.88	62	24 h	Jan. 1998 – Dec 1999	ICP-MS	
Queens County, NY	PS219	Urban	$PM_{2.5}$	1.5	72	24 h	Jul. – Sep. 2001	ICP-MS	Qureshi *et al.* (2006)
				1.9	62	24 h	Sep. – Oct. 2001	ICP-MS	
Atlanta, GA	Jefferson Street	Industrial and comercial	$PM_{2.5}$	3.4	529	24 h	Aug. 1998 – Aug. 2000	XRF	Kim *et al.* (2004)
Bronx County, NY	New York Botanical Garden	Urban	$PM_{2.5}$	6	422	24 h	Feb. 2000 – Dec. 2003	XRF	Qin *et al.* (2006)
Bronx County, NY	Intermediate School 52	Urban	$PM_{2.5}$	4	328	24 h	Jan. 2001 – Dec. 2003	XRF	
Queens County, NY	Queens College	Urban	$PM_{2.5}$	3	293	24 h	Apr. 2001 – Dec. 2003	XRF	
Union County, NJ	Elizabeth	Urban	$PM_{2.5}$	6	248	24 h	May 2001 – Dec. 2003	XRF	
Morris County, NJ	Chester	Suburban	$PM_{2.5}$	5	236	24 h	Jun. 2001 – Dec. 2003	XRF	
Brazil									
Santa Catarina	Capivari de Baixo [b)]	Urban	PM_2	0.10	105	24 h	Summer1999	ICP-MS	Godoy *et al.* (2005)
			$PM_{2\text{-}10}$	0.05	115	24 h	Summer1999	ICP-MS	
			PM_2	0.47	111	24 h	Winter 2000	ICP-MS	
			$PM_{2\text{-}10}$	0.21	113	24 h	Winter 2000	ICP-MS	
Argentina									
Buenos Aires	DINES	Downtown	PM_{10}	3.6	8	24 h	Jul. 2001	ICP-MS	Gómez *et al.* (2005)
	Hospital Alemán	Downtown	PM_{10}	5.7	8	24 h	Jul. 2001	ICP-MS	
	Casa Rapallini	Downtown	PM_{10}	5.3	8	24 h	Jul. 2001	ICP-MS	
	Casa Amarilla	Downtown	PM_{10}	4.0	8	24 h	Jul. 2001	ICP-MS	
	Metrogas	Downtown	PM_{10}	3.8	8	24 h	Jul. 2001	ICP-MS	
	Hospital Garrahan	Downtown	PM_{10}	6.0	8	24 h	Jul. 2001	ICP-MS	
	Jardín Japonés	Downtown	PM_{10}	4.8	8	24 h	Jul. 2001	ICP-MS	
	Biblioteca Nacional	Downtown	PM_{10}	5.2	8	24 h	Jul. 2001	ICP-MS	
	Aguas Argentinas	Downtown	PM_{10}	4.3	8	24 h	Jul. 2001	ICP-MS	
Asia/Oceania									
Japan									
Hokkaido	Sapporo	Large city	PM_{10}	2.4	268	25 days	Apr. 1974 – Mar. 1997	INAA	Japan Environmental Sanitation Center (1975-1998)
	Nohoro	Remote	PM_{10}	1.2	214	25 days	Apr. 1979 – Mar. 1997	INAA	
Miyagi	Sendai	Large city	PM_{10}	2.9	272	25 days	Apr. 1974 – Mar. 1997	INAA	
	Nonotake	Remote	PM_{10}	3.3	191	25 days	Apr. 1981 – Mar. 1997	INAA	
Niigata	Niigata	Industrial	PM_{10}	3.9	246	25 days	Oct. 1975 – Mar. 1997	INAA	
Tokyo	Tokyo	Large city	PM_{10}	10.6	270	25 days	Apr. 1974 – Mar. 1997	INAA	
Kanagawa	Kawasaki	Industrial	PM_{10}	12.2	271	25 days	Apr. 1974 – Mar. 1997	INAA	
Aichi	Nagoya	Large city	PM_{10}	8.8	242	25 days	Oct. 1976 – Mar. 1997	INAA	
Kyoto	Kyoto-hachiman	Small city	PM_{10}	10.2	188	25 days	Apr. 1981 – Mar. 1997	INAA	
Osaka	Osaka	Large city	PM_{10}	13.7	272	25 days	Apr. 1974 – Mar. 1997	INAA	
Hyogo	Amagasaki	Industrial	PM_{10}	10.6	240	25 days	Oct. 1976 – Mar. 1997	INAA	
Okayama	Kurashiki	Small city	PM_{10}	6.6	214	25 days	Apr. 1979 – Mar. 1997	INAA	
Shimane	Matsue	Small city	PM_{10}	2.0	269	25 days	Apr. 1974 – Mar. 1997	INAA	
Yamaguchi	Ube	Industrial	PM_{10}	4.6	185	25 days	Apr. 1981 – Mar. 1997	INAA	
Fukuoka	Omuta	Industrial	PM_{10}	5.9	260	25 days	Apr. 1974 – Mar. 1997	INAA	
	Chikugo-ogori	Small city	PM_{10}	4.6	210	25 days	Apr. 1979 – Mar. 1997	INAA	

Table 3. (continued)

Region / Nation / City	Site	Location	PM fraction	Sb concentration (ng/m^3)	Number of samples	Sampling / Measurement unit	Monitoring period	Analytical method [a]	Reference
Japan									
Tokyo	Shinjuku-ku, Hatsudai	Roadside	PM_2	14.0	1	2 weeks	Dec. 1997	INAA	Mizohata *et al.* (2000)
			$PM_{2\text{-}11}$	9.6	1	2 weeks	Dec. 1997	INAA	
			PM_2	5.5	1	2 weeks	Jun. 1998	INAA	
			$PM_{2\text{-}11}$	8.7	1	2 weeks	Jun. 1998	INAA	
	Ota-ku, Matsubara-bashi	Roadside	PM_2	9.6	1	2 weeks	Dec. 1997	INAA	
			$PM_{2\text{-}11}$	8.6	1	2 weeks	Dec. 1997	INAA	
	Kakano-ku, Wakamiya	Residential	PM_2	5.2	1	2 weeks	Jun. 1998	INAA	
			$PM_{2\text{-}11}$	1.4	1	2 weeks	Jun. 1998	INAA	
Tokyo	Bunkyo-ku, Chuo University	Urban	PM_2	4.93	93	25 days	May 1995 – Dec. 2004	ICP-MS	Furuta *et al.* (2005)
			$PM_{2\text{-}11}$	2.29	93	25 days	May 1995 – Dec. 2004	ICP-MS	
			$>PM_{11}$	0.23	93	25 days	May 1995 – Dec. 2004	ICP-MS	
Tokyo	Bunkyo-ku, Chuo University	Urban [c]	PM_2	2.47	36	25 days	Apr. 2003 – Mar. 2006	ICP-MS	Iijima *et al.* (2008b)
Gunma	Maebashi	Suburban	PM_2	2.17	36	2 weeks	Apr. 2003 – Mar. 2006	ICP-MS	
	Mt. Akagi	Mountainous	PM_2	0.50	36	2 weeks	Apr. 2003 – Mar. 2006	ICP-MS	
Kyoto	Yasaka	Mountainous	$PM_{2.5}$	0.109	42	24 h	Nov. 2001 – Jan. 2002	ICP-MS	Hioki *et al.* (2006)
			$PM_{2.5\text{-}10}$	<0.02	42	24 h	Nov. 2001 – Jan. 2002	ICP-MS	
			$PM_{2.5}$	0.607	30	12 h	Mar. 2002 – Apr. 2002 [d]	ICP-MS	
			$PM_{2.5\text{-}10}$	0.106	30	12 h	Mar. 2002 – Apr. 2002 [d]	ICP-MS	
China									
Beijing	Tsinghua University	Urban	PM_{10}	33	687	24 h	Mar. 2001 – Aug. 2003	LA-ICP-MS	Okuda *et al.* (2004)
Shanghai	Baoshan	Industrial (urban)	$PM_{2.5}$	33	13	24 h	Apr. 2004 – Apr. 2005	ICP-MS	Chen *et al.* (2008)
	Putuo	Residential (urban)	$PM_{2.5}$	22	13	24 h	Apr. 2004 – Apr. 2005	ICP-MS	
	Huangpu	Commercial (urban)	$PM_{2.5}$	21	13	24 h	Apr. 2004 – Apr. 2005	ICP-MS	
	Jiading	Control (suburban)	$PM_{2.5}$	22	13	24 h	Apr. 2004 – Apr. 2005	ICP-MS	
Hong Kong	Hong Kong University	Industrial and commercial	TSP	2.53	15	20 h	Jun. – Dec. 1998	ICP-MS	Zheng *et al.* (2007)
India									
Tirupati	Sri Venkateswara University	Urban	PM_{10}	0.51	52	24 h	Oct. 2001 – Sep. 2002	ICP-MS	Mouli *et al.* (2006)
Pakistan									
Changlagali	Changlagali	Rural	TSP	0.54	10	24 h	Jul. 1996	ICP-MS	Ghauri *et al.* (2001)
Jordan									
Al-Hashimya	Al-Hashimya	Urban	$PM_{2.2}$	9.95	62	24 h	Aug. 2000 – Aug. 2001	ICP-MS	Al-Momani *et al.* (2005)
			$>PM_{2.2}$	179.3	62	24 h	Aug. 2000 – Aug. 2001	ICP-MS	
Turkey									
Antalya	Antalya	Rural	PM_{10}	0.343	284	24 h	Feb. 1992 – Dec. 1993	INAA	Doğan *et al.* (2008)
Australia									
Port Kembla, NSW	Port Kembla [e]	Suburban	TSP	15.8	3	1 month	Feb. – Apr. 1978 [e]	INAA	Beavington *et al.* (2004)
			TSP	0.82	12	1 month	May 2001 – Apr. 2002 [e]	INAA	
Europe/Africa									
Spain									
Barcelona	Sagrera	Curb-side	PM_{10}	10	84 – 90 [f]	24 h	Jan. – Dec. 2001	ICP-MS	Querol *et al.* (2004)
Álava Province	Llodio	Urban background	PM_{10}	2	84 – 90 [f]	24 h	Jan. – Dec. 2001	ICP-MS	
Coruña Province	Bemantes	Rural background	PM_{10}	1	84 – 90 [f]	24 h	Jan. – Dec. 2001	ICP-MS	
Madrid	Alcobendas	Urban background	PM_{10}	7	84 – 90 [f]	24 h	Jan. – Dec. 2001	ICP-MS	
Huelva Province	Huelva	Urban background	PM_{10}	4	84 – 90 [f]	24 h	Jan. – Dec. 2001	ICP-MS	
Canary Islands	Canaries	Urban background	PM_{10}	3	84 – 90 [f]	24 h	Jan. – Dec. 2001	ICP-MS	
Tarragona Province	Tarragona	Urban background	PM_{10}	7	84 – 90 [f]	24 h	Jan. – Dec. 2001	ICP-MS	

Table 3. (continued)

Region / Nation / City	Site	Location	PM fraction	Sb concentration (ng/m^3)	Number of samples	Sampling / Measurement unit	Monitoring period	Analytical method [a]	Reference
Spain									
Ciudad Real Province	Puertollano [g]	Industrial	$PM_{2.5}$	3.2	111	24 h	Jan. 2004 – Mar. 2005	ICP-MS	Moreno *et al.* (2007)
			PM_{10}	4.2	110	24 h	Jan. 2004 – Mar. 2005	ICP-MS	
Germany									
Munich	Loc. I	Roadside	$PM_{2.5}$	2.40	8	2 weeks	May – Oct. 1994	ICP-MS	Dietl *et al.* (1997)
			$PM_{2.5-10}$	11.1	8	2 weeks	May – Oct. 1994	ICP-MS	
	Loc. II	Residential	$PM_{2.5}$	1.26	8	2 weeks	May – Oct. 1994	ICP-MS	
			$PM_{2.5-10}$	1.93	8	2 weeks	May – Oct. 1994	ICP-MS	
Greece									
Thessaloniki	S1	Residential and commercial	PM_{10}	8.3	52	24 h	summer 1997 – 1998	INAA	Voutsa *et al.* (2002)
	S2	Urban	PM_{10}	3.2	54	24 h		INAA	
	S3	Polluted residential	PM_{10}	5.9	47	24 h		INAA	
Finland									
Helsinki	Vallila	Urban	$PM_{2.3}$	0.77	53	24 h	Apr. 1996 – Jun. 1997	ICP-MS	Pakkanen *et al.* (2001)
			$PM_{2.3-15}$	0.78	16	24 h	Apr. 1996 – Jun. 1997	INAA	
	Luukki	Rural	$PM_{2.3}$	0.36	53	24 h	Apr. 1996 – Jun. 1997	ICP-MS	
			$PM_{2.3-15}$	0.036	16	24 h	Apr. 1996 – Jun. 1997	INAA	
Hungary									
Budapest	KFKI campus	Urban background	PM_2	0.69	33	24 h	Apr. – May 1996	INAA	Salma *et al.* (2001)
			PM_{2-10}	0.18	33	24 h	Apr. – May 1996	INAA	
	Széna Square	Downtown	PM_2	2.50	33	24 h	Apr. – May 1996	INAA	
			PM_{2-10}	6.60	33	24 h	Apr. – May 1996	INAA	
Switzerland									
Zurich	Zurich-Wiedikon	Curb-side	PM_{10}	8.8	59	24 h	Apr. 1998 – Mar. 1999	ICP-MS	Hueglin *et al.* (2005)
Zurich	Zurich-Kaserne	Urban	$PM_{2.5}$	0.57	79	24 h	Apr. 1998 – Mar. 1999	ICP-MS	
			PM_{10}	1.4	104	24 h	Apr. 1998 – Mar. 1999	ICP-MS	
Bern	Bern	Curb-side	$PM_{2.5}$	0.95	79	24 h	Apr. 1998 – Mar. 1999	ICP-MS	
			PM_{10}	5.6	103	24 h	Apr. 1998 – Mar. 1999	ICP-MS	
Basel	Basel	Suburban	$PM_{2.5}$	0.41	77	24 h	Apr. 1998 – Mar. 1999	ICP-MS	
			PM_{10}	0.66	103	24 h	Apr. 1998 – Mar. 1999	ICP-MS	
Payerne	Payerne	Rural	PM_{10}	0.26	17	24 h	Apr. 1998 – Mar. 1999	ICP-MS	
Chaumont	Chaumont	Rural	$PM_{2.5}$	0.16	78	24 h	Apr. 1998 – Mar. 1999	ICP-MS	
			PM_{10}	0.19	104	24 h	Apr. 1998 – Mar. 1999	ICP-MS	
Norway									
Spitsbergen	Ny Ålesund	Remote	$PM_{2.5}$	0.22	14	2 or 3 days	Winter 1983	INAA	Maenhaut & Cornille (1989)
			$PM_{2.5}$	0.048	18	2 or 3 days	Winter 1984	INAA	
			$PM_{2.5}$	0.0024	13	2 or 3 days	Summer 1984	INAA	
			$PM_{2.5}$	0.108	14	2 or 3 days	Winter 1986	INAA	
Finnmark County	Vardø	Remote	$PM_{2.5}$	0.125	13	2 or 3 days	Winter 1983	INAA	
			$PM_{2.5}$	0.093	14	2 or 3 days	Winter 1984	INAA	
			$PM_{2.5}$	0.0126	12	2 or 3 days	Summer 1984	INAA	
Zimbabwe									
Northern Zimbabwe	Rukomechi Research Station	Background	PM_2	0.03	312	24 h [h]	Dry season in Sep. 1994 – Jan. 2000	PIXE	Nyanganyura *et al.* (2007)
			PM_2	0.02	182	24 h [h]	Wet season in Sep. 1994 – Jan. 2000	PIXE	

[a] ICP-MS: Inductively Coupled Plasma Mass Spectrometry, XRF: X-Ray Fluorescence, LA-ICP-MS: Laser Ablation Inductively Coupled Plasma Mass Spectrometry, INAA: Instrumental Neutron Activation Analysis, ICP-AES: Inductively Coupled Plasma Atomic Emission Spectrometry, PIXE: Particle Induced X-ray Emission Spectrometry. [b] Twelve monitoring sites were located in the vicinity of coal fired power plant. Averaged data are shown. [c] Data were cited from Furuta *et al.* (2005) and Sato *et al.* (2008). [d] Monitoring was carried out during the Kosa events. [e] Monitoring site was located in the vicinity of copper smelter. Original data unit (ng/kg air) was corrected to ng/m^3 by multiplying with 1.226 kg/m^3 (at 15 °C, 760 mm Hg). The study was carried out before (in 1978) as well as after (during 2001 to 2002) the installation of modern precipitator. [f] Averaged data were calculated from the analysis of 84 to 90 samples per site, but exact sample nunbers were not shown. [g] Monitoring site was located in the vicinity of coal fired power plant. [h] Each data unit comprises samples collected during the daytime (12 h) for two consecutive days.

5. Automotive Brake Abrasion Dusts: Recent Revealed Predominant Source of Antimony in Airborne Particulates

According to the estimation by Pacyna & Pacyna (2001), stationary fossil fuel combustion (mainly coal combustion) significantly contributes to the global anthropogenic Sb inventory. In fact, coal combustion was often considered as the dominant source of airborne Sb in the previously reviewed literatures (Godoy *et al.*, 2005; Moreno *et al.*, 2007; Chen *et al.*, 2008). On the other hand, current air monitoring studies demonstrated that Sb is very often considered as a traffic-related element in urbanized areas. However, the emission amount of Sb from vehicular traffic was not considered in the estimation of anthropogenic Sb inventory by Pacyna & Pacyna (2001). Automobiles are gaining importance as the vital means of transport in highly populated cities. Therefore, we should enhance the scientific knowledge of the issues concerning traffic-related Sb in the near future. In this section, thus, automotive brake abrasion dust is highlighted as the predominant source of traffic-related Sb. The latest knowledge on the characteristics of brake abrasion dusts and their emission to the environment are also reviewed.

• Mechanisms Underlying the Functioning of Automotive Brake Systems and Types of Friction Materials

The brake is considered as one of the most important parts in a vehicle that contributes to safe control. According to their basic structures, automotive brakes are classified into disk brake systems and drum brake systems. Likewise, friction materials used for disk brakes and drum brakes are officially classified into brake pads and brake linings, respectively. However, in many researches, no distinction has been made between both parts, and friction materials have been referred to as brake linings only.

The mechanisms underlying the functioning of automotive brake systems and the different types of friction materials were well documented by some literatures (Kalton, 1995; Sasaki & Yano, 2007). A disk brake system generates the braking force by contacting the rotating disk with two brake pads. Since this system is exposed to air, high cooling efficiency that contributes to the high braking stability is expected. Hence, disk brake systems are commonly used in various types of passenger cars and small trucks. However, owing to its exposure to the atmosphere, this system directly emits brake abrasion dusts, which result in the traffic-related airborne particulates, to the atmosphere. On the other hand, a drum brake system generates the braking force by rubbing the inner surface of the rotating drum with brake linings. Since this system can generate a large braking force by a self-energizing effect (mechanism to generate larger braking force than input power by itself) without the use of any booster, they are commonly used in large trucks. Since the brake linings are enclosed in the drum, majority of the brake abrasion dusts are constrained inside the rotating drum and not directly released to the atmosphere.

Owing to its ideal braking characteristics such as homogeneity, moldability, and strength, asbestos has been used for the automotive friction materials over 100 years. However, ever since it was proved to be carcinogenic, its use has been strictly limited. In recent times, asbestos has been completely replaced by inorganic fibers (ceramic, grass, basalt, carbon, and potassium titanate fibers), organic fibers (aramid pulp, acrylic pulp, and cellulose fibers), and metallic fibers (steel, copper, and brass fibers). Commercially available automotive brake pads and linings are made of combined materials containing the above-mentioned fibers, inorganic fillers, metallic powders, and solid lubricants. Since these base materials are bound by phenolic resin, recent friction materials are known as resin-mold type. Depending on the mixing ratio of the steel fiber, resin-mold friction materials are further classified into three types; non-asbestos organic (NAO), low-steel, and semi-metallic. The main characteristics and market of each type are listed in Table 4.

Table 4. Characteristic differences among three classified brake pads

	NAO (Non-Asbestos Organic)	Low-steel	Semi-metallic
Major market	Japan / US	Europe	US / Europe
Fiber	Aramid / Ceramic / Copper / Potassium titanate	Steel (10–30%) / Copper	Steel (30–50%) / Copper
Binder	Phenolic resin	Phenolic resin	Phenolic resin
Additive	$BaSO_4$ / Sb_2S_3 / Graphite	$BaSO_4$ / Sb_2S_3 / Graphite	$BaSO_4$ / Sb_2S_3 / Graphite
Advantage	Low noise / Low aggression to rotor	High stability at high speed and temperature	High wear resistance at high load
Disadvantage	Low stability at high speed and temperature	High noise / High dust / High aggression to rotor	High vibration / Low stability / Rust

• Function and Composition of Antimony in Friction Materials

Sb_2S_3 acts as a lubricant and reduces excessive wear under high load conditions (Jang & Kim, 2000). Uexküll *et al.* (2005) reported that Sb_2S_3 could be partially oxidized by frictional heat to form Sb_2O_3, which is classified as a possible carcinogen. Therefore, the use of Sb compounds as friction materials would be associated with the increase in the concerns regarding the negative impact on the ecosystem such as pulmonary diseases.

The mixing ratios of various types of additives are decided depending on the braking characteristics required by different types of vehicles. For this reason, there is marked diversity between the elemental composition of metallic elements including Sb. Dietl *et al.* (1997) analyzed the elemental composition in various types of brake linings (it is not clear in this reference whether the authors distinguished between brake pads and brake linings) and found that the Sb concentration was 1–4%.

On the basis of published data, Weckwerth (2001) estimated that the Sb composition in brake linings (once again, it is not clear whether the authors distinguished between brake pads and brake linings) could be in the range of 1–5%

Uexküll *et al.* (2005) analyzed the elemental composition in the commercially available brake pads and brake linings and reported that the average Sb composition was 4.3% in the pads and 0.13% in the linings. Moreover, from the experiments by using 2 different brake dynamometers, they reported the Sb compositions in different brake abrasion dust samples to be 2.1% and 1.7%. These two dust samples were obtained from continuous experiments using 45 and 15 types of brake pads, respectively.

Iijima *et al.* (2007) analyzed the elemental composition in 3 types of brake pads and reported that the Sb composition was in the range of 0.71–1.7%.

From these results, it could be concluded that brake friction materials contain a significant amount of Sb. The actual Sb content in these friction materials is considerably higher than those found in chemical fablics or household wastes (several hundred μg/g order of Sb were found from them) which probably contain the Sb compounds as a flame retardant (Furuta *et al.*, 2005; Gómez *et al.*, 2005). Moreover, it is 10^5 times higher than that expected in the Earth's crust (0.2–0.3 μg/g), indicating that brake abrasion dusts dominantly contribute to the enrichment of Sb in airborne particulates.

Brake pads/linings contain various types of metallic elements. Weckwerth (2001) stated that the relationship between Sb and Cu concentrations is useful in identifying the brake abrasion dusts. Likewise, Iijima *et al.* (2007) mentioned that in addition to Sb and Cu, K, Ti, and Ba can be used to evaluate the contribution of brake abrasion dusts to airborne particulates.

Weckwerth (2001) calculated the Cu/Sb ratio in brake friction materials (4.6 ± 1.2) and compared it with that in the airborne particulates collected in Cologne, Germany. The Cu/Sb ratio in coarse particulates was 4.9 ± 1.7, which was very close to that in brake friction materials. From these results, Weckwerth concluded that the Sb content in coarse fractions is mainly affected by brake abrasion dusts. Likewise, Pakkanenn *et al.* (2001), Vousta *et al.* (2002), Gómez *et al.* (2005), and Hueglin *et al.* (2005) discussed the contribution of brake abrasion dusts to airborne Sb on the basis of the atmospheric behaviors of Sb and Cu.

• Particle Size Distributions of Brake Abrasion Dusts

Generally, mechanically abraded particulate matter is believed to be coarse (d = 2.5–10 μm) (Weckwerth, 2001). However, there are not many studies that have experimentally investigated the particle size distribution of automotive brake abrasion dusts.

Garg *et al.* (2000) examined the particle size distribution of brake abrasion dusts that were experimentally generated from six types of brake pads (two semi-metallic pads and four NAO pads) and one type of brake lining (semi-metallic lining). By using a brake dynamometer, brake abrasion dusts were emitted into the enclosed chamber and were subsequently collected by using a Micro-Orifice Uniform Deposit Impactor (MOUDI). The particle mass distributions of four types of pads showed bimodal profiles, whose peaks were found in the range less than 0.1 μm and greater than 10 μm. The other two pads and lining showed unimodal profiles, whose peaks were found in the range from 0.1 to 2.5 μm. Moreover, Garg *et al.* mentioned that the distributions varied with the disk temperature.

Sanders *et al.* (2003) have also used MOUDI to report the particle size distribution of brake abrasion dusts experimentally generated from three types of brake pads (a low-steel pad, a semi-metallic pad, and a NAO pad) by using a brake dynamometer. The particle mass distributions showed only unimodal profiles with peaks at approximately 6 μm, regardless of the type of brake pad.

Iijima *et al.* (2007, 2008a) examined the particle size distribution of brake abrasion dusts experimentally generated from a brake dynamometer by changing the braking parameters such as friction temperature, initial driving speed, and deceleration. The abrasion dusts were generated from three types of NAO pads, and the particle number distributions were measured by using an Aerodynamic Particle Sizer spectrometer. Figure 5(a) shows the particle number distributions obtained by changing the disk temperature, and Figure 5(b) shows the particle mass distributions derived from the number distributions. Only unimodal profiles were obtained in this experimental system, as in the case of the study by Sanders *et al.* (2003). It is noteworthy that the particle size distributions tended to shift to the coarser side with an increase in the disk temperature. This tendency was consistent with that observed by Garg *et al.* (2000). It is suggested that the friction surface becomes brittle under high friction temperature because of the thermal instability in the binding strength of the phenolic resin that holds the brake components together. At a disk temperature at 100 °C (typical braking condition in urban districts), the particle number and mass distribution peaks appeared at around 0.8 μm and 2 μm, respectively. This observation demonstrated that considerable amounts of fine particles as well as coarse particles would be emitted from automotive brakes.

The particle size distributions provided in the above-mentioned studies are not exactly in good agreement with one other. However, it is reasonable to understand that brake abrasion dusts can impact to both the fine and coarse fractions of airborne particulates. Additionally, it is known that the elemental composition in size-classified brake abrasion dusts retains the bulk pad composition (Iijima *et al.*, 2007). Therefore, automotive brake abrasion dusts will enhance Sb concentration in both the fine and coarse fractions of airborne particulates. These strong evidence confirm the increase in Sb concentration in airborne particulates in highly-trafficked areas.

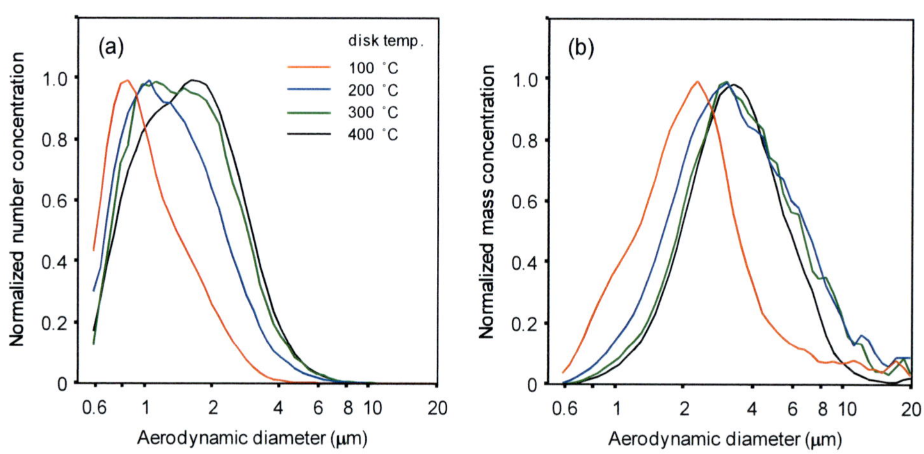

Figure 5.Normalized particle size distribution of automotive brake abrasion dusts. (a) normalized number distribution and (b) normalized mass distribution derived from (a). Observed signals are normalized by the maximum values.

• Estimation for Emission Amount of Brake-Derived Antimony

The emission factors of air pollutants emitted from automobiles such as exhaust gas components are generally defined as emission amount per driving distance (μg/km). Sternbeck *et al.* (2002) investigated the Sb concentrations in TSP collected at the inlet and outlet of two tunnels in Gothenburg, Sweden. The Sb concentrations were found to be higher at the outlets of the two tunnels (75 and 107 ng/m^3, respectively) than at the inlets (28 and 34 ng/m^3, respectively), indicating that Sb is certainly a traffic-related element. The Sb emission factors were calculated for each tunnel (31.8 and 50.8 μg/km, respectively) from the data of the concentration gradient between at the outlet and the inlet, ventilation flow rate through the tunnel, distance between the sampling locations, and the traffic flow rate. A similar study was carried out at the two tunnels in Milwaukee, US (Lough *et al.*, 2005), and the emission factors found to be of the same order of magnitude as those obtained for the tunnels in Gothenburg.

These data could be utilized in the estimation of emission inventories from traffic-related Sb. However, the distance-based emission factor is not really practical because brake abrasion dusts are emitted only during braking, whereas exhaust gases are continuously emitted. If the distance-based emission factor is used in the dispersion model, the predicted values for highways may be overestimated because the braking frequency is actually very low on highways. In contrast, the results obtained for intersections or urban district typically showing a stop-and-go driving pattern may be underestimated. With this in mind, Iijima *et al.* (2008a) determined the Sb emission factor in the course of individual vehicle braking (μg/braking/car). From the experimental data, the emission of brake abrasion dusts was regressed as a function of several braking parameters such as initial driving speed and deceleration. Using the regression function, the Sb emission factors under typical braking conditions in an urban district (initial driving speed: 50 km/h, deceleration: 1.0 m/s^2, and disk temperature: 100 °C) were calculated to be 32 μg/braking for PM_{10} and 22 μg/braking for $PM_{2.5}$. These data will contribute to the modeling of atmospheric Sb concentrations alongside roadways.

With regard to the emission inventory of brake-derived Sb, Iijima *et al.* (2007) and Tsunemi & Wada (2008) estimated annual emission amount in Japan. Iijima *et al.* (2007) estimated the annual emission amount to be 21 tons/yr on the basis of the data such as Sb composition in brake pads, number of passenger vehicles in Japan, and various assumptions regarding the weight and average lifetime of the brake pad. On the other hand, Tsunemi & Wada (2008) estimated the annual emission amount to be *ca.* 20 tons/yr on the basis of the substance flow analysis for Sb in Japan. These results were in good agreement with one another. In case of the estimation proposed by Iijima *et al.* (2007), the global annual emission amount of brake-derived Sb was calculated to be approximately 230 tons/yr (on the basis of the number of passenger vehicles in the world (605,500,000 vehicles, data were cited from Japan Automobile Manufacturers Association (JAMA), Inc., (2005)). However, it should be considered that the market shares of the three different types of brake pads vary considerably from one region to another, as shown in Table 4. The assumptions regarding the weight of the friction material, product lifetime, and wear characteristics are probably different for the three types of brake pads. Therefore, we must recognize that the extrapolation to the global scale may contain many uncertainties owing to the difference in these assumptions. Additionally, since the emission amount from trucks and buses were not counted in this estimation, the estimated value may be slightly underestimated. On the basis of these considerations, when we compare the contribution of brake-derived Sb to the other emission sources, the former would exceed the worldwide total natural emission inventories of Sb to the atmosphere (150 tons/yr) estimated by Shotyk *et al.* (2005b). Moreover, by considering the brake-derived Sb emission, worldwide anthropogenic emission inventory of Sb to the atmosphere (1,561 tons/yr) estimated by Pacyna & Pacyna (2001) would be increased by at least 15%, resulting in a value of approximately 1,800 tons/yr. Hence, automotive brake abrasion dusts should be recognized as one of the major anthropogenic sources of Sb emission to the atmosphere, particularly in urban cities.

• Summary

In summary, the emission of automotive brake abrasion dusts to the atmosphere could be characterized as follows.

1 Automotive brake friction materials generally contain a high percentage of Sb (1−5%), and thus, they contribute significantly to the enrichment of Sb in airborne particulates.
2 The relationship among the concentrations of Sb, Cu, K, Ti, and Ba in airborne particulates can be used to identify the contribution of automotive brake abrasion dusts to the airborne particulates.
3 The particle size of automotive brake abrasion dusts distributes from ultrafine ($d <$ 0.1 μm) to coarse ($d >$ 10 μm) ranges, explaining the enhancement of Sb concentrations in both fine and coarse fractions in highly-trafficked areas.
4 The emission amount of brake-derived Sb to the global atmosphere (230 tons/yr) is supposed to exceed that of the worldwide total natural inventories.

6. Future Prospectives and Final Remarks

Because of the recent rapid growth of the world economy, many trace metals in natural resources will not sufficiently supply the global demand. As a result, many kinds of trace metals are supplied by the recycling of industrial products. This situation holds true with Sb because its reserve-production ratio is short (16 years). In the United States, the recycling efficiency for Sb was estimated to be 89% of total scrap, and the recycling rate was about 20% of total Sb supply (Carlin, 2000). In the future, the recycle efficiency is supposed to increase because of the shortage of Sb mining reserve. Moreover, non-bromine flame retardants such as phosphorus flame retardant have been recently developed due to concerns for toxic compounds produced by halogenated products (Takeda, 2003). Tsunami & Wada (2008) estimated future usage of Sb on the assumption that annual Sb products such as flame retardants will decrease 10% from 2003 to 2015 and the recycle efficiency will increase up to 90% by 2015. In their assumptions, the total atmospheric emission of Sb into 2015 is estimated to decrease 30% compared to that in 2005. If the amount of Sb production is regulated around the world by legislative controls such as RoHS (Restriction of Hazardous Substances) and ELV (End of Life Vehicles), global Sb emission will drastically decrease and this aspect will reflect Sb concentration in environmental samples within chronologic records.

Various studies on the Sb chronology in environmental samples, which act as historical archives provide solid evidence to the fact that human activities have had a strong impact on the environmental. Moreover, monitoring studies on airborne Sb that have been recently carried out in the world show that human activities continue to significantly affect the environment. Identification of the dominant emission sources and an accurate estimation of the emission amounts from the respective Sb sources will be required for the determination of this human impact on the environment. Additionally, further studies on the fates of Sb compounds, including alteration of chemical species *via* ecological metabolisms, will help understand the environmental and geochemical cycles of Sb in a better manner. Furthermore, more in-depth knowledge of the toxicity of Sb compounds and the evaluation of their safety limits are indispensable for assessing the potential risk of Sb to ecosystems. On the basis of the above-mentioned criteria, the environmental significance of potentially toxic Sb and its compounds will be duly evaluated.

ACKNOWLEDGMENT

We are grateful for the financial support provided by the Ministry of Education, Science, Sports, and Culture, Japan through the Gants-in-Aid for scientific research (project number 19710022, 19750064, and 17550083), and we are also grateful to Kiyotaka Tsunemi (National Institute of Advanced Industrial Science and Technology), Toshimasa Ohara, Kiyoshi Tanabe (National Institute for Environmental Studies), Katsuyuki Takahashi (Japan Environmental Sanitation Center), Satoru Udagawa (Tokyo Dylec, Co.), Kiyoko Yano, Yasuo Takagi, Yosuke Sasaki (Akebono Brake Industry, Co., Ltd.), Hiroshi Tago, Kimiyo Kumagai, and Kunihisa Kozawa (Gunma Prefectural Institute of Public Health and Environmental Sciences) for helpful discussions.

REFERENCES

Al-Momani I. F., Daradkeh A. S., Haj-Hussein A. T., Yousef Y. A., Jaradat Q. M. & Momani K. A. (2005). Trace elements in daily collected aerosols in Al-Hashimya, central Jordan. *Atmos. Res.*, 73, 87–100.

Barrie L. A. & Hoff R. M. (1985). Five years of air chemistry observations in the Canadian Arctic. *Atmos. Environ.*, 19, 1995–2010.

Beavington F., Cawse P. A. & Wakenshaw A. (2004). Comparative studies of atmospheric trace elements: improvements in air quality near a copper smelter. *Sci. Total Environ.*, 332, 39–49.

Boeck M. D., Kirsch-Volders M. & Lison D. (2003). Cobalt and antimony: genotoxicity and carcinogenicity. *Mutat. Res.*, 533, 135–152.

Boyle R. W. & Jonasson I. R. (1984). The geochemistry of Sb and its use as an indicotor element ingeochmical prospecting. *J. Geochem. Explor.*, 20, 223–302.

Butterman W. C. & Carlin J. F. Jr. (2004). Antimony. In US Geological Survey ed. *Mineral Commodity Profiles*. Washington, DC U.S.A.: United States Government Printing Office.

Carlin J. F. Jr. (2000). Antimony Recycling in the United States in 2000. In US Geological Survey ed. *US Geological Survey Circular 1196-Q*. Washington, DC U.S.A.: United States Government Printing Office.

Carlin J. F. Jr. (2002). Antimony. In *Metals and Minerals, U.S. Geological Survey Minerals Yearbook 2001, volume I* (pp. 7.1-7.7). Washington, DC U.S.A.: United States Government Printing Office.

Cavallo D., Iavicoli I., Setini A., Marinaccio A., Perniconi B., Carelli G. & Iavicoli S. (2002). Genotoxic risk and oxidative DNA damage in workers exposed to antimony trioxide. *Environ. Mol. Mutagen.*, 40, 184–189.

Chen J., Tan M., Li Y., Zheng J., Zhang Y., Shan Z., Zhang G. & Li Y. (2008). Characteristics of trace elements and lead isotope ratios in $PM_{2.5}$ from four sites in Shanghai. *J. Hazard. Mater.*, 156, 36–43.

Cloy J. M., Farmer J. G., Graham M. C., MacKenzie A. B. & Cook G. T. (2005). A comparison of antimony and lead profiles over the past 2500 years in Flanders Moss ombrotrophic peat bog, Scotland. *J. Environ. Monit.*, 7, 1137–1147.

Council of the European Communities (1976). *Council Directive 76/464/EEC of 4 May 1976 on pollution caused by certain dangerous substances discharged into the aquatic environment of the Community*. Official Journal, L 129, 23–29.

Dietl C., Reifenhäuser W. & Peichl L. (1997). Association of antimony with traffic – occurrence in airborne dust, deposition and accumulation in standardized grass cultures. *Sci. Total Environ.*, 205, 235–244.

Doğan G., Güllü G. & Tuncel G. (2008). Sources and source regions effecting the aerosol composition of the Eastern Mediterranean. *Microchem. J.*, 88, 142–149.

Elliott B. M., Mackay J. M., Clay P. & Ashby J. (1998). An assessment of the genetic toxicology of antimony trioxide. *Mutat. Res.*, 415, 109–117.

Emsley J. (2001). Nature's Building Blocks: *An A–Z Guide to the Elements*. New York: Oxford Univ. Press.

Felicetti S. A., Thomas R. G. & McClellan R. O. (1974). Metabolism of two valence states of inhaled antimony in hamsters. *Am. Ind. Hyg. Assoc. J.*, 35, 292–300.

Filella M., Belzile N. & Chen Y. W. (2002a). Antimony in the environment: a review focused on natural waters I. Occurrence. *Earth-Sci. Rev.*, 57, 125–176.

Filella M., Belzile N. & Chen Y. W. (2002b). Antimony in the environment: a review focused on natural waters II. Relevant solution chemistry. *Earth-Sci. Rev.*, 59, 265–285.

Filella M., Belzile N. & Chen Y. W. (2007). Antimony in the environment: a review focused on natural waters III. Microbiota relevant interactions. *Earth-Sci. Rev.*, 80, 195–217.

Furuta N., Iijima A., Kambe A., Sakai K. & Sato K. (2005). Concentrations, enrichment and predominant sources of Sb and other trace elements in size classified airborne particulate matter collected in Tokyo from 1995 to 2004. *J. Environ. Monit.*, 7, 1155–1161.

Gao Y., Nelson E. D., Field M. P., Ding Q., Li H., Sherrell R. M., Gigliotti C. L., Van Ry D. A., Glenn T. R. & Eisenreich S. J. (2002). Characterization of atmospheric trace elements on $PM_{2.5}$ particulate matter over the New York–New Jersey harbor estuary. *Atmos. Environ.*, 36, 1077–1086.

Garg B. D., Cadle S. H., Mulawa P. A. & Groblicki P. J. (2000). Brake wear particulate matter emissions. *Environ. Sci. Technol.*, 34, 4463–4469.

Gebel T. (1997). Arsenic and antimony: comparative approach on mechanistic toxicology. *Chem.–Biol. Interact.*, 107, 131–144.

Gebel T., Christensen S. & Dunkelberg H. (1997). Comparative and environmental genotoxicity of antimony and arsenic. *Anticancer Res.*, 17, 2603–2607.

Gebel T. (1998). Suppression of arsenic-induced chromosome mutagenicity by antimony in V79 cells. *Mutat. Res.*, 412, 213–218.

Gerhardsson L. & Nordberg G. F. (1993). Lung cancer in smelter workers-interactions of metals as indicated by tissue levels. *Scand. J. Work Environ. Health*, 19, Suppl. 1, 90–94.

Ghauri B. M., Mirza M. I., Richter R., Dutkiewicz V. A., Rusheed A., Khan A. R. & Husain L. (2001). Composition of aerosols and cloud water at a remote mountain site (2.8 kms) in Pakistan. *Chemosphere–Global Change Sci.*, 3, 51–63.

Godoy M. L. D. P., Godoy J. M. & Artaxo P. (2005). Aerosol source apportionment around a large coal fired power plant–Thermoelectric Complex Jorge Lacerda, Santa Catarina, Brazil. *Atmos. Environ.*, 39, 5307–5324.

Gómez D. R., Giné M. F., Bellato A. C. S. & Smichowski P. (2005). Antimony: a traffic-related element in the atmosphere of Buenos Aires, Argentina. *J. Environ. Monit.*, 7, 1162–1168.

Green J. (2004). An overview of the fire retardant chemicals industry, past - present - future. *Fire Mater.*, 19, 197–204.

Groth D. H., Stettler L. E., Burg J. R., Busey W. M., Grant G. C. & Wong L. (1986). Carcinogenic effects of antimony trioxide and antimony ore concentrate in rats. *J. Toxicol. Environ. Health.*, 18, 607–626.

Hinkley T. K., Lamothe P. J., Wilson S. A., Finnegan D. L. & Gerlach T. M. (1999). Metal emissions from Kilauea, and a suggested revision of the estimated worldwide metal output by quiescent degassing of volcanoes. *Earth Planet. Sci. Lett.*, 170, 315–325.

Hioki T., Nakanishi S., Mukai H. & Murano K. (2006). Analysis of long-range transport of aerosols with water-soluble ionic species and trace metal components, collected continuously with particle size segregation in the coastal area of the Sea of Japan–focusing on the Kosa event in spring 2002. *J. Aerosol Res.*, 21, 160–175.

Hueglin C., Gehrig R., Baltensperger U., Gysel M., Monn C. & Vonmont H. (2005). Chemical characterisation of $PM_{2.5}$, PM_{10} and coarse particles at urban, near-city and rural sites in Switzerland. *Atmos. Environ.*, 39, 637–651.

Iijima A., Sato K., Yano K., Tago H., Kato M., Kimura H. & Furuta N. (2007). Particle size and composition distribution analysis of automotive brake abrasion dusts for the evaluation of antimony sources of airborne particulate matter. *Atmos. Environ.*, 41, 4908–4919.

Iijima A., Sato K., Yano K., Kato M., Kozawa K. & Furuta N. (2008a). Emission factor for antimony in brake abrasion dusts as one of the major atmospheric antimony sources. *Environ. Sci. Technol.*, 42, 2937–2942.

Iijima A., Tago H., Kumagai K., Kato M., Kozawa K., Sato K. & Furuta N. (2008b). Regional and seasonal characteristics of emission sources of fine airborne particulate matter collected in the center and suburbs of Tokyo, Japan as determined by multielement analysis and source receptor models. *J. Environ. Monit.*. 10, 1025–1032.

International Agency for Research on Cancer (IARC) (2001). Antimony trioxide and antimony trisulfide. In IARC Ed., *Monographs on the evaluation of carcinogenic risk to human* (pp. 291–305). Lyon: World Health Organization.

Jang H. & Kim S. J. (2000). The effects of antimony trisulfide (Sb_2S_3) and zirconium silicate ($ZrSiO_4$) in the automotive brake friction material on friction characteristics. *Wear*, 239, 229–236.

Japan Automobile Manufacturers Association (JAMA), Inc. (2005). *The Motor Industry of Japan 2005* (p. 42). Tokyo: JAMA.

Japan Environmental Sanitation Center (1975–1998). *Annual Report on Analytical Results of Airborne Particulate Matter-National Air Surveillance Network (NASN). Japan.* Kawasaki: Japan Environmental Sanitation Center.

Japan Oil, Gas, and Metals National Corporation (JOGMEC) (2007). *Antimony In Annual report of material flow database in 2007* (pp.123-126). Kawasaki, Japan: JOGMEC.

Jones R. D. (1994). Survey of antimony workers: mortality 1961-1992. *Occup. Environ. Med.*, 51, 772–776.

Kalton C. L. (1995). *Automotive Brakes and Antilock Braking Systems*. New York: Delmar Cengage Learning.

Kim E. Hopke P. K. & Edgerton E. S. (2004). Improving source identification of Atlanta aerosol using temperature resolved carbon fractions in positive matrix factorization. *Atmos. Environ.*, 38, 3349–3362.

Knasmuller S., Gottmann E., Steinkellner H., Fomin A., Pickl C., Paschke A., Göd R. & Kundi M. (1998). Detection of genotoxic effects of heavy metal contaminated soils with plant bioassays. *Mutat. Res.*, 420, 37–48.

Krachler M., Zheng J., Koerner R., Zdanowicz C., Fisher D. & Shotyk W. (2005). Increasing atmospheric antimony contamination in the northern hemisphere: snow and ice evidence from Devon Island, Arctic Canada. *J. Environ. Monit.*, 7, 1169–1176.

Krachler M., Zheng J., Fisher D. & Shotyk W. (2008). Atmospheric Sb in the Arctic during the past 16,000 years: Responses to climate change and human impacts. *Global Biogeochem. Cycles*, 22, GB1015.

Kuroda K., Endo G., Okamoto A., Yoo Y. S. & Horiguchi S. (1991). Genotoxicity of beryllium, gallium and antimony in short-term assays, *Mutat. Res.*, 264, 163–170.

Lantzsch L. & Gebel T. (1997). Genotoxicity of selected metal compounds in the SOS chromotest, *Mutat. Res.*, 389, 191–197.

Léonard A. & Gerber G. B. (1996). Mutagenicity, carcinogenicity and teratogenicity of antimony compounds. *Mutat. Res.*, 336, 1–8.

Li T., Archer G. F. & Carapella S.C. Jr. (1992). Antimony and antimony alloys. In *Kirk-Othmer encyclopedia of chemical technology 4th edition, Volume 3* (pp.367-381). New York, NY U.S.A.: John Wiley & Sons, Inc.

Lough G. C., Schauer J. J., Park J. S., Shafer M. M., Deminter J. T. & Weinstein J. P. (2005). Emissions of metals associated with motor vehicle roadways. *Environ. Sci. Technol.*, 39, 826–836.

Maenhaut W. & Cornille P. (1989). Trace element composition and origin of the atmospheric aerosol in the Norwegian Arctic. *Atmos. Environ.*, 23, 2551–2569.

McCallum R. I. (2005). Occupational exposure to antimony compounds, *J. Environ. Monit.*, 7, 1245–1250.

Mizohata A., Ito N. & Kusuya Y. (2000). Characterization of the atmospheric particulate matter observed at the sites near the heavy traffic road in Tokyo. *J. Jpn. Soc. Atmos. Environ.*, 35, 77–102.

Moreno T., Alastuey A., Querol X., Font O. & Gibbons W. (2007). The identification of metallic elements in airborne particulate matter derived from fossil fuels at Puertollano, Spain., *Int. J. Coal Geol.*, 71, 122–128.

Mote R. H. (1948) Antimony. In *Commodity reviews, U.S. Bureau of Mines Minerals Yearbook 1946, part II* (pp. 128-136). Washington, DC U.S.A.: United States Government Printing Office.

Mouli P. C., Mohan S. V., Balaram V., Kumar M. P. & Reddy S. J. (2006). A study on trace elemental composition of atmospheric aerosols at a semi-arid urban site using ICP-MS technique. *Atmos. Environ.*, 40, 136–146.

Newton P. E., Bolte H. F., Daly I. W., Pillsbury B. D., Terrill J. B., Drew R. T., Ben-Dyke R., Sheldon A. W. & Rubin L. F. (1994). Subchronic and chronic inhalation toxicity of antimony trioxide in the rat. *Fandam. Appl. Toxicol.*, 22, 561–576.

Nriagu J. (1983). *Lead and Lead poisoning in Antiquity*. New York: John Wiley.

Nriagu J. (1989). A global assessment of natural sources of atmospheric trace metals. *Nature*, 338, 47–49.

Nyanganyura D., Maenhaut W., Mathuthu M., Makarau A. & Meixner F. X. (2007). The chemical composition of tropospheric aerosols and their contributiong sources to a continental background site in northern Zimbabwe from 1994 to 2000. *Atmos. Environ.*, 41, 2644–2659.

Okuda T., Kato J., Mori J., Tenmoku M., Suda Y., Tanaka S., He K., Ma Y., Yang F., Yu X., Duan F. & Lei Y. (2004). Daily concentrations of trace metals in aerosols in Beijing, China, determined by using inductively coupled plasma mass spectrometry equipped with laser ablation analysis, and source identification of aerosols. *Sci. Total Environ.*, 330, 145–158.

Pacyna J. M. & Pacyna E. G. (2001). An assessment of global and regional emissions of trace metals to the atmosphere from anthropogenic sources worldwide. *Environ. Rev.*, 9, 269–298.

Pakkanen T. A., Loukkola K., Korhonen C. H., Aurela M., Mäkelä T., Hillamo R. E., Aarnio P., Koskentalo T., Kousa A. & Maenhaut W. (2001). Sources and chemical composition of atmospheric fine and coarse particles in the Helsinki area. *Atmos. Environ.*, 35, 5381–5391.

Patterson K. J. & Settle D. M. (1987). Magnitude of the lead flux to the atmosphere from volcanoes. *Geochim. Cosmochim. Acta*, 51, 675–681.

Qi C., Liu G., Chou C. L. & Zheng L. (2008). Environmental geochemistry of antimony in Chinese coals. *Sci. Total Environ.*, 389, 225–234.

Qin Y., Kim E. & Hopke P. K. (2006). The concentrations and sources of $PM_{2.5}$ in metropolitan New York City. *Atmos. Environ.*, 40, S312–S332.

Querol X., Alastuey A., Viana M. M., Rodriguez S., Artiñano B., Salvador P., Garcia do Santos S., Patier R. F., Ruiz C. R., de la Rosa J., Sanchez de la Campa A., Menendez M. & Gil J. I. (2004). Speciation and origin of PM_{10} and $PM_{2.5}$ in Sapin. *Aerosol Sci.*, 35, 1151–1172.

Qureshi S., Dutkiewicz V. A., Khan A. R., Swami K., Yang K. X., Husain L., Schwab J. J. & Demerjian K. L. (2006). Elemental composition of $PM_{2.5}$ in Queens, New York : Solubility and trmporal trends. *Atmos. Environ.*, 40, S238–S251.

Salma I., Maenhaut W., Zemplén-Papp É. & Záray G. (2001). Comprehensive characterisation of atmospheric aerosols in Budapest, Hungary: physicochemical properties of inorganic species. *Atmos. Environ.*, 35, 4367–4378.

Sanders P. G., Xu N., Dalka T. M. & Maricq M. M. (2003). Airborne brake wear debris: size distributions, composition, and a comparison of dynamometer and vehicle tests. *Environ. Sci. Technol.*, 37, 4060–4069.

Sasaki Y. & Yano K. (2007). Friction materials and inorganic fibers for automotive brakes. *Ceramics Japan*, 42, 380–384.

Sato K., Tamura T. & Furuta N. (2008). Partitioning between soluble and insoluble fractions of major and trace elements in size-classified airborne particulate matter collected in Tokyo. *J. Environ. Monit.*, 10, 211–218.

Schrader F. C. (1922). Antimony in 1919. In *Metals Part I* (pp.286-311). Denver, CO U.S.A.: U.S. Geological Survey Mineral Resources of the United States.

Schnorr T. M., Steenland K., Thun M. J. & Rinsky R. A. (1995). Mortality in a cohort of antimony smelter workers. *Am. J. Ind. Med.*, 27, 759–770.

Shotyk W., Krachler M. & Chen B. (2004a). Antimony in recent, ombrotrophic peat from Switzerland and Scotland: Comparison with natural background values (5,320 to 8,020 ^{14}C yr BP) and implications for the global atmospheric Sb cycle. *Global Biogeochem. Cycles*, 18, GB1016.

Shotyk W., Krachler M. & Chen B. (2004b). Anthropogenic Impacts on the Biogeochemistry and Cycling of Antimony. In Sigel A., Sigel H. & Sigel R. K. O. Eds., *Biogeochemistry Availability and Transport of Metals in the Environment*, vol. 44 of *Metal Ions in Biological Systems* (pp. 172–203). New York: Taylor & Francis.

Shotyk W., Krachler M. & Chen B. (2005a). Antimony: global environmental contaminant. *J. Environ. Monit.*, 7, 1135–1136.

Shotyk W., Chen B. & Krachler M. (2005b). Lithogenic, oceanic and anthropogenic sources of atmospheric Sb to a maritime blanket bog, Myrarnar, Faroe Islands. *J. Environ. Monit.*, 7, 1148–1154.

Shotyk W., Krachler M. & Chen B. (2006). Contamination of Canadian and European bottled waters with antimony from PET containers. *J. Environ. Monit.*, 8, 288–292.

Smichowski P. (2008). Antimony in the environment as a global pollutant: A review on analytical methodologies for its determination in atmospheric aerosols. *Talanta*, 75, 2–14.

Sternbeck J., Sjödin Å. & Andréasson K. (2002). Metal emissions from road traffic and the influence of resuspension–results from two tunnel studies. *Atmos. Environ.*, 36, 4735–4744.

Takeda K. (2003). *Data Book of Non-Bromine Flame Retardant Materials*, Tokyo, Japan New Energy and Industrial Technology Development Organization.

Taylor A. R. (1964). Abundance of chemical elements in the continental crust: a new table. *Geochim. et Cosmochim. Acta*, 28, 1273–1285.

Touval I. (1988). Antimony oxide. In *Handbook Of Fillers For Plastics*, Katz H.S. & Mileski J.V. ed. (pp. 279-291). Berlin Germany: Springer.

Tsunemi K. & Wada H. (2008). Substance flow analysis of antimony for risk assessment of antimony and antimony compounds in Japan. *J. Japan Inst. Metals*, 72, 91–98.

Uexküll O., Skerfving S., Doyle R. & Graungart M. (2005). Antimony in brake pads-a carcinogenic component? *J. Clean. Prod.*, 13, 19–31.

U.S. Environmental Protection Agency (USEPA) (1979). *Water Related Fate of the 129 Priority Pollutants, Vol. 1.* USEPA. Washington, DC: USEPA.

U.S. Geological Survey (USGS) (2008). Antimony. In *Mineral Commodity Summaries 2008* (pp. 24–25). Washington, DC: United States Government Printing Office.

Var F., Narita Y. & Tanaka S. (2000). The concentration, trend and seasonal variation of metals in the atmosphere in 16 Japanese cities shown by the results of National Air Surveillance Network (NASN) from 1974 to 1996. *Atmos. Environ.*, 34, 2755–2770.

Voutsa D., Samara C., Kouimtzis Th. & Ochsenkühn K. (2002). Elemental composition of airborne particulate matter in the multi-impacted urban area of Thessaloniki, Greece. *Atmos. Environ.*, 36, 4453–4462.

Watt W. D. (1983). Chronic inhalation toxicity of antimony trioxide: Validation of the threshold limit value. *Doctoral Dissertation*. Detroit, MI: Wayne State University.

Weckwerth G. (2001). Verification of traffic emitted aerosol components in the ambient air of Cologne (Germany). *Atmos. Environ.*, 35, 5525–5536.

Wedepohl K. H. (1995). The composition of the continental crust. *Geochim. Cosmochim. Acta*, 59, 1217–1232.

Wingren G. & Axelson O. (1993). Epidemiologic studies of occupational cancer as related to complex mixtures of trace elements in the art glass industry. *Scand. J. Work, Environ. Health*, 19, Suppl. 1, 95–100.

Zheng J., Tan M., Shibata Y., Tanaka A., Li Y., Zhang G., Zhang Y. & Shan Z. (2004). Characteristics of lead isotope ratios and elemental concentrations in PM_{10} fraction of airborne particulate matter in Shanghai after the phase-out of leaded gasoline. *Atmos. Environ.*, 38, 1191–1200.

Zheng M., Guo Z., Fang M. & Kester D. R. (2007). The influence of meteorology on the organic and inorganic properties of aerosols in Hong Kong. *Atmos. Res.*, 84, 323–336.

In: Airborne Particulates
Editor: Ming Cheng and Wen Liu

ISBN: 978-1-60692-907-0

Chapter 4

AIRBORNE PARTICULATE MATTER IN URBAN AREAS: A STUDY OF THE ROLE OF REACTIVE OXYGEN SPECIES ON THE TOXIC AND CARCINOGENIC EFFECTS AND IMPLICATIONS TO HUMAN HEALTH

Athanasios Valavanidis*[1]*, Konstantinos Fiotakis and Thomais Vlahogianni

Department of Chemistry, University of Athens,
University Campus Zografou,
15784 Athens, Greece

ABSTRACT

Airborne particulate matter (PM) from various combustion emission sources is considered as the most important pollutant for adverse health effects in the respiratory system. There is a large number of toxic and carcinogenic substances from airborne particulate matter entering the lung's alveoli which cause oxidative stress, inflammation and cytotoxic damage. These conditions lead progressively to increased morbidity and mortality as well as to increased risk to lung cancer. Studies investigated their size and Physicochemical characteristics, focusing especially to fine particulate matter with aerodynamic diameter below 10 μm (PM_{10} and $PM_{2,5}$), their transition metals, polycyclic aromatic hydrocarbons (PAH), stable quinoids and carcinogenic nitropyrenes. Fine particles enter the lungs and because of their small size are trapped in the alveoli and pulmonary tissues, where toxic substances in the presence of oxygen and redox reactions promote the production of reactive oxygen species (ROS) especially hydroxyl radicals, which are linked to the pathogenesis of pulmonary and cardiovascular adverse effects causing lipid peroxidation, damages to peptides and cellular DNA. In this study we investigated the problem of ROS generation from airborne PM, exhaust soot (diesel and

[1] Corresponding author: Assoc. Prof. A. Valavanidis, Tel. 00-30-210-7274479, Fax: 00-30-210-7274761, E-mail : valavanidis@chem.uoa.gr

gasoline) and soot from burned wood. Additionally, using Electron Paramagnetic Resonance (EPR) we examined the persistent free radicals in various samples, characteristic to a mixture of semiquinone radicals. Also, we studied the direct production of superoxide anion ($O_2^{\bullet -}$), hydrogen peroxide (H_2O_2) and the damaging hydroxyl radicals ($HO^{\bullet}$) by various extracts of particulate matter. Finally, we examined by EPR the formation of oxidative damage to guanosine nucleobase by PM in aqueous phosphate buffer (pH 7.4). From these results, it was suggested that the cytotoxic and carcinogenic potential of PM is the result of redox cycling of persistent quinoid radicals and transition metal ions on the surface of particles in the generation of ROS with potential for oxidant injurious effects in pulmonary tissues.

Keywords: airborne particulates, particulate matter, reactive oxygen species, free radicals, toxic effects, carcinogenic potential, pulmonary injury, human health

INTRODUCTION

In the last two decades numerous epidemiological investigations have established an association between exposure to airborne particulate matter (PM) from various air pollution sources premature mortality and various degenerative diseases [1, 2]. Case-control studies provided the evidence of a link between air pollution from traffic and lung cancer for non-smokers [3]. Short-term studies in the USA investigated the 20 largest metropolitan areas (50 million inhabitants) in the USA during 1987-1994. All-cause mortality increased by 0.5% for each 10 μg/m3 of PM10 [4]. These findings are very close to the European results from the APHEA programme with studies covering 43 million people living in 29 European cities. The combined effect estimate showed that all-cause daily mortality increased by 0.6% for each 10 μg/m3 increase in PM10 [5].

Although inhaled air particulate matter and other air pollutants have been linked epidemiologically with increasing respiratory and cardiovascular morbidity and mortality, there is still a fundamental lack of understanding of the underlying biological mechanisms [6]. Suspended air particulates, O3 and NOx share the same properties as potent oxidants, oxidizing important biological molecules, either through direct attack in the cellular compartments or indirectly through the activation of intracellular pathways [7, 8, 9]. Experimental observations in vivo and in vitro showed that particulate air pollutants cause pulmonary toxicity by retention and direct interaction in airways and alveoli, interference with either mucociliary or phagocytic clearance of inhalable particles, transport of toxic chemicals which are adsorbed onto the surface of the particle and generation of oxygen free radicals which produce inflammatory responses [10].

Experiments in animals and in human beings, have revealed that inhaled particulate matter cause potent proinflammatory effects involving lung epithelial cells [11] and impairment of the phagocytic function of human alveolar macrophages [12]. Also, particulate matter, directly or through uptake into epithelial cells and macrophages, activate oxidant pathways releasing cytotoxic oxygen free radicals [13, 14]. As a consequence of these processes there is cytokine and mediator releases, resulting in extensive neutrophil migration and T-lymphocyte recruitment and activation [15, 16, 17].

Inhaled Particle Penetration, Deposition, Clearance and Injury to the Respiratory Airways of Human Lungs

The aerodynamic diameter of the airborne particles plays an important role for the penetration into the lungs and their cytotoxicity. Practically there are four distinct fractions: the coarse fraction (>10 μm in diameter, formed by attrition of larger particles), the PM10, "thoracic" particles (smaller than 10 μm in diameter, penetrating into the lower respiratory system), the PM2.5 ("respirable" particles smaller than 2.5 μm, that can penetrate into the gas-exchange region of the lungs), and the ultrafine particles (smaller than 0.1 μm, which are abundant in numbers with very large surface area, but negligible particle mass). Recently, it has been recognized that ultrafine particles, with mass median diameter <0.1 μm, are more toxic when inhaled than PM10, because of their ability to be absorbed into tissues and the circulation. The greatly increased surface area of ultrafine particles is an important factor in their cardiopulmonary toxicity [18,19].

The larger particles (>10 μm) are filtered by the nose, oropharynx and larynx, but particles in the range of 10-1.0 μm penetrate deeper into airways and are deposited in the tracheobronchial tree or the respiratory bronchioles and the alveoli [20]. It has been calculated that during a pollution episode each lung acinus could receive on average 30 million particles and each alveolus about 1500 particles every 24 hours, about 50% being deposited. In chronic lung disease, some alveoli would receive much larger doses [21]. Deposition of inhaled particles can be predicted by mathematical modeling techniques based on various assumptions regarding fluid dynamics, transport through branching airway system, morphology of the lungs and deposition in airway bifurcations and alveoli [22]. Autopsy of lungs of residents from Mexico City, a location with relatively high ambient particle concentrations, indicated that the airway mucosa retained relatively large numbers of chain-aggregated masses of ultrafine carbonaceous spheres, some of which appeared to be combustion products [23]. The lung airways retain mostly PM2.5 rather than PM10, a finding supporting the observation, in some studies, that adverse health effects appear to correlate better with PM2.5 than with PM10 concentrations [24]. Analytical electron microscopy measurements in never-smoking long-term residents of Vancouver (low air pollution) showed that 96% of the effectively retained particles in the lung parenchyma were PM2.5 and only 5% were ultrafine particles [25].

Particle-Induced Pulmonary Oxidative Damage: Mechanisms

Although the actual mechanisms by which the particles cause acute and chronic pulmonary damage remains unresolved, it is widely accepted that inhalable particulate matter causes oxidative damage to the tissues of the lungs through the generation of oxygen free radicals. Experiments with metals and metal complexes associated with PM surfaces were found to cause acute lung injury in rats, following the instillation of metal-rich residual oil fly ash [26]. Bioavailable metals in the surface of PM, which can be removed in a water-soluble fraction, are commonly hypothesized to have a causative role in the lung tissue damage, through generation of oxygen free radicals [27, 28, 29]. Iron release from airborne particulate matter or metals which are present on their surfaces play a catalytic role in the generation of reactive oxygen species, especially hydroxyl radicals ($HO^{\bullet}$) causing extensive damage to cellular biological molecules [30,31].

Particulate air pollutants are mainly the result of combustion of fuels (gasoline, petrol, diesel) and other organic materials. All these particulates are chain-aggregated masses of fine carbonaceous spheres, which contain similar persistent organic free radicals, as in case of tobacco tar and diesel exhaust particles [32,33]. Generation of reactive oxygen free radicals from PM can be considered very important, because it can be linked to inflammatory reaction in the lungs as well as to pulmonary and cardiovascular injury [34].

The solid state spectra of most types of PM by Electron Paramagnetic Resonance presents a single broad unstructured signal with g-value in the range of 2.003-2.0045, which corresponds to the g-value that is characteristic for a family of semiquinone radicals [35,36]. These radicals are at high concentrations in PM, remarkably persistent for months and have the ability to generate oxygen free radicals (including $HO^{\bullet}$) by quinoid redox cycling. During redox cycling, the semiquinone intermediate reduces oxygen into the superoxide anion ($O_2^{\bullet-}$), and the cycle can be sustained by biological reducing equivalents, like NAD(P)H or ascorbate [35,37]. The redox cycling of semiquinones has been proposed as a mechanism of toxicological importance in the lungs for DEP (diesel exhaust particles)[38,39]. Numerous studies showed that the carcinogenicity of quinones results from the formation of ROS, through the redox cycling reactions (shuffling electron enzymatically or nonenzymatically among their reduced and oxidized forms and/or their semiquinone radicals) and with the involvement of metal ions and NADH [40,41,42].

Polycyclic aromatic hydrocarbons (PAH) are mainly products of incomplete combustion and can be found at high concentrations in PM, require metabolic activation to electrophiles, catalyzed by various enzymes through free radical mechanisms, to exert their carcinogenic effects [43]. The three principal pathways of metabolic activation of PAH are: a) generation of diol epoxides, catalyzed by several cytochrome P450s [44], leading to DNA adduct formation and considered to be essential to PAH carcinogenesis [45]; b) formation of radical cations, catalyzed by P450 peroxidases [46], and c) formation of reactive and redox active o-quinones, catalyzed by dihydrodiol dehydrogenases, which can contribute to PAH carcinogenesis and they have the potential to act as tumor promoters [47]. In addition to DNA adduct formation, benzo[a]pyrene-7,8-dione, another metabolite of BaP, can cause DNA strand scission through the formation of ROS, in the presence of NADH and Cu(II) [48,49].

The interaction of transition metals, including iron, manganese and copper, with various organic xenobiotics which can be found in PM (hydroquinone, phenolic compounds etc) via a metal-redox mechanism, results in their activation to radical intermediates, leading to reactive oxygen species and/or electrophiles which may cause a spectrum of lesions in DNA [50,51].

In this study, we used a series of specimens of particulate matter from various pollution sources: a) Total Suspended Particulates (TSP), PM_{10} and $PM_{2,5}$ from a commercial area with heavy vehicle traffic (center of Athens, Greece), b) fresh soot produced from diesel and gasoline vehicles and soot of diesel and gasoline deposited in the exhaust pipes, c) soot deposited on chimney walls from wood combustion. Our aim was to explore the possible mechanisms, interactions and parameters influencing the formation of reactive oxygen species of toxicological and carcinogenic interest. We investigated the generation of stable free radicals by Electron Paramagnetic Resonance (EPR), the formation of superoxide anion

($O_2^{\bullet-}$) and hydroxyl radicals ($HO^{\bullet}$) through the involvement of the transition metals and stable radicals of quinoids, present in the airborne particulate matter. Also, we investigated the radical damage to nucleobases of DNA by hydroxyl radicals produced from PM in aqueous buffer solutions.

Materials and Methods

Chemicals

Spin-traps DMPO (5,5-dimethyl-1- pyrroline-N-oxide), MNP (2-methyl-2-nitroso-propane, dimer), dG (2-deoxyguanosine), 8-HOdG standard (8-hydroxy-2'-deoxyguanosine) and deferoxamine mesylate were purchased from Sigma. EDTA-Na2 (ethylenediamine tetraacetic acid, disodium salt), phosphate buffer (pH 7.4), hydrogen peroxide (30%),were purchased from Merck. All other chemicals used were of analytical quality.

Samples of Airborne Suspended Particulates

(a) TSP in the period 2003–2004, in the centre of Athens (Division of Atmospheric Pollution and Noise Control of the Ministry of Environment, Regional Planning and Public Works), by high volume pumps (80m3h21) on a 24 h basis and collected using pre-weighed glass microfibre filters (Whatman GF/A 20.3 × 25.4 cm) with 99% collection efficiency for particles with 0.3 mm diameter. The location of the sampling site was in the centre of Athens mainly influenced by local traffic.

(b) Samples of soot from diesel vehicles were collected on annealed glass–fibre filters (Ø 5 cm), using a low volume air sampler, at a distance of 0.5m from the exhaust pipe of a diesel taxi (2.0 L engine, GM Astra model, 5 years old). Vehicle exhaust emissions sampling lasted for about 30 min in a test laboratory and used as "fresh" soot to distinguish from soot deposited in the exhaust pipes. Vehicle exhaust emissions sampling represent warm-engine operation at 800 (idle mode) and 2500 rpm (simulating urban driving conditions).

(c) Samples of gasoline exhaust particles were collected from a gasoline on annealed glass–fibre filters (Ø 5 cm), using a low volume air sampler, at a distance of 0.5m from a gasoline exhaust engine (1.6 L, passenger vehicle Ford, 10 years old, without catalytic converter). Vehicle exhaust emissions sampling lasted 1 h and was used as "fresh" soot in our experiments.

(d) Samples of soot from wood combustion (pine, olive oil tree, beech) were collected on a glass microfibre filters from a domestic fireplace exhaust chimney at a distance of 1m with a low volume air sampler.

(e) PM10 were collected in the same site as in the case of TSP with a high volume sampling system equipped with a six-stage slit cascade impactor (Andersen) with back-up filter, which effectively separates the PM into seven fractions, with the following cut-off diameters at 50% efficiency: >10.2, 10.2–4.2, 4.2–2.1, 2.1–1.4, 1.4–0.73, 0.73–0.41, <0.41 μm. Unloaded and loaded filters were kept in a darkened desiccator for 48 h before weighing for gravimetric analysis. Loaded filters were wrapped in solvent-rinsed aluminium foils and kept refrigerated until extraction (within 3–7 days).

(f) PM2.5 were collected with the same high-volume cascade impactor (Andersen) system. We used the fraction with aerodynamic diameter 2.1 mm and lower (we used the notation PM2.5 which is common in most studies) All samples were kept at -20°C for 3–7 days until used in experiments.

Electron Paramagnetic Resonance (EPR) Characterization of Particulate Matter

The EPR spectra of all samples were measured using a Varian E-4 EPR spectrometer. Typical parameters: 100 kHz, X-band; microwave frequency 9.4 GHz; attenuation power 20 mW; modulation amplitude 1-2 G; scan range 100 G; time constant 1 sec; scan time 8 min; receiver gain 2.5×10^2-5×10^3. The g-values of EPR spectra were calculated from the g-value (2.0036) of the stable free radical of 2,2-diphenyl-1-picrylhydrazyl (DPPH). The radical spins/g values of representative samples of soot (which represent stable free radicals) were determined by comparison with DPPH. Pre-weighed samples (mg) of soot were inserted in quartz EPR tubes and their spectrum was recorded. The area under the EPR single-broad peak was cut out and weighted, representing a quantitative measure of radical spins. Comparison was achieved with the EPR spectrum of pre-weighed sample of DPPH (which is known to be 95% in a stable free radical state).

Extraction Of PM From Filters

Filters were cut into smaller pieces and added to a conical flask with 50 ml of deionised water. PM were extracted, first by vigorous shaking (20 min) and subsequently by sonication in a waterbath at 20°C for another 20 min. The cleaned filters were removed carefully and because the obtained particle suspension was partly contaminated by small filter fibres (3–5% on a mass basis) a blank filter was treated the same way and was used as a control in all experiments. The suspension was centrifuged for 20 min and the precipitate was dried with a speed vacuum-dryer; under these conditions, 0.01–0.05 g of solid (TSP) was obtained.

Separation, Analysis and Identification of Quinoid Compounds

The EPR spectral characteristics of *o*- and *p*-semiquinone radical anions were observed in aqueous extracts of PM following the method by Zang *et al*[52]. Aqueous extracts of PM samples (sonication for 30 min) were incubated in air-saturated 50 mM carbonate-buffered solution (pH 10.0). The EPR spectral profile showed the presence of *o*- and *p*-semiquinone radical anions with *g* values of 2.0035–2.0040. To confirm the generation of *o*- and *p*-semiquinone radical anions, we compared the EPR spectra of freshly prepared, aerated buffered solutions (pH 10.0) of 1,2-and 1,4-naphthoquinone, 1,4-benzoquinone, catechol and 1,4-dihydroxybenzene under similar alkaline conditions. Also, after extraction of various PM, the remaining black solid dust was dried under reduced pressure and its EPR solid spectra recorded to investigate the shape and characteristics of EPR spectra as well as the presence of stable carbon-centred radicals (Fig. 2).

Spin Trapping and Measurement of Superoxide Anion and Hydroxyl Radicals

Pre-weighed samples of soot (approximately 50-100 mg) were placed in a conical flask with 50 ml phosphate buffer (pH 7.4) and aqueous extraction was performed under sonication for 30 min. Extracts were centrifuged at 25000 rpm for 5 min and the supernatant solution was used for EPR experiments. The first set of experiments concerned the generation of superoxide anion ($O_2^{\bullet -}$) spin-trapped by DMPO, without addition of hydrogen peroxide of the final volume of the mixture. The mixture consisted of 1 ml of soot extract, 2 ml of buffer and 2 ml of DMPO (50 mM) agitated gently by a mechanical shaker for 3 min and transferred in a quartz flat-cell for EPR measurements, because of the instability of the spin-trapped species DMPO-OOH. The positive standard for $O_2^{\bullet -}$ is based on the spin-trapping of DMPO-OOH generated by KO_2 in DMSO with the addition of 18-crown-6 ether to complex K^+. The mixture of 18-crown-6 ether/potassium superoxide (1:1, 5 μM) was dissolved in DMSO [53]. We tested samples of three different qualities: fresh (within of 1 h of collection), after one day and after 5 days, in order to check if volatile and semivolatile compounds are playing a role for the generation of $O_2^{\bullet -}$. The basic reactions involve the reduction of oxygen by the semiquinone radical ($QH^\bullet$) to superoxide anion, which leads to the formation of hydrogen peroxide. Measurements of the EPR spectra was performed with a flat-type quartz cell in a Varian E- 4 spectrometer operating with 1 G (0.1 mT) modulation amplitude, 20 mW microwave power, 3×10^3- 1.25×10^4 receiver gain, 100 G scan range, 1 s time constant and 4 min scan time.

Measurements of Superoxide Anion and Hydroxyl Radicals from Soot

Samples of soot from various air pollution sources were pre-weighted and extracted with dichloromethane under sonication (100 mg of soot extracted twice with 100 ml dichloromethane), and then extracts were evaporated to dryness, residues were dissolved in 10 ml DMSO [66]. EPR measurements of $O_2^{\bullet -}$ and $HO^\bullet$ produced from soot were tested with the EPR spectra produced by a DMSO solution of 50 mM DMPO and 18-crown-6/potassium superoxide (KO_2) (1:1, 5 μM), recorded for 4 min at room temperature after the last addition. The reaction mixture contained 1 ml of 50 mM phosphate buffer (pH 7.4) and 1 ml of DMSO solution of soot extracts (representing 20 mg equivalent) which was prepared as above. The reaction mixture was incubated at 37° C for 30 min and then 1 ml of 50 mM of DMPO in water was added to the mixture. The DMPO-OOH EPR signal disappears by the addition of 0.5 ml of [2mg/ml] of superoxide dismutase (SOD) to the mixture described earlier. Measurements of the EPR spectra was performed with a flat-type quartz cell in a Varian E- 4 spectrometer operating with 1 G (0.1 mT) modulation amplitude, 20 mW microwave power, 3×10^3- 1.25×10^4 receiver gain, 100 G scan range, 1 s time constant and 4 min scan time.

Spin Trapping and Measurement of $O_2^{\bullet-}$ And $HO^{\bullet}$ Generated in Aqueous Buffer Solution

Filters were cut into small pieces and immersed in 30 ml aqueous buffer solution and shaken for 10 min with subsequent ultrasonication for 10 min in a waterbath to release a suspension of PM. The suspension contains soluble and insoluble carbon particles, metal ions, inorganic salts, polycyclic aromatic hydrocarbons (PAHs) and quinoid compounds. The final particle concentration of PM preparations was estimated by comparative turbidometry at 405 nm using carbon black as a standard to ensure use of equal concentrations of particles (method of Knaapen *et al.*[54]). The first set of experiments concerned the generation of $O_2^{\bullet-}$ spin-trapped by DMPO, without addition of hydrogen peroxide in the final volume of the mixture. The mixture consisted of soot suspended in a total of 5 ml of buffer solution, containing 50 mM DMPO. The suspension was not degassed, and was agitated gently on a mechanical shaker for 2 min and transferred quickly to a quartz flat-cell for EPR measurements, because of the instability of the spin trapped species DMPO-OOH. The positive standard for the EPR spectrum of $O_2^{\bullet-}$ is based on the spin-trapping of DMPO-OOH (generated by KO_2 in DMSO with the addition of 18-crown-6 ether to complex K^+ dissolved in DMSO).52 Measurement of the EPR spectra was performed with a flat-type quartz cell in a Varian E-4 spectrometer operating with 1 G (0.1 mT) modulation amplitude, 20 mW microwave power, 3 x 10^3 to 1.25 x 10^4 receiver gain, 100 G scan range, 1 s time constant and 4 min scan time. The original spin-trapped radical DMPO-OOH was converted in 7–8 min to the DMPO-OH radical (1:2:2:1 quartet pattern). Complimentary experiments were carried out with the supernatant solution after centrifugation (20 min) and filtration (0.1 μm filter), containing soluble metal salts and other inorganic salts, in order to test the production of radicals without H_2O_2. The supernatant solution did not generate spin-trapped radical signals, but in the presence of H_2O_2 the DMPO-OH (1:2:2:1 quartet) EPR spectra were recorded, as expected (results not shown). In experiments with the whole suspension (after extraction of PM without centrifugation), the addition of 0.1 M H_2O_2 increased substantially the DMPO-OH signal; however, the DMPO-OOH adduct was not observed.

Spin-Trapping of Hydroxyl Radical Oxidative Damage to Guanosine-5'-Phosphate Nucleobase

Samples of particulate matter from various pollution sources were extracted by dichloromethane then evaporated to dryness and re-suspended in phosphate buffer (pH 7.4). These samples were tested for their ability to produce oxidative damage to the nucleobase, 5'-guanosine monophosphate, sodium salt (G-5'-P) through the production of hydroxyl radicals. As a standard method of hydroxyl radicals it was used the Fe^{2+}/EDTA(sodium salt)-H_2O_2 couple in phosphate buffer (pH 7.4) and the produced radical of the nucleobase was spin-trapped by the MNP (2-methyl-2-nitroso-propane, dimmer, Me_3C-N=O). The MNP solution contained 10-15% acetonitrile (having been left to stir for 2-5 hours to bring about the dissolution of the spin-trap). The EPR spectra support the formation of a mixture of C5 and C4-hydroxyl adducts [55]. To test the effect of chelators on the radical generating activity of PM samples, we used EDTA-Na_2 and desferrioxamine. Ferrous ions (Fe^{2+}) form complexes with EDTA, which increase the ability to produce hydroxyl radicals by lowering the redox potential of iron via chelation. Desferrioxamine blocks all available valences of iron thus reducing substantially its redox potential. Concentrations of the reactants in the final sample were typically 15-30 mg cm^{-3} (substrate, nucleobase), MNP 10^{-2} mol dm^{-3}, 25-50 mg cm^{-3} sample of particulate matter, hydrogen peroxide, H_2O_2, 1×10^{-2} mol dm^{-3}, phosphate buffer (pH 7.4). Before and during mixing, solutions for EPR experiments were purged with dry nitrogen for 2 min to expel oxygen and then transferred in a flat-type quartz cell. EPR instrument parameters (Varian E-4): modulation amplitude 1 G (0.1 mT); scan range 100 G; microwave power 20 mW; scan time 8 min; time constant 1.0 s; and receiver gain 5×10^3-1.25×10^4.

RESULTS

EPR MEASUREMENTS OF PERSISTENT QUINOID RADICALS

The EPR spectra of solid samples of particulate matter from a series of air pollution sources: A. diesel, B. gasoline exhaust particulate matter, C. wood burning soot and D. TSP from the Athens area, are shown in traces A-D in Figure 1. For comparison, trace E depicts the EPR spectrum of a sample of cigarette tar (in solid form) from smoking of 5 cigarettes retained on a Cambridge filter, extracted with benzene and evaporated to dryness. All samples have similar EPR spectra with a single broad unstructured signal with g-value in the range of 2.0035-2.0040, corresponding to a signal from semiquinone ($QH^\bullet$) radical. The broadness, (ΔHp-p = 9.5-10 G) of the EPR signal indicates that is the result of a group of semiquinone, but it may due to the non homogeneous samples and various interactions of the semiquinones with metal ions [36].

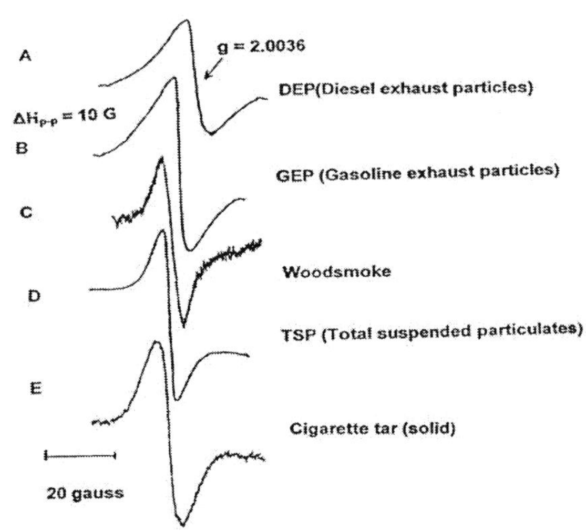

Figure 1. Electron Paramagnetic Resonance (EPR) spectra for the particulate matter samples from 5 different air pollution sources. Trace A, diesel exhaust soot from the center of Athens area; trace B, gasoline exhaust soot; trace C, wood burning soot; trace D, TSP (total suspended particulate); trace E, cigarette tar.

The radicals observed in solid particulate matter of all samples are persistent and are observed even in stored samples for several months. As in the case of cigarette tar, these semiquinone radicals is a dynamic mixture of hydroquinone (QH_2), semiquinone ($QH^\bullet$) and quinone (Q) (Zang, et al., 1995) with characteristic chemical stability [56].

Table 1. Concentrations of persistent radicals in samples of PM from various air pollution sources (radicals/g). Calculations were performed with stable radicals of pure DPPH crystals ($2{\times}10^{21}$ radicals/g). These concentrations are similar with cigarette tar radicals (10^{16}-10^{17} radicals/g) previously reported (Pryor et al, 1983).

PM samples from various pollution sources	Spins of free radicals per g of sample	Measurements in quartz EPR tubes
DPPH	2×10^{21}/ g	Crystalline, 95 % purity
TSP (total suspended particulates)	1-3 $\times$ 10^{15} /g	Extracts with organic solvent
Diesel exhaust particles	2-7 $\times$ 10^{17} /g	Extracts with organic solvent
Gasoline exhaust particles	1-3 $\times 10^{17}$ /g	Extracts with organic solvent
Soot of burned wood	1-2 $\times$ 10^{15} /g	Collected with brush
Tobacco tar (filter)	10^{16}- 10^{17} /g	collected on Cambridge filter, extracted with benzene

Measurements Of Superoxide Anion Radicals From PM Extracts

The production of superoxide anion from DMSO solutions of organic solution extracts of PM is considered as a very important mechanism for the generation of hydrogen peroxide and then the damaging hydroxyl radicals ($HO^\bullet$). We experimented with whole suspensions of particulate matter (both soluble and insoluble) in DMSO with the spin trap DMPO (concentrated, 110 mM). The production of $O_2^{\bullet -}$ depends on the incubation time. The incubation time (at 37°C) plays an important role. Superoxide anion adducts [DMPO-OOH] were formed after 10 min incubation at room temperature and in subsequent EPR spectra were still present after 20 min (Fig. 2).

Also, the EPR spectrum contains signals for another two radicals, namely adducts of hydroxyl radicals $HO^\bullet$ [DMPO-OH] and $^\bullet CH_3$ radicals [DMPO-CH_3], generated from the DMSO [$(CH_3)_2SO$] by the oxidative attack of the hydroxyl radical. The relative contribution of each of these spin adducts to the simulated spectrum was approximately 1:1:1. Fresh diesel exhaust particles (DEP) is the most active, compared to other PM samples, in the production of $O_2^{\bullet -}$. These results are similar to *in vitro* experiments with DEP in DMSO (DEP was extracted twice with dichloromethane, evaporated to dryness and residues were dissolved in DMSO) by Sagai and coworkers[57]. In order to show that the production of $O_2^{\bullet -}$ is the result of quinoid substances in PM, we washed DEP with methanol (removing most of the adsorbed compounds) and the remaining black soot (which is similar to Norit charcoal), after drying, produced under similar conditions very little $O_2^{\bullet -}$ (results not shown).

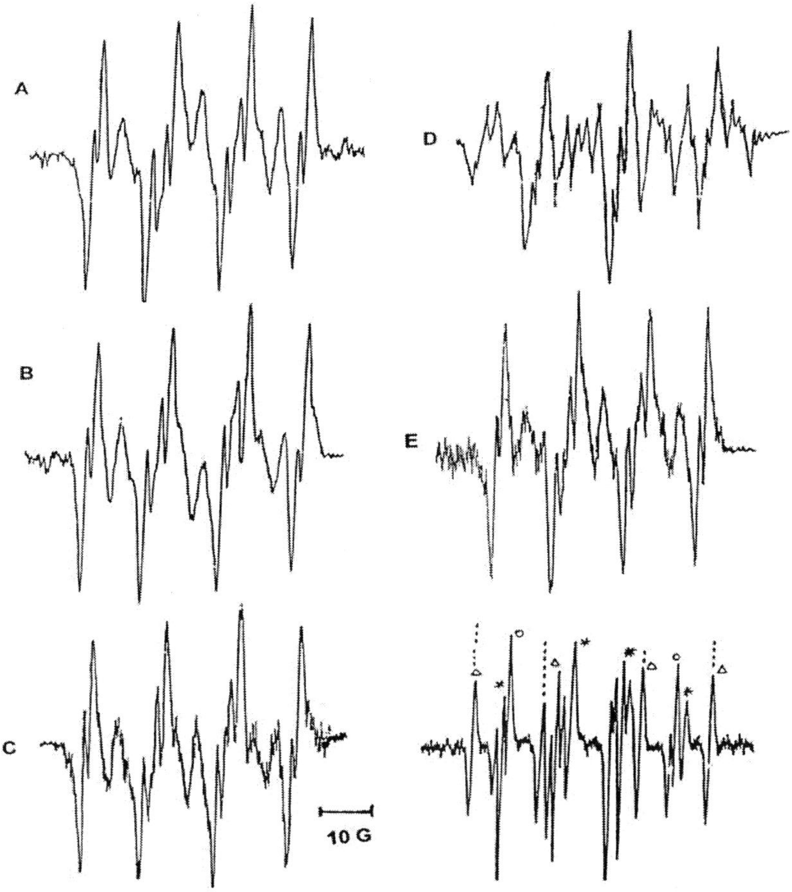

F (simulation)

Figure 2. EPR spectra of particulate matter in DMSO. The multiple spectrum represents three spin adducts: DMPO-OH, DMPO-OOH and DMPO-CH3 (produced from the attack of HO• radical to DMSO [(CH3)2SO]. A, DEP; B, GEP; C, TSP; D, woodsmoke soot; E, cigarette tar; F, simulation of three spin adducts with EPR parameters (hyperfine coupling constants):

Measurements of Superoxide Anion from PM Extracts in Aqueous (pH 7.4)

In another series of experiments, we used the supernatant solution (after centrifugation) and the whole suspensions of PM in phosphate buffer (pH 7.4) without H_2O_2 and in the presence of concentrated DMPO (100 mM; Fig. 3).

$O_2^{\bullet -}$ was formed immediately in the suspension (EPR spectra of spin adduct DMPO-OOH, 2 min after final mixing), but it was replaced very quickly (7–8 min) by hydroxyl radical adducts [DMPO-OH] due to the spontaneous thermal decay of superoxide adduct. The EPR spectrum of $HO^{\bullet}$ has a stronger EPR signal than that from $O_2^{\bullet -}$.

The supernatant solution (centrifuged and filtered, containing mainly soluble metal ions and inorganic salts) without H_2O_2 did not produce the spin adduct DMPO-OH (spectra not shown). However, when H_2O_2 was added to the supernatant solution, only a strong signal of DMPO-OH appeared (Fig. 4).

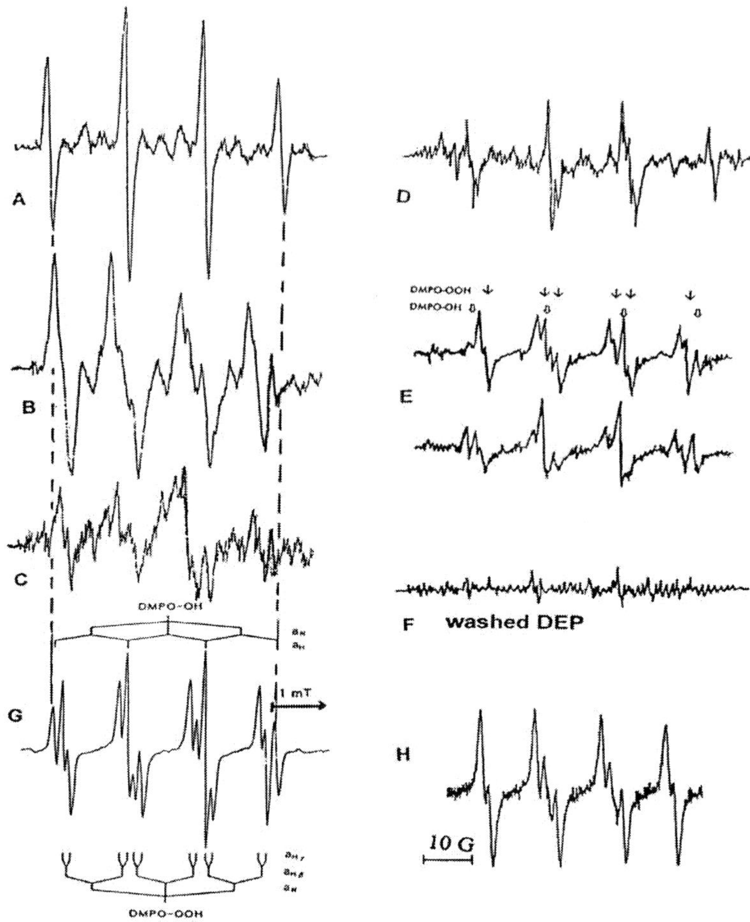

Figure 3. EPR measurements of O2•– and HO• generated from PM samples (suspensions) in phosphate buffer (pH 7.4) by various PM. A, TSP; B, DEP; C, GEP; D, wood smoke soot; E, cigarette tar; at two different time intervals, 2 min and 6 min; F, representative EPR spectrum of washed (with methanol) DEP; G, simulation of spin adducts of DMPO-OH and DMPO-OOH; H, EPR spectrum of DMPO-OOH spin adduct, produced in DMSO with 50 mM DMPO and mixture of KO2/18-crown-6 ether (1:1.5 μM).

EPR Measurements of Hydroxyl Radicals from PM with the Addition of H_2O_2

This investigation showed that $HO^{\bullet}$ can be generated by suspensions of PM without H_2O_2 through the redox cycling of quinoids and subsequent reactions. The addition of hydrogen peroxide (0.02–0.2 M) in the PM suspension enhanced the production of $HO^{\bullet}$, as indicated by the stronger EPR signal of the DMPO-OH adduct (Fig. 4). It is well known that $HO^{\bullet}$ production is the result of ferrous ion catalysed degradation of H_2O_2. Ferrous ions have been found in fresh diesel and gasoline particles, as well as in TSP samples from urban areas.[28, 29] Also, soluble ferric ions (but not Fe_2O_3) are more abundant in airborne particulate matter in urban areas and were observed in TSP. Another important step in the mechanism of $HO^{\bullet}$ generation is the reduction of soluble ferric ions to ferrous ions by $O_2^{\bullet-}$[58].

$$Fe^{3+} + O_2^{\bullet-} \longrightarrow Fe^{2+} + O_2 \quad \text{Eq. 1}$$

As iron appeared to be the most prominent transition metal present in PM suspensions, its role was assessedusing the iron chelators EDTA and desferrioxamine. The importance of ferrous and ferric ions in PM for the production of $HO^{\bullet}$ was established by the addition of EDTA-Na_2 which increased the signal of DMPO-OH. The addition of desferrioxamine inhibited $HO^{\bullet}$ formation totally. Under physiological conditions, iron is only slightly soluble, tending to form precipitates with anions such as HO^- and inorganic phosphate. However, a variety of chelating agents (such as EDTA) greatly increase the solubility of iron. The addition of EDTA to a generating system of $O_2^{\bullet-}$ in the presence of Fe^{3+} markedly increased the formation of $HO^{\bullet}$ by increasing the effective concentration of Fe^{3+} (the catalytic activity of iron requires at least one co-ordination site that is open or occupied by a readily dissociated ligand such as water) at neutral pH.

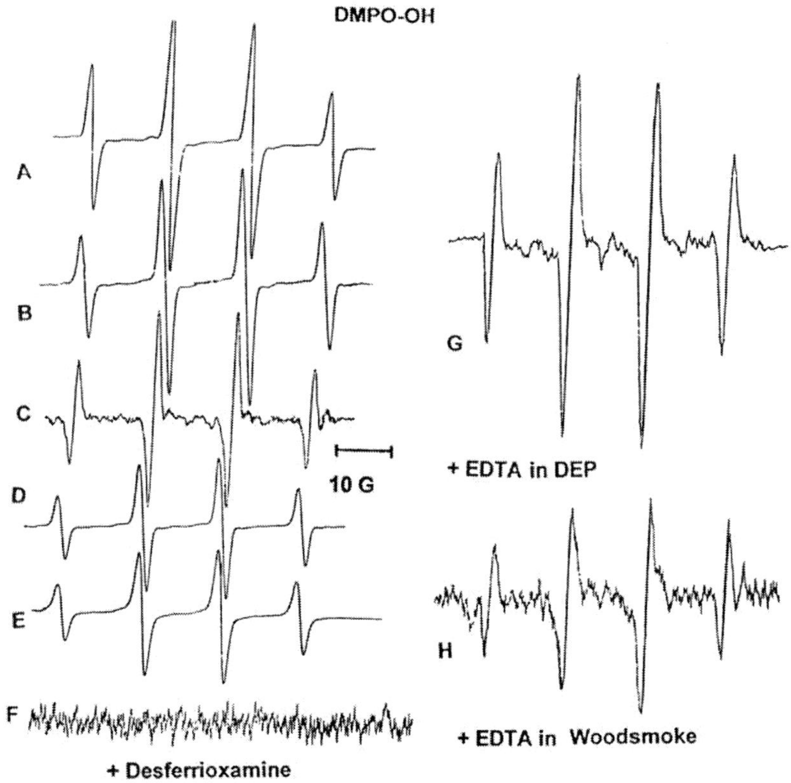

Figure 4. EPR spectra of DMPO-OH. Production of hydroxyl radicals by various PM samples (addition of H2O2) by a Fenton-type reaction. Addition of EDTA (disodium salt) in the mixture increased the generation of hydroxyl radical, whereas addition of desferrioxamine reduces substantially the production of hydroxyl radicals because it blocks effectively all co-ordination sites. A, DEP; B, GEP; C, TSP; D, wood smoke; E, cigarette tar; F, addition of desferrioxamine; G and H, addition of EDTA in DEP and wood smoke.

EPR Measurements of Hydroxy Adducts to Guanosine Nucleobase

Spin-trapping experiments confirm that hydroxyl radicals produced by various samples of PM in the presence of hydrogen peroxide induce damage in the form of hydroxy-adducts to guanosine 5'-monophosphate nucleobase in phosphate buffer solutions (pH 7.4). The EPR spectra observed are similar to the ones observed with the Fe^{2+}-EDTA/H_2O_2 couple, producing hydroxyl radicals. The hyperfine splittings: $\alpha(\alpha\text{-N})=1.54$ mT (or 15.4 G) is the dominant signal triplet of triplets (1:1:1), and $\alpha(\text{b-N}) = 0.3$ mT is assigned to with the spin trapping of the C_4-OH radical [55,59]. A weaker splitting is from the other two radical adducts C_5-OH or C_8-OH.

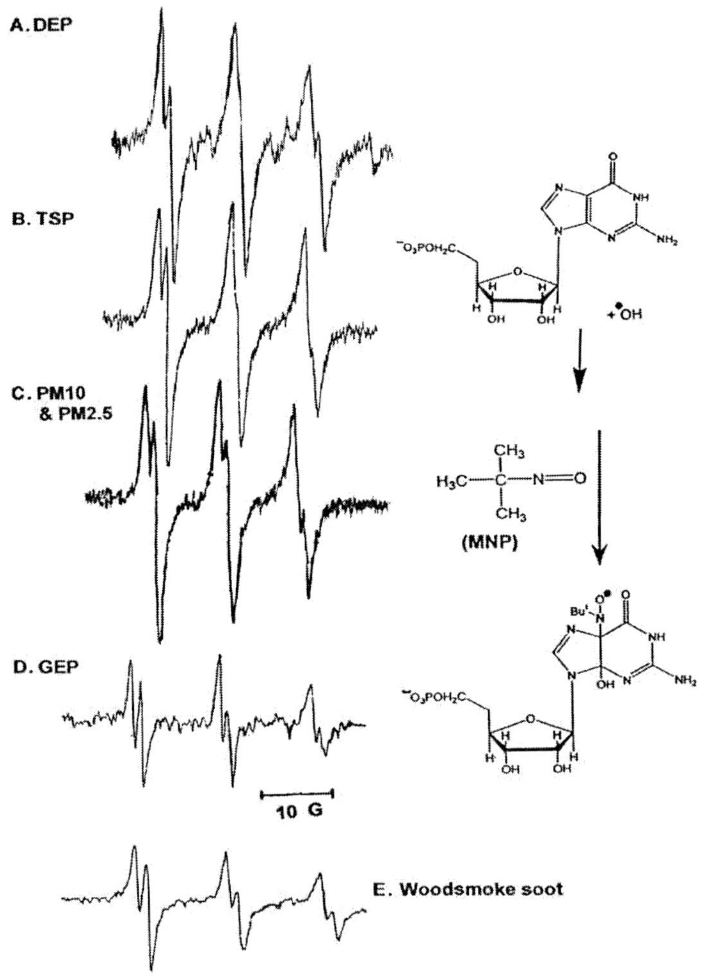

Figure 5. EPR spectra obtained from the reaction of HO• with 2' - deoxyguanosine at pH 7.4 in the presence of spin-trapping MNP. The hydroxyl radical was generated from various suspensions of airborne particulates with H2O2 in buffered solutions (pH 7.4). (A) diesel exhaust particles (DEP), (B) total suspended particulates (TSP), (C) PM10 and PM2.5 (particulate matter), (D), gasoline exhaust particles (GEP), (E) woodsmoke.

Discussion

The presence of quinones/semiquinones in particulate matter is the result of combustion processes. Quinoid compounds (hydroquinone, oxy-PAH, quinones) have been found in fireplace soot [60], in gasoline vehicle exhaust emissions [61] and as diones of PAH in particulate matter [62]. These active quinoid substances are chemisorbed to the surface of the particles, may not be easily leach out and constitute the persistent part of the semiquinones. Others which are adsorbed on the particles may be held with weaker forces and can be leached out into the aqueous phase of the alveoli. Similar radicals have been found in $PM_{2.5}$, are remarkably persistent and can be observed by EPR in samples that had been stored for months [63].

Various mechanisms have been proposed for the generation of superoxide anion by the dynamic mixture $QH_2/QH^{\bullet}/Q$ of quinoids in cigarette tar and similarly for particulate matter. Superoxide anion can be transformed easily into hydrogen peroxide and the biologically damaging hydroxyl radical in a metal-dependent reaction. Also, hydroxyl radicals can be produced by the reaction of $O_2^{\bullet-}$ and H_2O_2, as shown in the following reactions [14, 47].

$$QH^{\bullet} + O_2 \rightarrow Q + O_2^{\bullet-} + H^+ \qquad (1)$$

$$O_2^{\bullet-} + 2H^+ \rightarrow H_2O_2 + O_2 \qquad (2)$$

$$Q + e^- + H^+ \rightarrow QH^{\bullet} \qquad (3)$$

$$QH^{\bullet} + e^- + H^+ \rightarrow QH_2 \qquad (4)$$

$$H_2O_2 + Fe^{2+} \rightarrow Fe^{3+} + HO^{\bullet} + HO^- \qquad (5)$$

$$H_2O_2 + O_2^{\bullet-} \rightarrow HO^{\bullet} + HO^- + O_2 \qquad (6)$$

These reactions establish a redox cycle, where the dynamic quinoid mixture in the presence of oxygen produces cytotoxic hydroxyl radicals which are well known for damages to proteins, lipids and DNA. Transition metals play the usual role, in the Fenton like reaction, (5) for the generation of $HO^{\bullet}$. Reaction (6) can take place after the formation of hydrogen peroxide (reaction 2). Hydrogen peroxide can be synthesized endogenously by certain cell types as a response to activation by specific cytokines or growth factors, also exists physiologically at low concentrations in the lung tissues or is SOD converts $O_2^{\bullet-}$ to H_2O_2 or from inflammatory cells [64,65]. It has been reported that $O_2^{\bullet-}$ was produced during the metabolism of benzo[a]pyrene and 1-nitropyrene (both substances contained in PM in appreciable amounts) in the presence of lung microsomes [66,67].

Twenty five years ago, Pryor *et al.*[56] published their pioneering investigation into the presence of stable free radicals in cigarette tar, and formulated in subsequent years the underlying mechanisms concerning oxidative damage and carcinogenesis by tobacco tar substances. [32,34] Ambient suspended air particulate matter, produced from burned organic fuels or other materials, has many similarities to tobacco tar particles. Coarse (10–2.5 μm), fine (2.5–0.1 μm) and ultrafine (< 0.1 μm) particulate matter, which is produced by combustion of fuels (especially from motor vehicle exhausts) contains acid aerosols, organic compounds (quinones, PAHs, nitroPAHs, *etc.*), hydrocarbons, metals, crustal material and secondary particles formed by chemical reactions with gaseous air pollutants, adsorbed or attached to a typically carbonaceous core. The composition, size of particles and characteristics of PM depend on combustion sources, type of fuel, season and local geography. An important characteristic of all these types of PM is that the solid EPR spectra showed a strong and broad signal of persistent free radicals, suggesting a group of oxygen- and carbon-centred radicals adsorbed into a polymeric matrix. [14] The nature and chemical composition of various PMs (DEP, wood smoke) was investigated by numerous researchers. [8, 68, 57]

In our experiments, all stored samples of PM showed the same broad EPR pattern of persistent free radicals for several months. The stabilization of the semiquinone radicals is probably the result of their adsorption into a polymeric carbonaceous core. The broad EPR signal of PM samples covers a variety of radicals in a range of g values from 2.0028 to g = 2.0035–2.0040. Because of the carbonaceous core of the particles, we assumed that the main radicals are stable semiquinone radicals ($QH^{\bullet}$) or radical anions ($Q^{\bullet -}$) in a polymeric tarry matrix and carbon-centred radicals ($R^{\bullet}$) because of their ability to form ROS.

To compare our results, we draw information from previous investigations concerning cigarette tar and diesel exhaust particles (DEP). In the case of cigarette tar, Pryor and co-workers showed that the group of persistent radicals (g = 2.0038–2.0040) are a mixture of quinone, semiquinone, and hydroquinone with characteristic chemical stability and redox potential. [32,34,57] Quinoid compounds (hydroquinone, oxy-PAH, quinones) have been found in DEP, in NIST Standard Reference Material (SRM) and $PM_{2.5}$, [69] in gasoline vehicle exhaust emissions,[60] and as diones of PAH in particulate matter. [61, 62]

In a previous study [70], 1,2- and 1,4-naphthoquinones and 1,4-benzoquinone were identified in benzene and methanol extracts of DEP. A variety of PM samples from urban areas, such as PM_{10} and $PM_{2.5}$, showed a broad EPR signal of persistent semiquinone radicals.[35] These quinoid substances are direct products of combustion, or oxidative degradation products of PAH and phenols, chemisorbed or adsorbed on the particles and held with weaker forces, so that they can be leached out into the aqueous phase of the alveoli [35].

A number of studies showed that various PM samples have the ability to produce ROS (especially $HO^{\bullet}$) in vitro, but always in the presence of H_2O_2 [6,47]. The production of $HO^{\bullet}$ is an indirect indication of the presence of transition metals, especially Fe^{2+}, or other metals with a redox potential [71]. In a previous investigation in our laboratory, we showed that PM suspensions in aqueous phosphate buffer (pH 7.4), in the presence of H_2O_2, produced $HO^{\bullet}$ by a Fenton-type reaction. The amount of $HO^{\bullet}$ produced was proportional to the Fe^{2+} content of TSP.[29] In the present investigation, we focused on the role of stable quinoid radicals as the main protagonists in the toxic and carcinogenic properties of PM. If these quinoids are embedded deep into the alveolar lung tissues and in the presence of oxygen, they can continuously establish the formation of $O_2^{\bullet -}$ which leads in the aqueous environment to the production of H_2O_2 and, subsequently, in the generation of damaging $HO^{\bullet}$ through a Fenton type-reaction. Superoxide anion was produced by suspensions of PM samples in the absence of H_2O_2, but not by washed PM (with benzene or methanol) which lost most of its quinoid compounds. The remaining carbonaceous polymeric core has the ability to generate hydroxyl radicals (spin-trapped by DMPO) only in the presence of H_2O_2, but lower amount than suspensions of PM (data not shown). This is an indication that carbon-centred radicals ($R^{\bullet}$)

can react with H_2O_2 to produce hydroxyl radicals, or that remnants of transition metals in the polymeric matrix catalyse Fenton-type reactions.

Cellular reducing agents in biological systems, like NAD(P)H and ascorbate, can provide the electrons to reduce quinones back to semiquinones and further to hydroquinone species thus sustaining the redox cycle [35]. Furthermore, quinoids and other organic compounds in diesel soot might be metabolised by enzymatic reduction in biological systems to produce large amounts of $O_2^{\bullet-}$ [71]. A recent study found that ultrafine carbon black particles and transition metals interact by chemical potentiation or synergistic effect in a cell-free environment to generate damaging ROS [72].

Although in cell-free experiments we observed the production of ROS without the presence H_2O_2, it is well known that H_2O_2 exists physiologically in various cell compartments, or is generated endogenously by certain cell types as a response to activation by specific cytokines or growth factors from inflammatory cells [73,74]. Also, polycyclic aromatic hydrocarbons (PAHs) can be a source of superoxide anion in the lungs. It has been reported that $O_2^{\bullet-}$ was produced during the metabolism of benzo[a]pyrene and 1-nitropyrene (both substances contained in PM in appreciable amounts) in the presence of lung microsomes.[64,66]

The size of inhalable PM is also a very important factor for its cytotoxic potential. Fine particles can access subcellular compartments where they can deposit their toxic substances which are very difficult to clear by the mucociliary escalator. In these sites, they can effectively generate ROS near important biomolecules and cause oxidative damage.[18.19,24] The $PM_{2.5}$ fraction is known to be retained in the lung parenchyma and in the respiratory bronchioles.[25] Various biomarkers of free radical damage that have been developed recently (including 8-oxodo,2- aminoadipic semialdehyde in plasma proteins and malondialdehyde) found a significant association with $PM_{2.5}$ and oxidative stress in a group of 40 subjects.[67]

Our experiments with metal chelates, such as EDTA, showed that chelated metals, especially iron, without excluding other metals like Cu and V, can be more reactive in the production of hydroxyl radicals (Fig. 4). It is known that under physiological conditions iron is only slightly soluble, tending to form precipitates with anions such as HO^- and inorganic phosphate, whereas EDTA greatly increases its solubility. An explanation is that the addition of EDTA to a generating system of $O_2^{\bullet-}$ in the presence of Fe^{3+} markedly increased the formation of $HO^{\bullet}$ by increasing the effective concentration of Fe^{3+} (the catalytic activity of iron requires at least one coordination site that is open or occupied by a readily dissociated ligand such as water) at neutral pH [57]. Other studies showed that mobilization of soluble iron from particulate matter, in the ferrous or ferric condition, is very important for its catalytic redox cycling. Citrate, ascorbate and other chelators can mobilize iron in the lung lining fluid before phagocytosis of the particulates[75]. Also, chelating agents like EDTA at physiological pH lower the redox potential of Fe^{3+}/Fe^{2+} thereby favouring the redox reaction. Additional evidence for the role of iron in particulate matter is the inactivation by desferrioxamine, which totally inactivate Fe^{3+} by complexing all co-ordination sites (Fig. 4)[76].

The concentrations of metals in TSP, DEP, GEP and wood smoke established in our measurements were in the range found by other studies. TSP with the highest concentrations of iron and copper in soluble form generated higher amounts of $HO^{\bullet}$ compared to other PM samples. Fresh diesel also produces high amounts of $HO^{\bullet}$ radicals because of the presence of soluble ferrous ions. It has been shown experimentally that water-soluble metals in PM have the potential to cause damage to cellular DNA and other biomolecules, whereas insoluble metal oxides, such as Fe_2O_3, do not have this ability. Under aerobic conditions, the water-soluble metals in PM increase the production of oxygen free radicals enhancing pulmonary injury.[77]

The insoluble part of PM (carbonaceous aggregates of carbon-centred particles) showed a stable broad EPR signal at g = 2.0028; after separation and drying from the aqueous extracts, it was able to produce $HO^{\bullet}$ in the presence of H_2O_2, but not $O_2^{\bullet-}$ (without the presence of H_2O_2). Similarly methanol-washed DEP and other PM produce very small amounts of $O_2^{\bullet-}$ at room temperature. This is a good indication that quinoids are extracted by water or methanol from the polymeric carbonaceous core of PM. Similar results were observed in wood smoke and in methanol-washed DEP, as well as in pure charcoal (Norit A) by Sagai et al.[57] A recent study with DEP found that extractable components and the particle itself play a significant role in the generation of ROS and the catalysis of DNA strand breaks via an oxygen- and reductant-dependent process.[78]

Recent investigations with cell cultures showed that redox-cycling quinones extracted from airborne particulate matter or added to the medium induced mitochondrial dysfunction and apoptosis, through ROS generating mechanisms. Cellular analysis showed that the quinine enriched polar fraction of PM was more potent than the PAH-enriched fraction [79,80]

Also, there is a strong association between exposure to PM and oxidative stress mediated by ROS as an important mechanism of PM-induced lung inflammation. Through metals, quinoids, hydrocarbons and ultrafine constituents, PM exert oxidative stress on cells in the lungs by direct mechanisms or by stimulating the cells in the lungs to produce ROS [73,81, 82]. The role of ROS in pulmonary inflammation by ambient particulates is reviewed recently [83]. Also, increased particulate air pollution triggers adaptive responses to experimental animals. Exposure of a rat model to concentrated ambient particles produced increases of ROS in the lungs and heart, associated with Fe, Mn, Cu and Zn, and led to tissue-specific increases in the activities of SOD and catalase. [84].

The attack of $HO^{\bullet}$, mainly, on the C8-OH position of guanine and secondary on the C_4- and C_5- positions derived from incubation of guanine and their ribose and deoxyribose derivatives with PM/EDTA/H_2O_2 and MNP, at pH 7.4, is represented in EPR spectra in fig.5. Similar spectra were presented by Davies et al [55] derived from incubation of guanine and their ribose and deoxyribose derivatives, with Fe^{2+}/EDTA/H_2O_2 and MNP at pH 7.

CONCLUSIONS

The results of this study show that particulate matter from various pollution sources in urban areas, traffic-related diesel and gasoline exhaust particles and wood smoke, contains a mixture of stable semiquinone radical, metals and PAHs. The dynamic quinoid mixture in the presence of oxygen can produce superoxide anion and hydroxyl radicals without the presence of H_2O_2 in aqueous solution. It was found that the smaller of the aerodynamic diameter of the particles the higher the potential for the production of ROS, especially the damaging hydroxyl radical. The presence of persistent quinoid radicals and trace metal ions, especially iron, play an important role in mechanisms for the production of reactive oxygen species (ROS), leading to oxidative stress, pulmonary tissue injuries and DNA damage. Thus, traffic-related particulate pollution can be very damaging to human health.

References

[1] Brunekreef, B; Holgate, ST. Air pollution and health. *Lancet*, 2002, 360, 1233-1242.

[2] Pope, III AC; Burnett, RT; Thun, JM; Cale, EE; Krewski, D; Ito, K; Thurston, DG. Lung cancer, cardiopulmonary mortality and long-term exposure to fine particulate air pollution. *Journal of American Medical Association*, 2002, 287, 1132-1141.

[3] Nyberg, F; Gustavsson, P; Japur, L et al. Urban air pollution and lung cancer in Stockholm. *Epidemiology*, 2000, 11, 487-495.

[4] Samet, JM; Dominici, F; Curriero, CF; Coursac, I; Zeger, LS. Fine particulate air pollution and mortality in 20 cities, 1987-1994. *New England Journal of Medicine*, 2000, 343, 1742-1749.

[5] Katsouyanni, K; Zmirou, D; Spix, C et al. Short-term effects of air pollution on health: a European approach using epidemiological time-series data. The APHEA project: background, objectives, design. *European Respiratory Journal*, 1995, 8, 1030-1038.

[6] Li, N; Kim S; Wang M; Froines, J; Sioutas, C; Nel, A. Use of a stratified oxidative stress model to study the biological effects of ambient concentrated and diesel exhaust particulate matter. *Inhalation Toxicology* . 2002, 14, 459-486.

[7] Martin, DL; Krunkosky, MT; Dye, AJ; Fischer, BM; Jiang, FN; Rochelle, GL; Akley, JN; Dreber, KL. The role of reactive oxygen and nitrogen species in the response of airway epithelium to particles. *Environmental Health Perspectives* 1997, 105, 1301-1307.

[8] Bai, Y; Suzuki, A; Sagai, M. The cytotoxic effects of diesel exhaust particles on human pulmonary artery endothelial cells in vitro: role of active oxygen species. *Free Radical in Biology and Medicine,* 2001, 30, 555-562.

[9] Li, N; Venkatesan, I; Migue, A; Kaplan, R; Gujuluva, C; Alam, J; Nel, A. Induction of heme oxygenase-1 expression in macrophages by diesel exhaust particle chemicals and quinines via the antioxidant responsive element. *Journal of Immunology* 2000, 165, 3393-3401.

[10] Samet, J.M; Pepelko, EW; Sonawane, B; Hatch, EG; Driscoll, EK; Oberdorster, G. Risk assessment of oxidant gases and particulate air pollutants: uncertainties and research needs. *Environmental Health Perspectives,* 1994, 102(Suppl.10), 209-214.

[11] Fujii, T; Hayashi, S; Hogg, CJ; Vincent, R; Van Eeden, FS. Particulate matter induces cytokine expression in human bronchial epithelial cells. American Journal of Respiratory and *Cell Molecular Biology*, 2001, 25, 265-271.

[12] Lundborg, M; Johard, U; Lastbom, L; Gerde, P; Cammer, P. Human alveolar macrophage phagocytic function is impaired by aggregates of ultrafine carbon particles. *Environmental Research*, 2001, 86, 244-253.

[13] Renwick, CL; Donaldson, K; Clouter, A. Impairment of alveolar macrophage phagocytosis by ultrafine particles. *Toxicology and Applied Pharmacology*, 2001, 172, 119-127.

[14] Dellinger, B; Pryor, AW; Cueto, R; Squadrito, GL; Hedge, V; Deutsch, W.A. Role of free radicals in the toxicity of airborne fine particulate matter. *Chemical Research Toxicology*, 2001, 14, 1371-1377.

[15] Sun, G; Crissman, K; Norwood, J et al. Oxidative interactions and synthetic epithelial lining fluid with metal-containing particulate matter. *American Journal of Physiology Lung Cell Molecular Physiology*, 2001, 281, L807-L815.

[16] Van Eeden, FS; Tan, CW; Suwa, T et al. Cytokines involved in the systemic inflammatory response induced by exposure to particulate matter air pollutants PM(10). *American Journal Respiratory and Critical Care Medicine* 2001, 164, 826-830.

[17] Salvi, S; Blomberg, A; Rudell, B et al.,. Acute inflammatory responses in the airways and peripheral blood after short-term exposure to diesel exhaust in healthy human volunteers. *American Journal of Respiratory and Critical Care Medicine* 1999, 159, 702-709.

[18] Ferin, J; Oberdorster, G; Penney, DP. Pulmonary retention of ultra-fine and fine particles in rats. *American Journal of Respiration and Cell Molecular Biology*, 1992, 6, 535-542.

[19] Brown, DM; Wilson, MR; MacNee, W; Stone, V; Donaldson K. Size-dependent proinflammatory effects of ultrafine polystyrene particles: a role for surface area and oxidative stress in the enhanced activity of ultrafines. *Toxicology and Applied Pharmacology,* 2001, 175, 191-199.

[20] Muir, CDF; Particle deposition. In: Crystal, RG; West, BJ. Barres, JP; Cherniack, NS; Weibel, E.R. *The Lung. Vol. 2*. New York, Raven Press, 1991, 1839-1843.

[21] Seaton, A; MacNee, W; Donaldson, K; Godden D. Particulate air pollution and acute health effects. *Lancet* 1995, 345, 176-178.

[22] Hofmann, W. Modeling techniques for inhaled particle deposition: the state of the art. *Journal of Aerosol Medicine*, 1996, 9, 369-388.

[23] Brauer, M; Avila-Casado, C; Fortoul, TI; Vedal, S; Stevens, B., Churg, A. Air pollution and retained particles in the lung. Environmental Health Perspectives, 2001, 109, 1039-1043.

[24] Churg, A; Brauer, M. Ambient atmospheric particles in the airways of human lungs. *Ultrastructure Pathology*, 2000, 24, 353-361.

[25] Churg, A; Brauer, M; Human lung parenchyma retains PM2.5. *American Journal of Respiratory and Critical Care Medicine,* 1997, 155, 2109-2111.

[26] Dreher, K; Jaskot, R; Kodavanti, U; Lehmann, J; Winsett, D; Costa, D; 1996. Soluble transition metals mediate the acute pulmonary injury and airways hyperreactivity by residual oil fly ash particles. *Chest*, 109, 541-554.

[27] Ghio, A.J; Kennedy, PT; Whorton, AR; Crumbliss, AL; Hatch, EG; Hoidal, J.R;. Role of surface complexed iron in oxidant generation and lung inflammation induced by silicates, *American Journal of Physiology* (Lung Cell Molecular Physiology 7) 1992, 263, L511-L518.

[28] Donaldson, K; Brown, D.M; Mitchell, C; Dineva, M; Beswick, HP; Gilmour, P; MacNee, W. Free radical activity of PM10: iron-mediated generation of hydroxyl radicals. *Environmental Health Perspectives*, 1997, 105, 1285-1289.

[29] Valavanidis, A; Salika, A; Theodoropoulou, A. Generation of hydroxyl radicals by urban suspended particulate air matter. *The role of iron ions. Atmospheric Environment,* 2000, 34, 2379-2386.

[30] Pritchard, JR; Ghio, AJ; Lehmann, JR; Winsett, DW; Tepper, JS; Park, P; Gilmour, MI; Dreher, KL; Costa, DL. Oxidant generation and lung injury after exposure to particulate air pollutants are associated with concentrations of complexed iron. *Inhalation Toxicology*, 1996, 8, 457-477.

[31] Van Maaren, JMS; Borm, AJP; Knapen, A; van Herwijnen, M; Schilderman, AELP; Smith, KP; Aust, A; Tomatis, M; Fubini, B. In vitro effects of coal fly ashs: hydroxyl radical generation, iron release, and DNA damage and toxicity in rat lung epithelial cells. *Inhalation Toxicology*, 1999, 11, 1123-1141.

[32] Zang, L.-Y; Stone, K; Pryor, AW Detection of free radicals in aqueous extracts of cigarette tar by electron spin resonance. *Free Radical Biology and Medicine*, 1995, 19, 161-167.

[33] Ross, MM; Chedekel, MR; Risby, TH. Electron paramagnetic resonance spectrometry of diesel particulate matter. *Environmental International*, 1982, 7, 325-329.

[34] Church, DF; Pryor, AW; Free-radical chemistry of cigarette smoke and its toxicological implications. Environmental Health Perspectives 1985, 64, 111-126.

[35] Squadrito, G.L., Cueto, R., Dellinger, B., Pryor, W.A.,. Quinoid redox cycling as a mechanism for sustained free radical generation by inhaled airborne particulate matter. *Free Radical Biology and Medicine* 2001, 31, 1132-1138.

[36] Dellinger, B; Pryor, AW; Cueto, R; Squadrito, GL; Hedge, V; Deutsch, AW. Role of free radicals in the toxicity of airborne fine particulate matter. *Chemical Research Toxicology*, 2001, 14, 1371-1377.

[37] Roginsky, AV; Barsukova, KT; Stegmann, BH. Kinetics of redox interaction between substituted quinones and ascorbate under aerobic conditions. *Chemico-Biological Interactions*, 1999, 121, 177-197.

[38] Sagai, M; Lim, BH; Ichinose, T. Lung carcinogenesis by diesel exhaust particles and the carcinogenic mechanism via active oxygen. *Inhalation Toxicology*, 2000, 12, 215-223.

[39] Bai Y; Suzuki A; Sagai M. The cytotoxic effects of diesel exhaust particles on human pulmonary artery endothelial cells in vitro: role of active oxygen species. Free *Radical in Biology and Medicine*, 2001, 30, 555-562.

[40] Hirakawa, K; Oikawa, S; Hiraku, Y; Hirosawa, I; Kawanishi, S. Catechol and hydroquinone have different redox properties responsible for their differential DNA-damaging ability. *Chemical Research in Toxicology*, 2002, 15, 76-82.

[41] Bolton, JL; Trush, AM Penning, MT; Dryhurst, G; Monks, TJ. Role of quinines in toxicology. *Chemical Research in Toxicology*, 2000, 13, 135-160.

[42] Qui, BX; Cadenas, E. The role of NAD(P)H: quinone oxidoreductase in quinone-mediated p21 induction in humanl colon carcinoma cells. *Archives of Biochemistry and Biophysics,* 1997, 346, 241-251.

[43] Gelboin, HV. Benzo[α]pyrene metabolism, activation and carcinogenesis: Role and regulation of mixed function oxidases and related enzymes. *Physiology Review*, 1980, 60, 1107-1166.

[44] Thakker, DR; Yagi, H; Lu, AHY; Levin, W; Conney, AH; Jerina, DM. Metabolism of benzo[α]pyrene: Conversion of (±)-trans-7,8-dihydroxy-7,8-dihydrobenzo[α]pyrene to highly mutagenic 7,8-diol-9,10-epoxides. *Proceedings of National Academy of Sciences, USA,* 1976, 73, 3381-3385.

[45] Dennisenko, M.F., Pao, A., Tang, M.-S., Pfeifer, G.P. Preferential formation of benzo[a]pyrene adducts at lung cancer mutational hotspots in p53. *Carcinogenesis*, 1996, 14, 475-482.

[46] Cavalieri, EL; Rogan, EG. Central role of radical cations in metabolic activation of polycyclic aromatic hydrocarbons. *Xenobiotica*, 1995, 25, 677-688.

[47] Penning, MT; Burczynski, EM; Hung, C-H; McCoull, DK; Palackat, NT; Tsuruda, LS. Dihydrodiol dehydrogenases and polycyclic aromatic hydrocarbon activities: generation of reactive and redox active o-quinones. *Chemical Research in Toxicology* 1999, 12, 1-18.

[48] Flowers, L; Bleczinski, FW; Burczynski, EM; Harvey, GR; Penning, MT. Disposition and biological activity of benzo[a]pyrene-7,8-dione: A genotoxic metabolite generated by dihydrodiol dehyrogenase. *Biochemistry*, 1996, 35, 13664-13672.

[49] Ohnishi, S; Kawanishi, S. Double base lesions of DNA by a metabolite of carcinogenic benzo[a]pyrene. *Biochemical and Biophysical Research Communications*, 2002, 290, 778-782.

[50] Li, Y; Trush, M.A. Reactive oxygen-dependent DNA damage resulting from the oxidation of phenolic compounds by a copper-redox cycle mechanism. *Cancer Research*, 1994, 54, 1895s-1898s.

[51] Li, Y; Kuppusamy, P; Zweier, J.L; Trush, AM. ESR evidence for the generation of reactive oxygen species from the copper-mediated oxidation of the benzene metabolite, hydroquinone: role in DNA damage. *Chemico-Biological Interactions*, 1995, 94, 101-120.

[52] Zang, L.-Y; Stone, K; Pryor, AW; Detection of free radicals in aqueous extracts of cigarette tar by electron spin resonance. *Free Radical Biology and Medicine*, 1995, 19, 161-167.

[53] Arudi, R.L; Allen, AO; Bielski, BH. Some observations on the chemistry of KO2-DMSO solutions. *FEBS Letters* 1, 1981, 35, 265-267.

[54] Knaapen AM, Shi T, Borm AJP, Schins FPR. Soluble metals as well as insoluble particle fraction are involved in cellular DNA damage induced by particulate matter. *Mol Cell Biochem,* 2002; 234/235: 317–326.

[55] Davies, JM; Gilbert, BC; Hazlewood, C; Polack, N. EPR spin-trapping studies of radical damage to DNA. *Journal of Chemical Society, Perkin Transactions* 1995, 2, 13-21.

[56] Pryor, AW; Hales, BJ; Premovic, IP; Church, DF. The radicals of cigarette tar: their nature and suggested physiological implications. *Science*, 1983, 220, 425-427.

[57] Sagai, M; Saito, H; Ichinose, T; Kodama, M; Mori, Y. Biological effects of diesel exhaust particles. I. In vitro production of superoxide and in vivo toxicity in mouse. *Free Radical in Biology and Medicine,* 1993, 14, 37-47.

[58] Graft, E; Mahoney, JR; Bryant, BG; Eaton, JW. Iron-catalyzed hydroxyl radical formation. *J Biol Chem* 1984, 259, 3620–3624.

[59] Veira, ACJS; Candeias, LP; Steenken, *S. Journal of Chemical Physics*, 1993, 90, 881-887.

[60] Purvis, CR; McCrillis, CR; Kariher, PH. Fine particulate matter (PM) and organic speciation of fireplace emissions. *Environmental Science and Technology*, 2000, 34, 1653-1658.

[61] Cadle, HS; Mulawa, P; Groblicki, P; Laroo, C; Ragazzi, AR; Nelson, K; Gallagher, G; Zielinska, B. In-use light-duty gasoline vehicle particulate matter emissions on three driving cycles. *Environmental Science and Technology*, 2001, 35, 26-32.

[62] Koeber, R; Bayona, MJ; Niessner, R. Determination of benzo[a]pyrene diones in air particulate matter with liquid chromatography mass spectrometry. *Environmental Science and Technology,* 1999, 33, 1552-1558.

[63] Squadrito, GL; Cueto, R; Dellinger, B; Pryor, AW. Quinoid redox cycling as a mechanism for sustained free radical generation by inhaled airborne particulate matter. *Free Radical Biology and Medicine*, 2001, 31, 1132-1138.

[64] Burton, HR. Superoxide and hydrogen peroxide in relation to mammalian cell proliferation. *Free Radical Biology and Medicine*, 1985, 18, 775-794.

[65] Hensley, K; Robinson, AK; Gabbita, PS; Salsman, S; Floyd, AR. Reactive oxygen species, cell signaling, and cell injury. *Free Radical Biology and Medicine* 2000, 28, 1456-1462.

[66] Sullivan, PD. Free radical of benzo(a)pyrene and derivatives. *Environmental Health Perspectives,* 1985, 64, 283-295.

[67] Nachtman, JP. Superoxide generation by 1-nitropyrene in rat lung microsomes. *Research Communication in Pathology and Pharmacology*, 1986, 51, 73-80.

[68] Fine PM, Cass GR, Simoneit BRT. Chemical characterization of fine particle emissions from the fireplace combustion of woods grown in the southern United States. *Environ Sci Technol* 2002; 36: 1442–1451.

[69] Cho AK, Di Stefano E, You Y et al. Determination of four quinones in diesel exhaust particles, SRM 1649a, and atmospheric PM2.5. *Aerosol Sci Technol* 2004; 38: 68–81.

[70] Kumagai Y, Taira J, Sagai M. Apparent inhibition of superoxide dismutase activity in vitro by diesel exhaust particles. *Free RadicBiol Me*d 1995; 18: 365–371.

[71] Shi T, Schins RPF, Knaapen AM et al. Hydroxyl radical generation by electron paramagnetic resonance as a new method to monitor ambient particulate matter composition. *J Environ Monitor* 2003; 5: 550–556.

[72] Kappus H. Overview of enzyme systems involved in bioreduction of drugs and in redox cycling. *Biochem Pharmacol* 1986; 35: 1–6.

[73] Wilson MR, Lightbody JH, Donaldson K, Sales J, Stone V. Interaction between ultrafine particles and transition metals in vivo and in vitro. *Toxicol Appl Pharmacol* 2002; 184: 172–179.

[74] Gavett SH, Haykal-Coates N, Copeland LB, Heinrich J, Gilmour MI. Metal composition of ambient PM2.5 influences severity of allergic airways disease in mice. *Environ Health Perspect* 2003;111: 1471–1477.

[75] Sorensen M, Autrup H, Moller P et al. Linking exposure to environmental pollutants with biological effects. *Mutat Res* 2003; 544: 255–271.

[76] Burkitt MJ, Gilbert BC. The autoxidation of iron(III) in aqueous systems: the effect of iron chelation by physiological, nonphysiological and therapeutic chelators on the generation of reactive oxygen species and the inducement of biomolecular damage. *Free Radic Res Commun* 1991; 14: 107–123.

[77] Gutteridge JMC, Richmond R, Halliwell B. Inhibition of the iron-catalysed formation of hydroxyl radicals from superoxide and of lipid peroxidation by desferrioxamine. *Biochem J* 1979; 184: 469–472.

[78] Pan C-JG, Schmitz DA, Cho AK, Froines J, Fukuto JM. Inherent redox properties of diesel exhaust particles: catalysis of the generation of reactive oxygen species by biological reductants. *Toxicol Sci* 2004; 81: 225–232.

[79] Xia T, Korge P, Weiss JN et al. Quinones and aromatic chemical compounds in particle matter induce mitochondrial dysfunction: implications for ultrafine particle toxicity. *Environ Health Perspect* 2004; 112: 1347–1358.

[80] Henry TR, Wallace KB. Differential mechanisms of cell killing by redox cycling and arylating quinines. *Arch Toxicol* 1996; 70: 482–489.

[81] Prahalad AK, Inmon J, Dailey LA, Madden MC, Ghio AJ, Gallagher JE. Air pollution particles mediated oxidative DNA base damage in a cell free system and in human airway epithelial cells in relation to particulate metal content and bioreactivity. *Chem Res Toxicol* 2001; 4: 879–887.

[82] Shi T, Knaapen AM, Begerow J, Birmili W, Borm PJ, Schins RP. Temporal variation of hydroxyl radical generation and 8- hydroxy-2'-deoxyguanosine formation by coarse and fine particulate matter. *Occup Environ Med* 2003; 60: 315–321.

[83] Upadhyay D, Panduri V, Ghio A, Kamp DW. Particulate matter induces alveolar epithelial cell DNA damage and apoptosis: role of free radicals and the mitochondria. Am J Respir *Cell Mol Biol* 2003; 29: 180–187.

[84] Toa F, Gonzalez-Flecha B, Korbzik L. Reactive oxygen species in pulmonary inflammation by ambient particulates. *Free Radic Biol Med* 2003; 35: 327–340.

In: Airborne Particulates
Editor: Ming Cheng and Wen Liu

ISBN: 978-1-60692-907-0

Chapter 5

MULTIPLE TECHNIQUES FOR RESEARCHING AIRBORNE PARTICULATES: A COMPREHENSIVE CASE STUDY OF FALLON, NEVADA

Paul R. Sheppard1[a], Gary Ridenour[b] and Mark L. Witten[c]

[a]Laboratory of Tree-Ring Research, University of Arizona, Tucson, Arizona USA 85721
[b]625 W. Williams Ave., Suite B, Fallon, Nevada USA 89406
[c]Department of Pediatrics, University of Arizona, Tucson, Arizona USA 85721

ABSTRACT

Airborne particulates are an important public health concern. Because of toxic effects of airborne particulates, they are monitored regularly for total loading and chemical constituents. Monitoring includes various techniques to measure airborne particulates. The primary objective of this paper is to describe, compare, and contrast airborne particulate techniques, both theoretically and in practice. Many of these techniques have been applied recently in a single location–Fallon, Nevada–where a cluster of childhood leukemia has been ongoing since 1997. This offers an opportunity to assess advantages and disadvantages of airborne particulate techniques.

Techniques: Total suspended particulate (TSP) chemistry is the filtering of air to capture airborne dust for measurement and interpretation of elements. Surface dust chemistry is the measurement and interpretation of elements in fine sediments that accumulate on outdoor surfaces. Lichen chemistry is the measurement and interpretation of elements in or on lichens. Leaf chemistry is the measurement and interpretation of elements in or on leaves of trees and other plants. Bark chemistry is the measurement and interpretation of elements on the surface of bark of trees and other plants. Dendrochemistry is the measurement and interpretation of chemical elements in tree rings.

Fallon: TSP chemistry showed elevated tungsten and cobalt in airborne particulates of Fallon compared to other towns of west central Nevada. Morphological and chemical

[1] Corresponding author: (520) 621-6474, fax: (520) 621-8229, sheppard@ltrr.arizona.edu

testing of Fallon airborne tungsten particles showed them to be anthropogenic in origin, not natural. Surface dust chemistry and leaf surface chemistry showed peaks of tungsten and cobalt just northwest of the center of town. Lichen tissue chemistry showed elevated airborne tungsten and cobalt within Fallon compared to outlying desert. Dendrochemistry showed environmental availability of tungsten increasing in Fallon during the mid 1990s, coinciding roughly with the onset of the cluster of childhood leukemia in that town.

No single technique for measuring airborne particulates is perfect. Consequently, it would be best to apply multiple techniques in order to capitalize on advantages and offset disadvantages, as illustrated by this comprehensive study of airborne particulates in Fallon.

INTRODUCTION

Airborne particulates are an important public health concern. Outdoor airborne particulates are omnipresent, even in rural areas where air might appear to be clean relative to urban settings (Holmes, 2001). Inhalation of airborne particulates can have many toxicological effects on humans. For example, excessive inhalation of dust can lead to various respiratory ailments (Lippman 2000). Additionally, inhalation of trace elements contained in airborne particulates potentially leads to other illnesses (Costa 2000). A notable example of this is inhalation of lead resulting in increased body burdens of lead (Nriagu 1980), causing a litany of health effects (Singhal and Thomas 1980). Because of toxic effects of airborne particulates, major programs exist to regularly monitor total loading of airborne particulates and their chemical constituents (e.g., U.S. EPA 2008).

As part of these monitoring programs, various techniques exist to measure airborne particulates. The most obvious technique is filtering dust directly from air using regulated vacuum devices (Baron and Willeke 2001). Many other techniques employ less direct means to establish at least relative loadings of elements in airborne dust. The primary objective of this paper is to describe, compare, and contrast airborne particulate techniques, both theoretically and in practice. Many of these techniques have been applied recently in a single location–Fallon, Nevada (Figure 1a)–offering an opportunity to assess their relative advantages and disadvantages for studying airborne particulates.

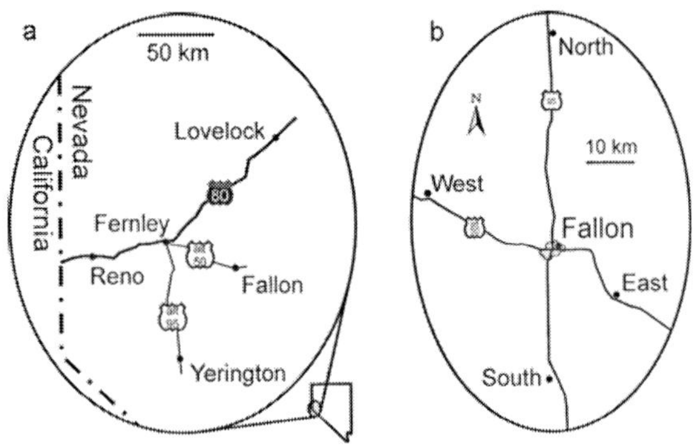

Figure 1.Maps of (a) west central Nevada showing Fallon and comparison towns where TSP samples were collected, and (b) Fallon and surrounding desert where lichens were collected. In (a), other towns besides Fallon were comparison towns for TSP and tree rings. In (b), locations marked outside of Fallon were sample sites for lichens in outlying desert.

Table 1. Characteristics of techniques for researching airborne particulates.

Technique	Field Work	Field Expense	Lab Procedures	Indication of Particulates	Temporal Resolution	Multiple Time Steps	Spatial Resolution	Characterize Particles
total suspended particulates	difficult	high	detailed	direct	hourly to daily	no	low	yes
surface dust	very easy	very low	detailed	direct	daily to weekly	no	high	yes
lichen tissues	moderate	moderate	moderate	indirect	annual to multidecadal	possibly	potentially high	no
lichen surfaces	moderate	moderate	moderate	intermediate	weekly to monthly	no	mixed	possibly
leaf tissues	easy	low	easy	indirect	weekly to seasonal	no	high	no
leaf surfaces	easy	low	easy	intermediate	weekly to seasonal	no	high	possibly
bark surfaces	moderate	moderate	moderate	intermediate	annual to multiannual	no	high	possibly
tree rings	difficult	moderate	very detailed	indirect	annual to centennial	yes	low	no

AIRBORNE PARTICULATE TECHNIQUES (TABLE 1)

Total Suspended Particulates

Total suspended particulate (TSP) chemistry is the filtering of air to capture airborne dust for measurement and interpretation of elements (Solomon et al. 2001). Advantages of TSP are (1) particulate chemistry is determined directly, i.e., without intermediate incorporation of elements by indirect monitors, which makes interpretation of results straightforward, and (2) collected particles can be removed from filters and characterized morphologically and chemically. Disadvantages are (1) field collectors are expensive and cumbersome to deploy, (2) laboratory procedures are detailed and include many steps, (3) individual filters represent only one time interval, and (4) spatial resolution is typically low, largely due to the expense of collectors. Temporal resolution of TSP ranges from hourly to daily, which is on the high-frequency end of the spectrum for airborne particulate techniques. Case studies exist worldwide using TSP chemistry to quantify atmospheric loading of heavy metals (e.g., Trindade et al. 1981; Foner and Ganor 1992; Karue et al. 1992; Pastuszka et al. 1993; Pirrone et al. 1995; Moreno-Grau et al. 2000; Ragosta et al. 2002; Guo et al. 2004; Rajput et al. 2005).

Surface Dust

Surface dust chemistry is the measurement and interpretation of elements in fine sediments that accumulate on outdoor surfaces (Duggan and Inskip 1985). Advantages of surface dust are (1) particulate chemistry is determined directly, (2) collected particles can be isolated and characterized morphologically and chemically, (3) field work is very easy and inexpensive, and (4) spatial resolution can be high, largely due to the ease of field collection. Disadvantages are (1) laboratory procedures are detailed and include many steps, and (2) individual samples represent only one time interval. Temporal resolution of surface dust ranges from daily to weekly depending on frequency and intensity of rain storms. Case studies exist worldwide using surface dust chemistry to quantify atmospheric loading of heavy metals and/or identify spatial patterns of deposition of airborne particulates (e.g., Harrison 1979; Duggan 1984; Fergusson and Ryan 1984; Thornton et al. 1985; Rapsomanikis and Donard 1985; Davies et al. 1987; Tam et al. 1987; Wong and Mak 1997; Benin et al. 1999; Reid et al. 2003; Clark et al. 2005).

Lichens

Lichen chemistry is the measurement and interpretation of elements in or on lichens (Garty 1993), which are symbiotic associations of fungi and algae (Baron 1999). Lichens lack true roots, thereby diminishing the influence of growth substrates on their chemical composition (Szczepaniak and Biziuk 2003; Wolterbeek 2002). Lichens also lack a cuticle, allowing easy incorporation of atmospheric constituents directly into their tissues (Conti and Cecchetti 2001; Falla et al. 2000). An advantage of lichen chemistry is its potentially high spatial resolution: Lichens are widespread geographically (Richardson 1992, 1995), colonizing tree bark or rock surfaces practically everywhere (Wolterbeek and Bode 1995). Even if lichens happen not to be growing in a study site of interest, they can be transplanted to serve as biomonitors of airborne particulates (Conti et al. 2004). It is also theoretically possible to measure multiple times steps from a single lichen, as inner parts of lichens are older than outer edges (Hale 1983), though this feature of lichen growth is not exploited much in lichen chemistry. Disadvantages of lichens are that field and laboratory work is moderately detailed and/or expensive. The indication of airborne particulates from lichen tissue is indirect, but rinsing tissues and analyzing rinse solutions is a more direct version of lichen surface chemistry. Similarly, particles are not easily characterized morphologically and chemically from lichen tissue, but characterization might be possible from lichen surface rinse samples. Temporal resolution of lichen tissue chemistry ranges from annual to multidecadal (Sensen and Richardson 2002), as lichens can live for decades (Bull et al. 1994). Temporal resolution of lichen surface chemistry is weekly to monthly depending on frequency and intensity of rainfall. Case studies exist worldwide using lichen chemistry to monitor or assess airborne elements (e.g., Jeran et al. 1996; Scerbo et al. 1999; Chiarenzelli et al. 2001; Zhang et al. 2002; Zschau et al. 2003; Bennett and Wetmore 2003; Loppi and Pirintsos 2003; Nash et al. 2003; Yun et al. 2003; Yenisoy-Karakaş and Tuncel 2004; Cuny et al. 2004; Helena et al. 2004; Freitas and Pacheco 2004).

Leaves

Leaf chemistry is the measurement and interpretation of elements in or on leaves of trees and other plants (Bargagli 1993). Advantages of leaf chemistry are (1) spatial resolution is high, especially in urban areas where trees are typically heavily planted for landscaping, and (2) field and laboratory work is easy and inexpensive. A disadvantage is that samples represent only one time interval. The indication of airborne particulates from leaf tissue is indirect, but rinsing leaves and analyzing rinse solutions is a more direct version of leaf surface chemistry. Particles are not easily characterized morphologically and chemically from leaf tissue, but characterization might be possible from leaf surface rinse samples. Temporal resolution of leaf samples ranges from weekly to seasonal to perhaps multiannual, depending on rainfall frequency and whether the tree species studied are deciduous or evergreen. Many case studies exist worldwide using leaf chemistry to quantify atmospheric loading of heavy metals and/or identify their spatial patterns (e.g., Ward et al. 1977; Dasch 1987; Aksoy and Öztürk 1997; Alfani et al. 1996; Aksoy et al. 1999; Gupta et al. 2004; Salve et al. 2006; Rossini Oliva and Mingorance 2006).

Bark

Bark chemistry is the measurement and interpretation of elements on the surface of bark of trees and other plants (Walkenhorst et al. 1993). Because of the nature of bark growth, airborne particulates are found mostly on bark surfaces, less so in bark tissue itself (Hampp and Höll 1974). An advantage of bark chemistry is high spatial resolution, especially in urban areas that are landscaped. Disadvantages are (1) field and laboratory work is moderately difficult and expensive, and (2) bark samples represent only one time interval. Temporal resolution of bark surface chemistry ranges from annual to multiannual depending on how fast bark sloughs off from the tree species studied. Case studies exist worldwide using bark chemistry to quantify airborne particulates (e.g., Lötschert and Köhm 1978; Kakulu 2003; Mandiwana et al. 2006; Ward et al. 2006).

Tree Rings

Dendrochemistry is a subdiscipline of dendrochronology, the dating and interpretation of tree rings to reconstruct and analyze past environmental conditions (Fritts 1976). In dendrochemistry, the measurement of primary interest is chemical elements, or suites of multiple elements, contained in individual growth rings or groups of multiple rings (Amato 1988; Smith and Shortle 1996). The principal advantage of dendrochemistry is that relative environmental availability of elements can be reconstructed for multiple intervals back in time, making tree rings unique amongst techniques for studying airborne particulates. Temporal resolution can range from annual to centennial, depending on ages of trees sampled. Disadvantages are (1) fieldwork is difficult and expensive, (2) laboratory preparations are very detailed, including overcoming potential contamination from metal sampling tools (Sheppard and Witten 2005), (3) determination of airborne particulate chemistry is indirect, (4) characterizing particle morphology and chemistry is not possible, and (5) spatial resolution is usually low, largely due to the difficulty of field sampling. Even with these disadvantages, dendrochemistry has been done in a wide array of applications in environmental science, including for heavy metals such as Pb (e.g., Hagemeyer and Weinand 1996), Ni (e.g., Yanosky and Vroblesky 1992), Cd (e.g., Guyette et al. 1991), and Hg (e.g., Zhang et al. 1995).

FALLON, NEVADA

Fallon is a small, rural, farming community (Greater Fallon Area Chamber of Commerce 2008) located in west central Nevada (Figure 1a). Its climate is cool to mild and dry, with a mean annual temperature and precipitation of 10.7° C and 127 mm, respectively, as typified from meteorological data from Fallon (monthly data from 1931 to 2008 obtained on-line from the National Climatic Data Center, NOAA 2008).

A cluster of childhood leukemia has been ongoing in Fallon since 1997. Officially, 16 cases of childhood leukemia were diagnosed from 1997 to 2002 inclusive (Expert Panel 2004), and one additional case was reported in December 2004 (Nevada State Health Division 2004). Given Fallon's pediatric population of ~2500 children up to 19 years in age (U.S. Census 2000) and a national expected rate of childhood leukemia of 4.1 cases per 100,000 children up to 19 years in age per year (U.S. NCI 2003), the expected rate of childhood leukemia for Fallon should be only one case every ten years.

This cluster, deemed "one of the most unique ever reported" (Steinmaus et al. 2004), has prompted multiple investigations to determine if an environmental cause might be responsible. Prior research has focused on drinking water (Moore et al. 2002), jet fuel (U.S.

ATSDR 2002), pesticides (U.S. CDC 2003; Rubin et al. 2007), surface water (U.S. ATSDR 2003a), outdoor air (U.S. ATSDR 2003b), surface soil and indoor dust (U.S. ATSDR 2003c), potential lingering effects of underground nuclear bomb testing in the area (Seiler 2004), and groundwater (Seiler et al. 2005).

Inhalation of dust (airborne particulates) can be a particular health concern for children, who often come in close contact with dust and soil when playing (U.S. ATSDR 2003c). Consequently, TSP chemistry of outdoor air has also been tested in and around Fallon (U.S. ATSDR 2003b). However, early air testing deployed just two high-volume samplers, one in the center of town and one located several kilometers west of town, exemplifying one of the disadvantages of TSP, i.e., low spatial resolution. That was not enough sampling to determine spatial patterns of airborne particulates, and ultimately no association was found between airborne particulate chemistry in Fallon and the childhood leukemia cluster (U.S. ATSDR 2003b, page 25). Because that sampling regimen was sparse, outdoor airborne particulates of Fallon were studied again with a more comprehensive, multi-technique approach. To date, TSP, surface dust, lichens, leaf surfaces, and tree rings have been analyzed in Fallon as well as in nearby comparison towns and from surrounding outlying desert. The following section compares results from these airborne particulate techniques used in Fallon.

Table 2. Summary of results from airborne particulate techniques used in Fallon, Nevada.

Technique	Number of Samples	Number of Days Sampling	Unit of Measurement	Range of Values W	Range of Values Co	Source
total suspended particulates	83	29	$\frac{\text{ng metal}}{\text{m}^3 \text{ air}}$	<1 – ~40	<1 – ~8	Sheppard et al. 2006
surface dust	125	3	$\frac{\mu\text{g metal}}{\text{g dust}}$	<1 – 934	<1 – 98	Sheppard et al. 2007a
lichen tissues	30	3	$\frac{\mu\text{g metal}}{\text{g lichen}}$	<1 – ~40	<1 – ~9	Sheppard et al. 2007b
leaf surfaces	95	2	$\frac{\mu\text{g metal}}{\text{g leaf}}$	<1 – ~17	<1 – ~6	Sheppard et al., in press
tree rings	100s	10	$\frac{\text{ng metal}}{\text{g wood}}$	~40 – ~170	~80 – ~120	Sheppard et al. 2007c

AIRBORNE PARTICULATES IN FALLON (TABLE 2)

Total Suspended Particulates

Detailed methodology of our TSP study in Fallon is described elsewhere (Sheppard et al. 2006). Based on an extensive collection of 83 samples spanning 29 sampling days in Fallon as well as in nearby comparison towns (Figure 1a), tungsten and cobalt can be elevated in airborne particulates of Fallon compared to other towns of west central Nevada. Tungsten and cobalt varied in Fallon airborne particulates both temporally and spatially, with maximum values many times higher than minimum values. Capitalizing on the advantage of TSP that particles can be removed from filters and characterized morphologically and chemically, airborne tungsten particles of Fallon were mostly small (< 5 μm in diameter) and evenly sized, and they were composed mostly of pure tungsten or a mixture of tungsten with cobalt and/or other metals (Sheppard et al. 2007d). As such, airborne tungsten particles of Fallon were anthropogenic in origin, not natural.

The high variability in TSP chemistry results provoked the use of other techniques for studying airborne particulates to better understand airborne tungsten and cobalt in Fallon.

Surface Dust

Detailed methodology of our surface dust study in Fallon is described elsewhere (Sheppard et al. 2007a). Based on an extensive collection of 125 samples located throughout and around Fallon, tungsten and cobalt showed major peaks in surface dust just northwest of the center of town. Peak tungsten and cobalt values were much higher than background values measured from the outskirts of Fallon. Based on this technique, the source area of airborne tungsten and cobalt was fine-tuned to just northwest of the center of town.

Lichen Tissues

Detailed methodology of our lichen tissue study in Fallon is described elsewhere (Sheppard et al. 2007b). Based on a collection of 10 separate lichens growing at one site in Fallon and 20 separate lichens growing in four different locations outside of town (Figure 1b), airborne tungsten and cobalt were elevated within Fallon compared to outlying desert. Tungsten and cobalt were not elevated in the rock substrates of the sampled lichens, so the source of these two metals to lichens within Fallon must have been airborne. Furthermore, natural desert of west central Nevada did not seem to be the source of elevated airborne tungsten and cobalt within Fallon.

Leaf Surfaces

Detailed methodology of our leaf surface study in Fallon is described elsewhere (Sheppard et al., in press). Based on an extensive collection of 95 samples located throughout and around Fallon, airborne tungsten and cobalt concentrations showed major peaks just northwest of the center of town. Peak tungsten and cobalt values were much higher than background values measured from the outskirts of Fallon. Based on this technique, the source area of airborne tungsten and cobalt was confirmed to just northwest of the center of town.

Tree Rings

Detailed methodology of our tree-ring study in Fallon is described elsewhere (Sheppard et al. 2007c). Based on an extensive collection of many trees sampled in Fallon as well as in nearby comparison towns, with each tree having tens of growth rings, environmental availability of tungsten began increasing in Fallon during the mid 1990s, coinciding roughly with the onset of the cluster of childhood leukemia in that town. Cobalt has been elevated in tree rings of Fallon relative to trees in nearby comparison towns throughout this time period.

Conclusion

No single technique for measuring airborne particulates is perfect. Each technique has advantages and disadvantages. Because of this, applying only one technique in research of a particular location might not be fully indicative of and/or convincing about airborne particulates in the study area. Consequently, it would be best to apply multiple techniques in order to capitalize on advantages and offset disadvantages. In general, corroboration across multiple techniques would be more convincing than relying on results of any one technique (Reid and Thompson 1996; Reimann and de Caritat 2005).

All five airborne particulate techniques used so far in Fallon have corroborated one another. Tungsten and cobalt were elevated in airborne particulates of Fallon compared to nearby towns (total suspended particulates) and to outlying desert (lichen tissues). Airborne tungsten and cobalt showed peaks near the center of Fallon compared to the outskirts of town (surface dust and leaf surfaces). Environmental availability of tungsten increased in Fallon during the mid 1990s, coinciding roughly with the onset of the cluster of childhood leukemia (tree rings). Morphological analysis of actual tungsten particles from TSP filters showed them to be anthropogenic in origin, not natural (particle analysis).

This comprehensive study of airborne particulates in Fallon capitalized on the advantages of each technique used. TSP gave a direct loading of airborne tungsten and cobalt (mass of metal per volume of air sample), and it allowed for morphological and chemical characterization of airborne tungsten particles. Lichen chemistry allowed for comparing airborne particulates between Fallon and outlying desert. Surface dust and leaf surface chemistry allowed for fine-scale spatial mapping of airborne tungsten and cobalt within Fallon. Dendrochemistry gave a time series of relative environmental availability of tungsten and cobalt.

As has been stated in other published environmental research of ours pertaining to Fallon, it cannot be concluded from only environmental data that elevated airborne tungsten and/or cobalt cause childhood leukemia. Such a connection requires direct biomedical testing. Nonetheless, given that childhood leukemia in Fallon is the "most unique cluster ever reported" (Steinmaus et al. 2003) and that Fallon is distinctive environmentally by its elevated airborne tungsten and cobalt as determined using multiple airborne particulate techniques, it stands to reason that additional biomedical research is warranted to assess the leukogenicity of airborne tungsten and cobalt (e.g., Miller et al., 2001; Sun et al. 2003; Kalinich et al. 2005; Steinberg et al. 2007).

Acknowledgments

Our environmental research in Fallon was funded in part by the Gerber Foundation and the Cancer Research and Prevention Foundation. This comprehensive review was funded in part by U.S. Environmental Protection Agency. This article does not necessarily reflect the views of the U.S. Environmental Protection Agency and no official endorsement should be inferred.

References

Aksoy, A.; Hale, W. H. G.; Dixon, J. M. Capsella bursa-pastoris (L.) Medic. as a biomonitor of heavy metals. *Sci. Tot. Environ.* 1999, 226, 177-186.

Aksoy, A.; Öztürk, M. A. Nerium oleander L. as a biomonitor of lead and other heavy metal pollution in Mediterranean environments. *Sci. Tot. Environ.* 1997, 205, 145-150.

Alfani, A.; Maisto, G.; Iovieno, P.; Rutigliano, F. A.; Bartoli, G. Lead contamination by atmospheric pollutants as assessed by elemental analysis of leaf tissue, leaf surface deposit and soil. *J. Plant Physiol.* 1996, 148, 243-248.

Amato, I. Tapping tree rings for the environmental tales they tell. *Analyt. Chem.* 1988, 60, 1103A-1107A.

Bargagli, R. Plant leaves and lichens as biomonitors of natural or anthropogenic emissions of mercury. In *Plants as Biomonitors: Indicators of Heavy Metals in the Terrestrial Environment*; Markert, B. A.; VCH, New York, NY, 1993; pp 461-484.

Baron, G. *Understanding Lichens*. Richmond Publishing Company, Slough, England, 1999.

Baron, P. A.; Willeke, K. *Aerosol Measurement: Principles, Techniques, and Applications*, 2nd Edition. Wiley, New York, NY, 2001.

Benin, A. L.; Sargent, J. D.; Dalton, M.; Roda, S. High concentrations of heavy metals in neighborhoods near ore smelters in northern Mexico. *Environ. Health Perspec.* 1999, 107, 279-284.

Bennett, J. P.; Wetmore, C. M. Elemental chemistry of four lichen species from the Apostle Islands, Wisconsin, 1987, 1995 and 2001. *Sci. Tot. Environ.* 2003, 305, 77-86.

Bull, W. B.; King, J.; Kong, F. C.; Moutoux, T.; Phillips, W. M. Lichen dating of coseismic landslide hazards in alpine mountains. *Geomorphology* 1994, 10, 253-264.

Chiarenzelli, J.; Aspler, L.; Dunn, C.; Cousens, B.; Ozarko, D.; Powis, K. Multi-element and rare earth element composition of lichens, mosses, and vascular plants from the Central Barrenlands, Nunavut, *Canada. Appl. Geochem.* 2001, 16, 245-270.

Clark, C. S.; Thuppil, V.; Clark, R.; Sinha, S.; Menezes, G.; D'Souza, H.; Nayak, N.; Kuruvilla, A.; Law, T.; Dave, R.; Shah, S. Lead in paint and soil in Karnataka and Gujarat, India. *J. Occupat. Environ. Hyg.* 2005, 2, 38-44.

Conti, M. E.; Cecchetti, G. Biological monitoring: lichens as bioindicators of air pollution assessment–a review. *Environ. Pollut.* 2001, 114, 471-492.

Conti, M. E.; Tudino, M.; Stripeikis, J.; Cecchetti, G. Heavy metal accumulation in the lichen Evernia prunastri transplanted at urban, rural and industrial sites in central Italy. *J. Atmo. Chem.* 2004, 49, 83-94.

Costa, M. Trace elements: Aluminum, arsenic, cadmium, and nickel. In *Environmental Toxicants: Human Exposures and Their Health Effects*, 2nd Edition; Lippman, M; John Wiley & Sons, New York, NY, 2000; pp 811-850.

Cuny, D.; Davranche, L.; Thomas, P.; Kempa, M.; Van Haluwyn, C. Spatial and temporal variations of trace element contents in Xanthoria parietina thalli collected in a highly industrialized area in northern France as an element for a future epidemiological study. *J. Atmo. Chem.* 2004, 49, 391-401.

Dasch, J. M. Measurement of dry deposition to surfaces in deciduous and pine canopies. *Environ. Pollut.* 1987, 44, 261-277.

Davies, D. J. A.; Watt, J. M.; Thornton, I. Lead levels in Birmingham dusts and soils. *Sci. Tot. Environ.* 1987, 67, 177-185.

Duggan, M. J. Temporal and spatial variations of lead in air and in surface dust–implications for monitoring. *Sci. Tot. Environ.* 1984, 33, 37-48.

Duggan, M. J.; Inskip, M. J. Childhood exposure to lead in surface dust and soil: a community health problem. *Public Health Rev.* 1985, 13, 1-54.

Expert Panel. Final Report and Recommendations to the Nevada State Health Division, Expert Panel on Childhood Leukemia in Churchill County, Nevada. Available: *http://health2k.state.nv.us/healthofficer/leukaemia/FALLONexpertpanel022304.pdf,* 2004.

Falla, J.; Laval-Gilly, P.; Henryon, M.; Morlot, D.; Ferard, J. F. Biological air quality monitoring: A review. *Environ. Monit. Assess.* 2000, 64, 627-644.

Fergusson, J. E.; Ryan, D. E. The elemental composition of street dust from large and small urban areas related to city type, source and particle size. *Sci. Tot. Environ.* 1984, 34, 101-116.

Foner, H. A.; Ganor, E. The chemical and mineralogical composition of some urban atmospheric aerosols in Israel. *Atmos. Environ.* 1992, 26B, 125-133.

Freitas, M. C.; Pacheco, A. M. G. Bioaccumulation of cobalt in Parmelia sulcata. *J. Atmo. Chem.* 2004, 49, 67-82.

Fritts, H. C. *Tree Rings and Climate.* Academic Press, New York, NY, 1976.

Garty, J. Lichens as biomonitors for heavy metal pollution. In *Plants as Biomonitors: Indicators of Heavy Metals in the Terrestrial Environment*; Markert, B. A.; VCH, New York, NY, 1993; pp 193-263.

Greater Fallon Area Chamber of Commerce. History of Fallon. Available: http://*www.fallonchamber.com*, 2008.

Guo, Z. G.; Feng, J. L.; Fang, M.; Chen, H. Y.; Lau, K. H. The elemental and organic characteristics of PM2.5 in Asian dust episodes in Qingdao, China, 2002. *Atmos. Environ.* 2004, 38, 909-919.

Gupta, A.; Kumar, R.; Maharaj Kumari, K.; Srivastava, S. S. Atmospheric dry deposition to leaf surfaces at a rural site of India. Chemosphere 2004, 55, 1097-1107.

Guyette, R. P.; Cutter, B. E.; Henderson, G. S. Long-term correlations between mining activity and levels of lead and cadmium in tree-rings of eastern red-cedar. *J. Environ. Qual.* 1991, 20, 146-150.

Hagemeyer, J.; Weinand, T. Radial distributions of Pb in stems of young Norway spruce trees grown in Pb-contaminated soil. *Tree Physiol.* 1996, 16, 591-594.

Hale, M. E. *The Biology of Lichens*, 3rd Edition. Edward Arnold, London, England, 1983.

Hampp, R.; Höll, W. Radial and axial gradients of lead concentration in bark and xylem of hardwoods. *Arch. Environ. Contam. Toxicol.* 1974, 2, 143-151.

Harrison, R. M. Toxic metals in street and household dusts. *Sci. Tot. Environ.* 1979, 11, 89-97.

Helena, P. N.; Franc, B.; Cvetka, R. L. Monitoring of short-term heavy metal deposition by accumulation in epiphytic lichens (Hypogymnia physodes (L.) Nyl.). *J. Atmo. Chem.* 2004, 49, 223-230.

Holmes, H. *The Secret Life of Dust*. John Wiley & Son, New York, NY, 2001.

Jeran, Z.; Jaćimović, R.; Batič, F.; Smodiš, B.; Wolterbeek, H. Th. Atmospheric heavy metal pollution in Slovenia derived from results for epiphytic lichens. *Fresenius J. Anal. Chem.* 1996, 354, 681-687.

Kakulu, S. E. Trace metal concentration in roadside surface soil and tree back [sic]: a measurement of local atmospheric pollution in Abuja, Nigeria. *Environ. Monit. Assess.* 2003, 89, 233-242.

Kalinich, J. F.; Emond, C. A.; Dalton, T. K.; Mog, S. R.; Coleman, G. D.; Kordell, J. E.; Miller, A. C.; McClain, D. E. Embedded weapons-grade tungsten alloy shrapnel rapidly induces metastatic high-grade rhabdomyosarcomas in F344 rats. *Environ. Health Perspec.* 2005, 113, 729-734.

Karue, J.; Kinyua, A. M.; El-Busaidy, A. H. S. Measured components in total suspended particulate matter in a Kenyan urban area. *Atmos. Environ.* 1992, 26B, 505-511.

Lippmann, M. Ambient particulate matter. In *Environmental Toxicants: Human Exposures and Their Health Effects*, 2nd Edition; Lippman, M; John Wiley & Sons, New York, NY, 2000; pp 31-63.

Loppi, S.; Pirintsos, S. A. Epiphytic lichens as sentinels for heavy metal pollution at forest ecosystems (central Italy). *Environ. Pollut.* 2003, 121, 327-332.

Lötschert, W.; Köhm, H. J. Characteristics of tree bark as an indicator of high-immission areas. Oecologia 1978, 37, 121-132.

Mandiwana, K. L.; Resane, T.; Panichev, N.; Ngobeni, P. The application of tree bark as bio-indicator for the assessment of Cr (VI) in air pollution. *J. Hazard. Mater.* 2006, B137, 1241-1245.

Miller, A. C.; Mog, S.; McKinney, L.; Luo, L.; Allen, J.; Xu, J. Q.; Page, N. Neoplastic transformation of human osteoblast cells to the tumorigenic phenotype by heavy metal-tungsten alloy particles: induction of toxic effects. *Carcinogenesis* 2001, 22, 115-125.

Moore, L. E.; Lu, M.; Smith, A. H. Childhood cancer incidence and arsenic exposure in drinking water in Nevada. *Arch. Environ. Health* 2002, 57, 201-206.

Moreno-Grau, S.; Pérez-Tornell, A.; Bayo, J.; Moreno, J.; Angosto, J. M.; Moreno-Clavel, J. Particulate matter and heavy metals in the atmospheric aerosol from Cartagena, Spain. *Atmos. Environ.* 2000, 34, 5161-5167.

Nash, T. H.; Gries, C.; Zschau, T.; Getty, S.; Ameron, Y.; Zambrano, A. Historical patterns of metal atmospheric deposition to the epilithic lichen Xanthoparmelia in Maricopa County, Arizona, USA. *J. Physique IV* 2003, 107(Part 2), 921-924.

Nevada State Health Division. New Childhood Leukemia Case Confirmed. News Release, 20 December 2004. Available: *http://health2k.state.nv.us/pio/releases/122004PressRelLeukemia.pdf*, 2004.

NOAA. National Climatic Data Center. National Oceanic and Atmospheric Administration. Available: *http://www.ncdc.noaa.gov/oa/ncdc.html,* 2008.

Nriagu, J. O. Lead in the atmosphere and its effects on lead in humans. In *Lead Toxicity*; Singhal, R. L.; Thomas, J. A.; Urban & Schwarzenberg, Baltimore, MD, 1980; pp 483-503.

Pastuszka, J.; Hławiczka, S.; Willeke, K. Particulate pollution levels in Katowice, a highly industrialized Polish city. *Atmos. Environ.* 1993, 27B, 59-65.

Pirrone, N.; Keeler, G. J.; Warner, P. O. Trends of ambient concentrations and deposition fluxes of particulate trace metals in Detroit from 1982 to 1992. *Sci. Tot. Environ.* 1995, 162, 43-61.

Ragosta, M.; Caggiano, R.; D'Emilio, M.; Macchiato, M. Source origin and parameters influencing levels of heavy metals in TSP, in an industrial background area of Southern Italy. *Atmos. Environ.* 2002, 36, 3071-3087.

Rajput, M. U.; Ahmad, S.; Ahmad, M.; Ahmad, W. Determination of elemental composition of atmospheric aerosol in the urban area of Islamabad, Pakistan. *J. Radioanal. Nucl. Chem.* 2005, 266, 343-348.

Rapsomanikis, S.; Donard, O. Lead and zinc in roadside dust from a suburb in Athens, Greece. *Environ. Tech. Lett.* 1985, 6, 145-148.

Reid, E. A.; Reid, J. S.; Meier, M. M.; Dunlap, M. R.; Cliff, S. S.; Broumas, A.; Perry, K.; Maring, H. Characterization of African dust transported to Puerto Rico by individual particle and size segregated bulk analysis. *J. Geophys. Res. Atmo.* 2003, 108(D19), No. 8591.

Reid, M.; Thompson, S. Ecological fieldwork methods. In *Essential Environmental Science: Methods & Techniques*; Watts, S.; Halliwell, L.; Routledge, London, England, 1996; pp 351-389.

Reimann, C.; de Caritat, P. Distinguishing between natural and anthropogenic sources for elements in the environment: regional geochemical surveys versus enrichment factors. *Sci. Tot. Environ.* 2005, 337, 91-107.

Richardson, D. H. S. Pollution Monitoring with Lichens. The Richmond Publishing Company, Slough, England, 1992.

Richardson, D. H. S. Metal uptake in lichens. *Symbiosis* 1995, 18, 119-127.

Rossini Oliva, S.; Mingorance, M. D. Assessment of airborne heavy metal pollution by aboveground plant parts. *Chemosphere* 2006, 65, 177-182.

Rubin, C. S.; Holmes, A. K.; Belson, M. G.; Jones, R. L.; Flanders, W. D.; Kieszak, S. M.; Osterloh, J.; Luber, G. E.; Blount, B. C.; Barr, D. B.; Steinberg, K. K.; Satten, G. A.; McGeehin, M. A.; Todd, R. L. Investigating childhood leukemia in Churchill County, Nevada. *Environ. Health Perspec.* 2007, 115, 151-157.

Salve, P. R.; Maurya, A.; Wate, S. R. Atmospheric dry deposition on leaves at an urban location. *Bull. Environ. Contam. Toxicol.* 2006, 77, 834-837.

Scerbo, R.; Possenti, L.; Lampugnani, L.; Ristori, T.; Barale, R.; Barghigiani, C. Lichen (Xanthoria parientina) biomonitoring of trace element contamination and air quality assessment in Livorno Province (Tuscany, Italy). *Sci. Tot. Environ.* 1999, 241, 91-106.

Seiler, R. L. Temporal changes in water quality at a childhood leukemia cluster. *Ground Water* 2004, 42, 446-455.

Seiler, R. L.; Stollenwerk, K. G.; Garbarino, J. R. Factors controlling tungsten concentrations in ground water, Carson Desert, Nevada. *Appl. Geochem.* 2005, 20, 423-441.

Sensen, M.; Richardson, D. H. S. Mercury levels in lichens from different post trees around a chlor-alkali plant in New Brunswick, Canada. *Sci. Tot. Environ.* 2002, 293, 31-45.

Sheppard, P. R.; Hallman, C. L.; Ridenour, G.; Witten, M. L. Spatial patterns of tungsten and cobalt on leaf surfaces of trees in Fallon, Nevada. *Land Contam. Reclam.*, in press, 2009, 17.

Sheppard, P. R.; Ridenour, G.; Speakman, R. J.; Witten, M. L. Elevated tungsten and cobalt in airborne particulates in Fallon, Nevada: possible implications for the childhood leukemia cluster. *Appl. Geochem.* 2006, 21, 152-165.

Sheppard, P. R.; Speakman, R. J.; Ridenour, G.; Glascock, M. D.; Farris, C.; Witten, M. L. Spatial patterns of tungsten and cobalt in surface dust of Fallon, Nevada. *Environ. Geochem. Health* 2007a, 29, 405-412.

Sheppard, P. R.; Speakman, R. J.; Ridenour, G.; Witten, M. L. Using lichen chemistry to assess airborne tungsten and cobalt in Fallon, Nevada. *Environ. Monit. Assess.* 2007b, 130, 511-518.

Sheppard, P. R.; Speakman, R. J.; Ridenour, G.; Witten, M. L. Temporal variability of tungsten and cobalt in Fallon, Nevada. *Environ. Health Perspec.* 2007c, 115, 715-719.

Sheppard, P. R.; Toepfer, P.; Schumacher, E.; Rhodes, K.; Ridenour, G.; Witten, M. L. Morphological and chemical characteristics of airborne tungsten particles of Fallon, Nevada. *Microsc. Microanal.* 2007d, 13, 296-303.

Sheppard, P. R.; Witten, M. L. Laser trimming tree-ring cores for dendrochemistry of metals. *Tree-Ring Res.* 2005, 61, 87-92.

Singhal, R. L.; Thomas, J. A. *Lead Toxicity*. Urban & Schwarzenberg, Baltimore, MD, 1980.

Smith, K. T.; Shortle, W. C. Tree biology and dendrochemistry. In *Tree Rings, Environment, and Humanity*; Dean, J. S.; Meko, D. M.; Swetnam, T. W.; Radiocarbon, Tucson, AZ, 1996; pp 629-635.

Solomon, P. A.; Norris, G.; Landis, M.; Tolocka, M. Chemical analysis methods for atmospheric aerosol components. In *Aerosol Measurement: Principles, Techniques, and Applications*, 2nd Edition; Baron, P. A.; Willeke, K.; Wiley, New York, NY, 2001; pp 261-293.

Steinberg, K. K.; Relling, M. V.; Gallagher, M. L.; Greene, C. N.; Rubin, C. S.; French, D.; Holmes, A. K.; Carroll, W. L.; Koontz, D. A.; Sampson, E. J.; Satten, G. A. Genetic studies of a cluster of acute lymphoblastic leukemia cases in Churchill County, Nevada. *Environ. Health Perspec*. 2007, 115, 158-164.

Steinmaus, C.; Lu, M.; Todd, R. L.; Smith, A. H. Probability estimates for the unique childhood leukemia cluster in Fallon, Nevada, and risks near other U.S. military aviation facilities. *Environ. Health Perspec*. 2004, 112, 766-771.

Sun, N. N.; Fastje, C. D.; Wong, S. S.; Sheppard, P. R.; Macdonald, S.; Ridenour, G.; Hyde, J. D.; Witten, M. L. Dose-dependent transcriptome changes by metal ores on a human acute lymphoblastic leukemia cell line. *Toxicol. Ind. Health* 2003, 19, 157-163.

Szczepaniak, K.; Biziuk, M. Aspects of the biomonitoring studies using mosses and lichens as indicators of metal pollution. *Environ. Res*. 2003, 93, 221-230.

Tam, N. F. Y.; Liu, W. K.; Wong, M. H.; Wong, Y. S. Heavy metal pollution in roadside urban parks and gardens in Hong Kong. *Sci. Tot. Environ*. 1987, 59, 325-328.

Thornton, I.; Culbard, E.; Moorcroft, S.; Watt, J.; Wheatley, M.; Thompson, M.; Thomas, J. F. A. Metals in urban dusts and soils. *Environ. Technol. Lett*. 1985, 6, 137-144.

Trindade, H. A.; Pfeiffer, W. C.; Londres, H.; Costa-Ribeiro, C. L. Atmospheric concentration of metals and total suspended particulates in Rio de Janeiro. *Environ. Sci. Technol*. 1981, 15, 84-89.

U.S. ATSDR. Evaluation of Potential Exposures from the Fallon JP-8 Fuel Pipeline. Agency for Toxic Substances and Disease Registry. Available: *http://www.atsdr.cdc.gov/HAC/PHA/fallonpipe/fallon_toc*.html, 2002.

U.S. ATSDR. Surface Water, Sediment, and Biota Human Exposure Pathway Analysis for Churchill County: Fallon Leukemia Project, Fallon, Churchill County, Nevada. Agency for Toxic Substances and Disease Registry. Available: *http://www.atsdr.cdc.gov/HAC/PHA/fallonwater/finalwater.pdf*, 2003a.

U.S. ATSDR. Air Exposure Pathway and Assessment: Fallon Leukemia Cluster Investigation. Agency for Toxic Substances and Disease Registry. Available: *http://www.atsdr.cdc.gov/HAC/PHA/fallonair/finalair.pdf*, 2003b.

U.S. ATSDR. Pathway Assessment for Churchill County Surface Soils and Residential Indoor Dust: Fallon Leukemia Project, Fallon, Churchill County, Nevada. Agency for Toxic Substances and Disease Registry. Available: *http://www.atsdr.cdc.gov/HAC/PHA/ fallonsoil/ finalsoil.pdf*, 2003c.

U.S. CDC. A Cross-Sectional Exposure Assessment of Environmental Exposures in Churchill County, Nevada. Centers for Disease Control and Prevention. Available: *http://www.cdc.gov/nceh/clusters/fallon*, 2003.

U.S. Census. United States Census 2000. Available: *http://www.census.gov/main/www/cen2000.html*, 2000.

U.S. EPA. National Air Monitoring Strategy Information. U.S. Environmental Protection Agency, Washington, DC. Available at: *http://www.epa.gov/ttn/amtic/monitor.html*, 2008.

U.S. NCI. Age-Adjusted SEER Incidence and the U.S. Death Rates and 5-Year Relative Survival Rates by Primary Cancer Sites, Sex, and Time Period. SEER Cancer Statistics Review, 1975-2000, Table XXVII-3: Childhood Cancers, U.S. National Cancer Institute. Available: *http://www.seer.cancer.gov,* 2003.

Walkenhorst, A.; Hagemeyer, J.; Breckle, S. W. Passive monitoring of airborne pollutants, particularly trace metals, with tree bark. In *Plants as Biomonitors: Indicators of Heavy Metals in the Terrestrial Environment*; Markert, B. A.; VCH, New York, NY, 1993; pp 523-540.

Ward, N. I.; Brooks, R. R.; Roberts, E. Silver in soils, stream sediments, waters and vegetation near a silver mine and treatment plant at Maratoto, New Zealand. *Environ. Pollut.* 1977, 13, 269-280.

Ward, T. J.; Spear, T.; Hart, J.; Noonan, C.; Holian, A.; Getman, M.; Webber, J. S. Trees as reservoirs for amphibole fibers in Libby, Montana. *Sci. Tot. Environ.* 2006, 367, 460-465.

Wolterbeek, B. Biomonitoring of trace element air pollution: principles, possibilities and perspectives. *Environ. Pollut.* 2002, 120, 11-21.

Wolterbeek, H. Th.; Bode, P. Strategies in sampling and sample handling in the context of large-scale plant biomonitoring surveys of trace element air pollution. *Sci. Tot. Environ.* 1995, 176, 33-43.

Wong, J. W. C.; Mak, N. K. Heavy metal pollution in children playgrounds in Hong Kong and its health implications. *Environ. Tech.* 1997, 18, 109-115.

Yanosky, T. M.; Vroblesky, D. A. Relation of nickel concentrations in tree rings to groundwater contamination. *Water Resources Res.* 1992, 28, 2077-2083.

Yenisoy-Karakaş, S.; Tuncel, S. G. Geographic patterns of elemental deposition in the Aegean region of Turkey indicated by the lichen, Xanthoria parietina (L.) Th. Fr. *Sci. Tot. Environ.* 2004, 329, 43-60.

Yun, M.; Longerich, H. P.; Wadleigh, M. A. The determination of 18 trace elements in lichens for atmospheric monitoring using inductively coupled plasma-mass spectrometry. *Can. J. Anal. Sci. Spectros.* 2003, 48, 171-180.

Zhang, L.; Qian, J. L.; Planas, D. Mercury concentrations in tree rings of black spruce (Picea mariana Mill. B.S.P.) in boreal Quebec, Canada. *Water Air Soil Pollut.* 1995, 81, 163-173.

Zhang, Zh. H.; Chai, Z. F.; Mao, X. Y.; Chen, J. B. Biomonitoring trace element atmospheric deposition using lichens in China. *Environ. Pollut.* 2002, 120, 157-161.

Zschau, T.; Getty, S.; Gries, C.; Ameron, Y.; Zambrano, A.; Nash III, T. H. Historical and current atmospheric deposition to the epilithic lichen Xanthoparmelia in Maricopa County, Arizona. *Environ. Pollut.* 2003, 125, 21-30.

In: Airborne Particulates
Editor: Ming Cheng and Wen Liu
ISBN: 978-1-60692-907-0

Chapter 6

CHEMICAL CHARACTERIZATION OF AMBIENT PARTICLE: PAHS AND CHLORINATED PAHS

Takeshi Ohura
Institute for Environmental Sciences, University of Shizuoka, Japan

ABSTRACT

Airborne particle matter contains a myriad of inorganic and organic species, many of which can adversely affect human health. Of these constituents, the class of polycyclic aromatic hydrocarbons (PAHs) is noteworthy because it is ubiquitous and includes several potent carcinogens. Most PAHs emitted to the ambient air are formed during the incomplete combustion of organic materials; that is, anthropogenic activities such as motor vehicle use and industrial activities become major sources of atmospheric PAHs associated with particles. PAH profiles based on specific PAH concentration showed differential behaviors for certain relatively heavy molecular weight PAHs, suggesting that the ratios give valuable information regarding the stability and emission sources of the PAHs. Such outdoor PAHs pollution is considerably impacted on the indoors; indoor PAH concentrations are lower than, and significantly correlated to outdoor PAH concentrations. Humans spend most of their lives indoors. Therefore any evaluation of the human health risks caused by exposure to air pollution requires a detailed understanding of pollutant levels not only indoors but also outdoors. Furthermore, as a series of structurally related aromatics, halogenated aromatic hydrocarbons (HAHs) such as polychlorinated dibenzo-*p*-dioxins/dibenzofurans and polychlorinated biphenyls have been also widespread in the environment, which have prompted great concern about their adverse human effects. Recently, chlorinated PAHs (ClPAHs), structural hybrids of PAHs and HAHs, have been found in the atmospheric particles in an urban city, Japan; 1-chloropyrene was detected at the highest concentration (7.5 pg/m^3 in annual mean), followed by 6-chlorobenzo[*a*]pyrene (5.6 pg/m^3) and 9,10-dichlorophenanthrene (5.1 pg/m^3). Those ClPAHs have been shown to elicit dioxin-like toxicity and mutagenic properties, especially the toxicities of exposure to ClPAHs based on aryl hydrocarbon receptor-mediated activity were estimated to be 30~50 times higher than that of dioxins. The estimation raises the problem that the adverse effects of exposure to ClPAHs deserve further attention.

INTRODUCTION

Particle matter (PM) air pollutants are a mixture of various sized ambient particles that have potentially toxic chemicals associated with them. These particles are distributed bimodally in two fractions with mean aerodynamic diameters of 0.5 μm (fine particles) and 20 μm (coarse particles) [1]. Fine and coarse particles differ in many aspects other than size, for example, origin, composition, and half-life [1-5]. Fine particles are produced primarily by combustion processes. They are composed largely of soot, diesel engine exhaust particles, and secondary reaction products of sulfur oxides, nitrogen oxides, and hydrocarbons [3, 6-8]. The half-life of fine particles has been defined as "days to weeks" [1]. Coarse particles are generated primarily from resuspended dust (soil dust and street dust), sea salt, and other crustal materials [9]. The half-life of coarse particles has been defined as "minutes to hours" [1].

Fine particles with diameter <2.5 μm (PM2.5) are of the greatest health concern because when inhaled, they can be deposited more deeply in the lungs than coarse particles [10-12]. Thus, the combination of toxic chemicals with fine particles can have an especially deleterious effect on human health, including causing a predisposition to respiratory disease. Several studies have also shown a relationship between PM2.5 and human health, up to and including mortality and morbidity [13, 14].

A number of hazardous compounds included organic and inorganic are present in the ambient particles. Polycyclic aromatic hydrocarbons (PAHs) have been known to be one of the toxicant since the early 1910s [15]. PAHs are produced by incomplete combustion of organic substances [16-18]. Anthropogenic sources of PAHs include the combustion of materials for energy supply such as coal, oil, gas, and wood [17, 18]. In particular, industrial plants such as coke and aluminum production, residential heating, and power generation as stationary and mobile sources comprise the majority of human-produced PAH sources [18]. Natural sources of PAHs include volcanic activity and forest fire [17].

PAHs showing high mutagenicity and carcinogenicity are in common relatively high molecular weights (>4-ring), that is, particulate PAHs [19]. Therefore, to investigate the occurrences of PAHs associated with (fine) particles will be important to assess human risks of the exposure.

One of interesting features of PAHs will be advanced for the diversity of the derivatives; isomers, congeners, and substitution product. Nitrated PAHs are well-known as one of potent direct mutagen in air pollutants [20]. Recently, chlorine-substituted PAHs (ClPAHs) have been also found in the environment [21]. Although the environmental and biological data are very limited in comparison to those of PAHs, some researchers have worked vigorously to address the problem. This chapter looks at the ambient distributions, fates, sources, and toxicities of PAHs and ClPAHs associated with particles.

AMBIENT LEVELS AND DISTRIBUTIONS OF PAHS.

PAHs are present ubiquitously in the environment as hazardous pollutants, and are often observed in small quantities of the order of ng/m^3 in the air. They are formed mainly as a result of pyrolytic processes, in particular the incomplete combustion of organic materials;

that is, anthropogenic activities such as motor vehicle use and industrial activities become major sources of atmospheric PAHs associated with particulates [16].

A lot of surveys concerning ambient PAHs have been carried out at several locations, locally as well as around the world. To compare the ambient PAH data obtained in around the world will make us show that the current status of PAH pollutions and differences of the sources, and impact on health risks of the exposure. However, we here need to be deeply impressed that the many studies cannot be simply compared among them because those collection methods (glass fiber filter or Teflon filter, etc.), particles size collected (PM2.5 or TSP, etc.), analytical conditions (recovery and sensitivity, etc.) may be different. In addition, we should consider that ambient PAH concentrations in certain country considerably vary in the city, local area, season, meteorological conditions, and the day-by-day.

Of the seasonal trends of PAHs associated with particles, there is universal fact that the concentrations generally trend to be elevated in colder seasons as compared with those in warmer seasons [22]. It is believed that the concentration is dependent on several seasonal factors and atmospheric loss mechanisms such as long-range transport, space heating, photolysis/photooxidation, and gas-particle partitioning, etc. Indeed, much of studies show that there is significant correlation between ambient PAH concentrations and the temperature, indicating that temperature (i.e. including seasonal sources and factors) is driven by ambient PAH concentrations [22].

Long-term trends of ambient PAHs have been investigated in many countries in which a decrease of the concentration have been observed. In an urban city, Shizuoka, Japan, the concentrations of benzo[*a*]pyrene trended moderately downward over the past decade, whereas the contributions of individual PAHs to 5PAHs (benzo[*a*]pyrene + benz[*a*]anthracene + pyrene + fluoranthene + phenanthrene) associated with particles remained static [23]. This fact means that the significant sources did not vary greatly during the period. Similarity, in 1970s and 1990s in northern area, Sapporo, Japan, the BaP concentrations and the indirect mutagenicity of the ambient particles also showed the decrease trends. The factor decreasing the concentrations could be due to a large decrease in coal consumption.

To investigate spatial distributions of ambient PAHs in certain regions will be further help to assess the exposure risks and impacts on the pollution. The local distributions could be significantly influenced by not only the possible local sources but also the seasonal and climate conditions such as wind direction and weather. Indeed, the characteristic local distributions for each PAH have been frequency observed in industrial cities, in which the distributions are different from each PAH specie [24]. It indicates to present the independent sources for the PAH in the surveyed region.

PAHS IN INDOORS.

Humans spend most of their lives indoors. Therefore, on evaluation of the human health risks caused by exposure to air pollutions, detailed surveys of the pollutant levels indoors will be required. In addition, this understanding must include an estimation of the contribution of outdoor pollutants to indoor pollution, because indoors PAH pollutions are often driven by the outdoors [25-29].

For particulate PAHs (>4-ring), the indoor levels trend to be frequently lower than the outdoor, and the both PAH concentrations showed significant correlations in each other [26-29]. The indoor levels could be affected by not only corresponding outdoor levels but also various indoor's features such as house structure and air exchange, and sources such as smoking and cooking, etc. Although such activities elevate temporarily the indoor PAH level, the significant differences of PAHs levels among indoor microenvironments (living, bedroom, kitchen, etc.) are typically not observed in the samples [29]. This fact would be due to high air exchange rate of room-to-room compared with room-to-outdoor.

Personal Exposure to PAHs.

Throughout the environmental surveys of PAHs, investigating the personal exposures rather than the indoors will be most acceptable procedure to evaluate the risks of exposures. To date, much of studies on the personal exposure have been performed, in which personal exposure levels of PAHs associated with particles have trended to be relatively higher than the indoor and outdoor levels [29]. This means that the personal activities could often associate to the production of PAHs.

Smoking is most potent source affecting personal exposure to PAHs in the living space. Ohura et al. showed that smoking activity significantly elevate the personal levels of not only PAHs but also fine particles, but not significant for indoors [29]. Therefore, if smoker's sample include when the surveys of personal and indoors PAHs were carried out, it will be difficult to suggest the factors affecting the personal levels from indoor sources.

As noted in above, investigating personal exposure to PAHs is important to assess the risks of exposure, whereas it takes extra effort to perform actually in comparison to survey of indoor and outdoor. So, predictions of personal exposure levels from the indoor and outdoor concentrations have been attempted. The predicted personal concentrations of BaP, i.e. calculated from the contributions of time-weighted indoor (living room and bedroom) and outdoor concentrations, were consistent with the measured concentrations of BaP, which accounted for approximately 90% of the personal levels measured [29]. On the other hand, personal exposure to PM2.5 calculated from the concentrations in various microenvironments was underestimated relative to the levels measured (approximately 50% of the personal levels measured) [29]. In spite of the high contribution of PAHs to PM2.5, no significant correlation was found between the concentration of PM2.5 and the associated PAH concentration either indoors or outdoors, and in summer or winter [28]. This showed that the amount of PAH associated with PM2.5 in air fluctuated according to the season and the sites regardless of the particle concentrations.

Sources of Ambient PAHs

For emission sources of PAHs in urban air, vehicle exhausts generally seem to be the dominant [16, 17]. Also, domestic heating and industrial activities should be concerned to the significant sources [17]. The identification and characterization of those possible sources have been done by various attempts; principal component analysis, receptor model, counter map, trajectory analysis, and PAH profiles. On the analysis of PAH sources, the individual PAH concentration profile analysis would be useful and simple technique, because there are much of reports investigating PAH concentration ratios as source indicator.

Benzo[e]pyrene is often used as a index for the profile analysis because of its photostability [30]. The certain PAH ratio to benzo[e]pyrene concentration gives valuable information regarding the stability and emission sources of the PAHs. For example, the PAH profiles of [benzo[ghi]perylene]/[benzo[e]pyrene] and [coronene]/[benzo[e]pyrene] are suggested to be indicators of traffic as a pollutant source [31]. To date, much of studies of further PAH profiles have been reported [32-40]. The selected PAH profiles of the significant sources are listed in Table 1.

Table 1. Ratios for the PAHs attributed to specific sources

PAH ratio[a]	Vehicular emissions	Gasoline	Diesel	Non-traffic sources	Wood combustion	Used motor oil	Incinerators	Coal
BaA/BaP		0.5	1.0		1.0			
BaA/Chry	0.40−0.63	0.28−1.2	0.17−0.36		0.93	0.5		
BaP/(BaP+Chry)		0.49	0.73					
BaP/BghiP	0.3−0.78	1.27	0.46−0.81				0.14−0.6	
BghiP/BeP	2.02			0.8				
BghiP/Cor		1.5	2.5		2.0			
BghiP/IP	0.9	3.5−3.8	1.1−1.2		0.8			
BeP/BaP		1.1−1.3	2.0−2.5		0.44			
BbF/BkF			>0.5					
BbF/Cor		0.3	2.9		2.9			
BkF/IP		0.4	0.5		0.6			
Chry/BeP		2.5	1.6		2.4			
Cor/BeP	1.54			0.3				
Fluor/BeP	3.5±0.5							
Fluor/(Fluor+Py)		0.43±0.08				0.36±0.08		
Fluor/Py	0.6							
IP/BghiP	0.40−0.63	0.28−1.2	0.17−0.36		0.93			
IP/(IP+BghiP)		0.18	0.35−0.70		0.62			0.56
Py/BaP		0.85±0.38	0.81±0.87		0.71			

[a] BaA; benz[*a*]anthracene, BaP; benzo[*a*]pyrene, BbF; benzo[*b*]fluoranthene, BeP; benzo[*e*]pyrene, BghiP; benzo[*ghi*]perylene, BkF; benzo[*k*]fluoranthene, Chry; chrysene, Cor; coronene, Fluor; fluoranthene, IP; indeno[1,2,3-*cd*]pyrene, Py; pyrene.

Note that the analytical approach to identify the sources should pay attention to use because it is often difficult to discriminate between some sources. Ambient PAHs are susceptible to atmospheric reactions such as oxidation and photolysis so that PAH concentrations measured may not reflect the true levels emitted from the related sources. This would relate the distance of sampling point to possible emission sources. In addition, to perform the profile analysis using either the particulate or gaseous PAHs data could be incorrect handling.

FATES OF AMBIENT PAHS.

Atmospheric fates of PAHs will be accounted for two significant atmospheric reactions: (i) heterogeneous processes such as photolysis/photooxidation and gas-particle interactions, which involve particle-associated compounds (including particulate PAHs with larger 4-ring), and (ii) homogeneous gas-phase reactions of volatile 2- and 3-ring and semivolatile 4-ring PAHs initiated by OH (daytime) and NO3 radicals (nighttime) and ozone (24 h) [22]. Therefore, it is now generally agreed that photodegradation is the major chemical pathway for the loss of 4- to 6-ring PAHs in ambient particles. The reaction rates, mechanism, and products of this overall phenomenon depend not only on the structures and UV-visible absorption spectra (on and in aerosol surfaces) of specific PAHs but also on the physical and chemical natures of the particles on/in which they are adsorbed/absorbed.

The photochemistry of PAHs sorbed to combustion particles is believed to occur in an organic layer surrounding the graphitic particle core [22]. So, to understand simply the mechanism of photodegradation PAH in combustion aerosols, McDow, and co-workers have developed the laboratory experimental system in which solutions containing various organics found in ambient aerosols are irradiated in photoreactors [41-44]. To date, the photodegradabilities of many PAH species have been investigated using the system in various complex organic solutions, resulting that the photochemical behaviors and reactions have been appeared.

OCCURRENCES OF CLPAHS.

Chlorinated polycyclic aromatic hydrocarbons (ClPAHs, Figure 1) with 3- to 5-ring, and substituted 1- to 3-chlorine have been found in the environment such as in air, tap water, and sediments, so that they are ubiquitous environmental contaminants [21]. In addition, the combustion of polyvinylchloride and plastic wrap made from polyvinylidene chloride result in the production of ClPAHs, suggesting that combustion of organic materials including chlorine is a possible source of environmental pollution [45-47]. To date, although the survey of ambient ClPAHs is very limited, Ohura and co-workers have recently investigated the occurrence of ClPAHs associated with particles [23, 48-51]. Those results show that 1-chloropyrene was detected at the highest concentration (7.5 pg/m^3 in annual mean) among targeted 12 ClPAHs in an urban city, Shizuoka, Japan, followed by 6-chlorobenzo[*a*]pyrene (5.6 pg/m^3) and 9,10-dichlorophenanthrene (5.1 pg/m^3) [51]. In the further study, all of targeted 26 species of ClPAHs with 3- to 5-ring were also detected in either gaseous or particulate phase, total concentration of which was 17 pg/m^3 (annual mean) [49]. The concentrations of particulate ClPAHs trended to be 2~100 times lower than those of gaseous ClPAHs, and were approximately 60~300 times as low as those of corresponding parent

PAHs. From the compositional analysis, relatively high molecular weight ClPAHs, i.e. particulate ClPAHs, dominated in colder seasons rather than in warmer seasons.

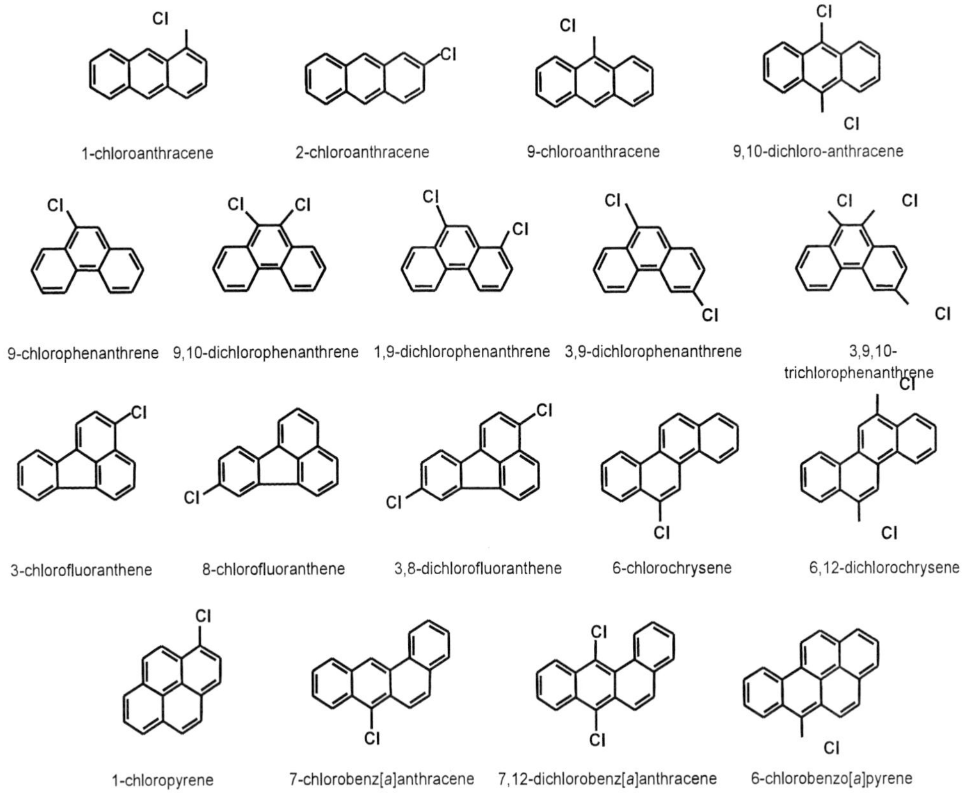

Figure 1. Structures of environmentally relevant ClPAHs.

The observed particulate ClPAHs concentrations tended to be high in colder seasons and low in warmer seasons as a typical trend in ambient PAHs [49, 51]. This means that ambient ClPAHs levels are strongly related to the atmospheric behaviors as shown in the case of PAHs of gas/particle partitioning, seasonal sources, and climatic conditions including the photodecay reactions and short/long-range transports, etc.

Next, photochemical degradation of ClPAHs have been also investigated to simulate the compound's fate on aerosol surfaces [50-52]. The photolysis rates of ClPAHs varied greatly according to the skeleton of PAHs, the stability is following order; 6-chlorobenzo[a]pyrene ~ 1-chloropyrene < 7-chlorobenz[a]anthracene < chlorophenanthrenes < chlorofluoranthenes. In addition, the photolysis rates of chlorophenanthrenes and 1-chloropyrene were higher than those of corresponding parent PAHs, whereas chlorofluoranthenes, 7-chlorobenz[a] anthracene and 6-chlorobenzo[a]pyrene were more stable under irradiation compared to respective parent PAH [52]. It suggests that certain PAHs would be stabilized upon chlorination. Indeed, such stabilization of chlorine-substituted aromatics has been observed in the case of chlorinated dioxins [53]. Furthermore, the predictions of direct photolysis half-lives and average quantum yields have been performed using the quantitative structure-property relationship models [54, 55]. Ohura et al. suggested that the significant factors affecting the half-lives and average quantum yields of ClPAHs in cyclohexane were ELUMO+1, total energy and surface area, and ELUMO, ELUMO-EHOMO and total energy, respectively [52].

TOXICITIES OF CLPAHS.

In relation to biological toxicity, mutagenicity with metabolic activation by a liver homogenate (S-9 mixture) using Salmonella test has been observed for some ClPAHs [56]. Thus, like PAHs, the ClPAHs could be metabolized to form oxidized metabolites with profile similar to that of the respective parent PAHs. In addition, metabolites of certain ClPAHs by fish liver microsomes have been identified [57]. The tumorigenicity has been also investigated in neonatal male mouse, showing that 7-chlorobenz[a]anthracene is potent carcinogen [58].

Exposure to PAH-containing substances increases the risk of cancer in humans to be mediated through aryl hydrocarbon receptor (AhR) activation [59, 60]. The ligand-bound AhR, a cytosolic ligand-activated transcription factor complex, activates the transcription of genes related to the drug-metabolizing enzymes such as cytochrome P450 (CYP) and glutathione S-transferase. The most characterized pathway involves translocation of the activated AhR to the nucleus, where it binds with the AhR nuclear translocator protein to form a heterodimer. Binding of the heterodimer leads to transcriptional modulation of genes that contain a xenobiotic responsive element [61]. To promote PAH toxicity, interaction with AhR seems to be an essential mechanism. Ohura et al. investigated the human AhR ligand-binding activity of environmentally relevant 18 ClPAHs, suggesting that the toxicities of exposure to ClPAHs based on the AhR activity were estimated to be approximately 30~50 times higher than that of dioxins [62]. The estimation raises the problem that the adverse effects of exposure to ClPAHs deserve further attention.

CONCLUSION

PAHs are members of a unique class of air pollutants relevant to many scientific and social issues having a variety of aspects: chemical, toxicological, engineering, technological, public health, economic and regulatory, and legislative. The ubiquitous nature of airborne PAHs is evident from the fact that the 16 U.S. Environmental Protection Agency "Priority Polycyclic Aromatic Hydrocarbon Pollutants" in urban airsheds throughout the world [22].

For monitoring of ambient PAHs, the analytical methods including the sampling, extraction, purification, and analytical devices had been already developed, so that the surveys of airborne PAH concentrations have been carried out in various sites in the world. Nevertheless, the reason why we continue to monitor the PAHs could be due to its high toxicities and being common in the environment. PAHs in the air have a number of isomers, environmental occurrences, behaviors and toxicities of which are significantly different nature among them; they are specific for PAHs.

Cancer is a leading cause of death worldwide. From a total of 58 million deaths worldwide in 2005, cancer accounts for 7.6 million (or 13%) of all deaths [63]. The main types of cancer leading to overall cancer mortality are:

- Lung (1.3 million deaths/year);
- Stomach (almost 1 million deaths/year);
- Liver (662,000 deaths/year);
- Colon (655,000 deaths/year) and
- Breast (502,000 deaths/year).

Lung cancer is the most common cause of death from cancer worldwide. That is, air quality is most important problem to our health. In addition, more than 70% of all cancer deaths in 2005 occurred in low and middle income countries. Recent air pollution accompanying rapid industrial developments like China and India has become a serious problem in the area. In the future, pollution suffered outside the area where it originates, by which health hazards could be also serious problem in the world.

References

[1] Wilson, WE. & Suh, HH. (1997) Fine particles and coarse particles: concentration relationships relevant to epidemiologic studies. *J. Air Waste Manag. Assoc.* 47, 1238-1249.

[2] Poster, DL. Hoff, RM. & Baker, JE. (1995) Measurement of the particle-size distributions of semivolatile organic contaminants in the atmosphere. *Environ. Sci. Technol.* 29, 1990-1997.

[3] Kleeman, MJ. & Cass, GR. (1998) Source contributions to the size and composition distribution of urban particulate air pollution. *Atmos. Environ.* 32, 2803-2816.

[4] Hughes, LS. Allen, JO. Kleeman, MJ. Johnson, RJ. Cass, GR. Gross, DS. Gard, EE. Galli, ME. Morrical, BD. Fergenson, DP. Dienes, T. Noble, CA. Liu, D. Silva, PJ. & Prather, KA. (1999) Size and composition distribution of atmospheric particles in Southern California. *Environ. Sci. Technol.* 33, 3506-3515.

[5] Hughes, LS. Allen, JO. Bhave, P. Kleeman, MJ. Cass, GR. Liu, D-Y. Fergenson, DP. Morrical, BD. & Prather, KA. (2000) Evolution of atmospheric particles along trajectories crossing the Los Angeles Basin. *Environ. Sci. Technol.* 34, 3058-3068.

[6] Magliano, KL. Hughes, VM. Chinkin, LR. Coe, DL. Haste, TL. Kumar, N. & Lurmann, FW. (1999) Spatial and temporal variations in PM10 and PM2.5 source contributions and comparison to emissions during the 1995 integrated monitoring study. *Atmos. Environ.* 33, 4757-4773.

[7] Pun, BK. & Seigneur, C. (2001) Sensitivity of particulate matter nitrate formation to precursor emissions in the California San Joaquin Valley. *Environ. Sci. Technol.* 35, 2979-2987.

[8] Lim, HJ. & Turpin, BJ. (2002) Origins of primary and secondary organic aerosol in Atlanta: results of time-resolved measurements during the Atlanta supersite experiment. *Environ. Sci. Technol.* 36, 4489-4496.

[9] Boon, KF. Kiefert, L. & McTainsh, GH. (1998) Organic matter content of rural dusts in Australia. *Atmos. Environ.* 32, 2817-2823.

[10] Pope III, CA. Bates, DV. & Raizenne, ME. (1995) Health effects of particulate air pollution: time for reassessment? *Environ. Health Perspect.* 103, 475-480.

[11] Schwartz, J. & Neas, LM. (2000) Fine particles are more strongly associated than coarse particles with acute respiratory health effects in schoolchildren. *Epidemiology*. 11, 6-10.

[12] Pope III, CA. (2007) Mortality effects of longer term exposures to fine particulate air pollution: review of recent epidemiological evidence. *Inhalat. Toxicol.* 19, 33-38.

[13] Pope III, CA. Thun, MJ. Namboodiri, MM. Dockery, DW. Evans, JS. Speizer, FE. & Heath, CWJ. (1995) Particulate air pollution as a predictor of mortality in a prospective study of U.S. adults. *Am. J. Respir. Crit. Care Med.* 151, 669-674.

[14] Schwartz, J. Dockery, DW. & Neas, LM. (1996) Is daily mortality associated specifically with fine particles? *J. Air Waste Manag. Assoc.* 46, 927-939.

[15] Phillips, DH. (1983) Fifty years of benzo(a)pyrene. *Nature*. 303, 468-472.

[16] Baek, SO. Field, RA. Goldstone, ME. Kirk, PW. Lester, JN. & Perry, R. (1991) A review of atmospheric polycyclic aromatic hydrocarbons: Sources, fate and behavior. *Water Air Soil Pollut.* 60, 279-300.

[17] Maliszewska-Kordybach, B. (1999) Sources, concentrations, fate and effects of polycyclic aromatic hydrocarbons (PAHs) in the environment. Part A: PAHs in air. *Polish Journal of Environmental Studies*. 8, 131-136.

[18] Mastral, AM. & Callen, MS. (2000) A review on polycyclic aromatic hydrocarbon (PAH) emission from energy generation. *Environ. Sci. Technol.* 34, 3051-3057.

[19] Bidleman, TF. (1988) Atmospheric processes. *Environ. Sci. Technol.* 22, 361-367.

[20] Durant, JL. Busby, WFJ. Lafleur, AL. Penman, BW. &C respi, CL. (1996) Human cell mutagenicity of oxygenated, nitrated and unsubstituted polycyclic aromatic hydrocarbons associated with urban aerosols. *Mutat. Res.* 371, 123-157.

[21] Ohura, T. (2007) Environmental behavior, sources, and effects of chlorinated polycyclic aromatic hydrocarbons. *ScientificWorldJournal*. 7, 372-380.

[22] Finlayson-Pitts, BJ. & Pitts, JNJ. (2000) Chemistry of the Upper and Lower Atmosphere. Anonymous Academic Press, San Diego

[23] Kitazawa, A. Amagai, T. & Ohura, T. (2006) Temporal trends and relationships of particulate chlorinated polycyclic aromatic hydrocarbons and their parent compounds in urban air. *Environ. Sci. Technol.* 40, 4592-4598.

[24] Ohura, T. Amagai, T. Fusaya, M. & Matsushita, H. (2004) Spatial distributions and profiles of atmospheric polycyclic aromatic hydrocarbons in two industrial cities in Japan. *Environ. Sci. Technol.* 38, 49-55.

[25] Liu, Y. Zhu, L. & Shen, X. (2001) Polycyclic aromatic hydrocarbons (PAHs) in indoor and outdoor air of Hangzhou, China. *Environ. Sci. Technol.* 35, 840-844.

[26] Naumova, YY. Eisenreich, SJ. Turpin, BJ. Weisel, CP. Morandi, MT. Colome, SD. Totten, LA. Stock, TH. Winer, AM. Alimokhtari, S. Kwon, J. Shendell, D. Jones, J. Maberti, S. & Wall, SJ. (2002) Polycyclic aromatic hydrocarbons in the indoor and outdoor air of three cities in the U.S. *Environ. Sci. Technol.* 36, 2552-2559.

[27] Levy, JI. Dumyahn, T. & Spengler, JD. (2002) Particulate matter and polycyclic aromatic hydrocarbon concentrations in indoor and outdoor microenvironments in Boston, Massachusetts. *J. Expo. Anal. Environ. Epidemiol.* 12, 104-114.

[28] Ohura, T. Amagai, T. Sugiyama, T. Fusaya, M. & Matsushita, H. (2004) Characteristics of particle matter and associated polycyclic aromatic hydrocarbons in indoor and outdoor air in two cities in Shizuoka, Japan. *Atmosph. Environ.* 38, 2045-2054.

[29] Ohura, T. Noda, T. Amagai, T. & Fusaya, M. (2005) Prediction of personal exposure to PM2.5 and carcinogenic polycyclic aromatic hydrocarbons by their concentrations in residential microenvironments. *Environ. Sci. Technol.* 39, 5592-5599.

[30] Nielsen, T. (1984) Reactivity of polycyclic aromatic hydrocarbons toward nitrating species. *Environ. Sci. Technol.* 18, 157-163.

[31] Nielsen, T. (1996) Traffic contribution of polycyclic aromatic hydrocarbons in the center of a large city. *Atmosph. Environ.* 30, 3481-3490.

[32] Simcik, MF. Eisenreich, SJ. & Lioy, PJ. (1999) Source apportionment and source/sink relationships of PAHs in the coastal atmosphere of Chicago and Lake Michigan. *Atmos. Environ.* 33, 5071-5079.

[33] Kavouras, IG. Lawrence, J. Koutrakis, P. Stephanou, EG. & Oyola, P. (1999) Measurement of particulate aliphatic and polynuclear aromatic hydrocarbons in Santiago de Chile: source reconciliation and evaluation of sampling artifacts. *Atmos. Environ.* 33, 4977-4986.

[34] Khalili, NR. Scheff, PA. & Holsen, TM. (1995) PAH source fingerprints for coke ovens, diesel and, gasoline engines, highway tunnels, and wood combustion emissions. *Atmos. Environ.* 29, 533-542.

[35] Li, CK. & Kamens, RM. (1993) The use of polycyclic aromatic hydrocarbons as source signatures in receptor modeling. *Atmos. Environ.* 27, 523-532.

[36] Guo, H. Lee, SC. Ho, KF. Wang, XM. & Zou, SC. (2003) Particle-associated polycyclic aromatic hydrocarbons in urban air of Hong Kong. *Atmos. Environ.* 37, 5307-5317.

[37] Park, SS. Kim, YJ. & Kang, CH. (2002) Atmospheric polycyclic aromatic hydrocarbons in Seoul, Korea. *Atmos. Environ.* 36, 2917-2924.

[38] Ravindra, K.;Bencs, L.;Wauters, E.;Hoog, J.;Deutsch, F.;Roekens, E.;Bleux, N.;Berghmans, P. &Grieken, RV. (2006) Seasonal and site-specific variation in vapour and aerosol phase PAHs over Flanders (Belgium) and their relation with anthropogenic activities. *Atmos. Environ.* 40, 771-785.

[39] Papageorgoulou, A. Manoli, E. Touloumi, E. & Samara, C. (1999) Polycyclic aromatic hydrocarbons in the ambient air of Greek towns in relation to other atmospheric pollutants. *Chemosphere*. 39, 2183-2199.

[40] Zhou, J. Wang, T. Huang, Y. Mao, T. & Zhong, N. (2005) Size distribution of polycyclic aromatic hydrocarbons in urban and suburban sites of Beijing, China. *Chemosphere*. 61, 792-799.

[41] Odum, JR. McDow, SR. & Kamens, RM. (1994) Mechanistic and kinetic studies of the photodegradation of benz[a]anthracene in the presence of methoxyphenols. *Environ. Sci. Technol.* 28, 1285-1290.

[42] McDow, SR. Sun, Q. Vartiainen, M. Hong, Y. Yao, Y. Fister, T. Yao, R. & Kamens, RM. (1994) Effect of composition and state of organic components on polycyclic aromatic hydrocarbon decay in atmospheric aerosols. *Environ. Sci. Technol.* 28, 2147-2153.

[43] Jang, M. & McDow, SR. (1995) Benz[a]anthracene photodegradation in the presence of known organic constituents of atmospheric aerosols. *Environ. Sci. Technol.* 29, 2654-2660.

[44] Jang, M. & McDow, SR. (1997) Products of benz[a]anthracene photodegradation in the presence of known organic constituents of atmospheric aerosols. *Environ. Sci. Technol.* 31, 1046-1053.

[45] Fujima, S. Ohura, T. & Amagai, T. (2006) Simultaneous determination of gaseous and particulate chlorinated polycyclic aromatic hydrocarbons in emissions from the scorching of polyvinylidene chloride film. *Chemosphere*. 65, 1983-1989.

[46] Wang, D. Piao, S. Chu, S. & Xu, X. (2001) Chlorinated polycyclic aromatic hydrocarbons from polyvinylchloride combustion. *Bull. Environ. Contam. Toxicol.* 66, 326-333.

[47] Wang, D. Xu, X. Chu, S. & Zhang, D. (2003) Analysis and structure prediction of chlorinated polycyclic aromatic hydrocarbons released from combustion of polyvinylchloride. *Chemosphere*. 53, 495-503.

[48] Horii, Y. Ok, G. Ohura, T. & Kannan, K. (2008) Occurrence and profiles of chlorinated and brominated polycyclic aromatic hydrocarbons in waste incinerators. *Environ. Sci. Technol.* 42, 1904-1909.

[49] Ohura, T. Fujima, S. Amagai, T. & Shinomiya, M. (2008) Chlorinated polycyclic aromatic hydrocarbons in the atmosphere: Seasonal levels, gas-particle partitioning, and origin. *Environ. Sci. Technol.* 42, 3296-3302.

[50] Ohura, T. Kitazawa, A. & Amagai, T. (2004) Seasonal variability of 1-chloropyrene on atmospheric particles and photostability in toluene. *Chemosphere*. 57, 831-837.

[51] Ohura, T. Kitazawa, A. Amagai, T. & Makino, M. (2005) Occurrence, profiles, and photostabilities of chlorinated polycyclic aromatic hydrocarbons associated with particulates in urban air. *Environ. Sci. Technol.* 39, 85-91.

[52] Ohura, T. Amagai, T. & Makino, M. (2008) Behavior and prediction of photochemical degradation of chlorinated polycyclic aromatic hydrocarbons in cyclohexane. *Chemosphere*. 70, 2110-2117.

[53] Kim, M. & O'Keefe, PW. (2000) Photodegradation of polychlorinated dibenzo-p-dioxins and dibenzofurans in aqueous solutions and in organic solvents. *Chemosphere*. 41, 793-800.

[54] Chen, J. Quan, X. Schramm, KW. Kettrup, A. & Yang, F. (2001) Quantitative structure?property relationships (QSPRs) on direct photolysis of PCDDs. *Chemosphere*. 45, 151-159.

[55] Niu, J. Wang, L. & Yang, Z. (2007) QSPRs on photodegradation half-lives of atmospheric chlorinated polycyclic aromatic hydrocarbons associated with particulates. *Ecotoxicol. Environ. Safety*. 66, 272-277.

[56] Colmsjo, A. Rannug, A. & Rannug, U. (1984) Some chloro derivatives of polynuclear aromatic hydrocarbons are potent mutagens in *Salmonella typhimurium*. *Mutat. Res.* 135, 21-29.

[57] Ueng, T-H. Ueng, Y-F. & Chou, MW. (1994) Regioselective metabolism of benzo[a]pyrene and 7-chlorobenz[a]anthracene by fish liver microsomes. *Toxicol. Lett.* 70, 89-99.

[58] Fu, PP. Von Tungeln, LS. Zhan, D-J. & Bucci, T. (1996) Potent tumorigenicity of 7-chlorobenz[a]anthracene and 7-bromobenz[a]anthracene in the neonatal B6C3F1 male mouse. *Cancer Lett.* 101, 37-42.

[59] Bostrom, C-E. Gerde, P. Hanberg, A. Jernstrom, B. Johansson, C. Kyrklund, T. Rannug, A. Tornqvist, M. Victorin, K. & Westerholm, R. (2002) Cancer risk assessment, indicators, and guidelines for polycyclic aromatic hydrocarbons in the ambient air. *Environ. Health Perspect.* 110, 451-488.

[60] Menzie, CA. Potocki, BB. & Santodonato, J. (1992) Exposure to carcinogenic PAHs in the environment. *Environ. Sci. Technol.* 26, 1278-1284.

[61] Poland, A. & Knutson, JC. (1982) 2,3,7,8-Tetrachlorodibenzo-p-dioxin and related halogenated aromatic hydrocarbons: examination of the mechanism of toxicity. *Annu. Rev. Pharmacol. Toxicol.* 22, 517-554.

[62] Ohura, T. Morita, M. Makino, M. Amagai, T. & Shimoi, K. (2007) Aryl hydrocarbon receptor-mediated effects of chlorinated polycyclic aromatic hydrocarbons. *Chem. Res. Toxicol.* 20, 1237-1241.

[63] World Health Organization. (2006) Fact sheet N°297.

In: Airborne Particulates
Editor: Ming Cheng and Wen Liu

ISBN: 978-1-60692-907-0

Chapter 7

CHARACTERISTIC OF ATMOSPHERIC PARTICULATES AND METALLIC ELEMENTS COMPOSITION STUDY IN CENTER TAIWAN*

Guor-Cheng Fang
Air Toxic and Environmental Analysis Laboratory
Hungkuang University, Sha-Lu, Taichung 433, Taiwan

ABSTRACT

Particulate matter (PM) is a major factor which affects the air quality of ambient environment in Taiwan region. The consecutive study of ambient air pollutants for different particulate seizes and their chemical compositions were conducted. Several character sampling sites were selected in this study. They are Taichung Harbor (TH) and WuChi traffic (WT) and Taichung airport (TA) sampling sites. As for TH and WT sampling sites, measured the concentrations of $PM_{2.5}$ (PM with aerodynamic diameter <2.5 μm, Fine particulate), $PM_{2.5-10}$ (PM with aerodynamic diameter 2.5-10 μm, Coarse particulate) and total suspended particulates (TSP) and analyzed their chemical compositions during March 2004 to January 2005. The measurement results obtained that average TSP concentrations (μg/m^3) were 149.50±29.71 and 106.67±32.04 at TH and WT sampling sites, respectively. Average concentrations (μg/m^3) of coarse particulate were 28.42±8.67 and 22.31±5.46 at TH and WT sampling sites, respectively, and average concentrations (μg/m^3) of fine particulate were 44.77±9.20 and 36.81±5.61 μg/m^3 at TH and WT sampling sites, respectively, during sampling periods. Additionally, the two primary components of metallic elements on TSP particulates were Fe (1151±236 ng/m^3) and Zn (793±104 ng/m^3), and Fe (1303±298 ng/m^3) and Zn (1081±277 μg/m^3) at the TH and WT sampling sites, respectively. The two primary components of metallic elements in coarse particulates were Fe (825±155 ng/m^3) and Mg (363±66 ng/m^3), and Fe (839±188 ng/m^3) and Mg (294±96 ng/m^3) at the TH and WT sampling sites, respectively. Furthermore, the two primary components of metallic elements in fine particulates were Fe (253±67 ng/m^3) and Mg (167±44 ng/m^3), and Zn (491±133 ng/m^3) and Fe (382±103

* A version of this chapter was also published in *Atmospheric Science Research Progress,* edited by Chih-Hao Yang published by Nova Science Publishers, Inc. It was submitted for appropriate modifications in an effort to encourage wider dissemination of research.

ng/m^3) at TH and WT sampling sites, respectively, during sampling periods. As for Taichung TA sampling site, measured the concentrations of different particulate sizes ($PM_{2.5}$, $PM_{2.5-10}$ and TSP) and analyzed their compositions during September to December of 2005. The measurement results obtained that average TSP, fine and coarse particulates were 120.6, 38.13 and 24.45 μg/m^3, respectively. Besides, two main compositions of average metallic element concentrations in atmospheric particulates were as follows: Fe, 301.67±89.68 ng/m^3 and Mg, 110.22±29.95 ng/m^3 in fine particulate; Fe, 67.07±29.74 ng/m^3 and Mg, 59.67±22.92 ng/m^3 in coarse particulate; Fe, 385.78±117.84 ng/m3 and Mg, 119.21±39.81 ng/m^3 in fine particulate.

Keywords: harbor; airport; traffic; particulate matter; metallic element; statistical analysis.

1. Introduction

Atmospheric pollution in the Taichung county near the Taiwan Strait, mainly due to aerosols (particulate matter (PM)), has become a significant issue for nearby residents. The population in and surrounding this area is approximately 1.5 million (Taichung County Government, 2007). According to previous studies (Baker *et al.*, 1997; Chester *et al.*, 1994; Eisenreich *et al.*, 1997; Gao *et al.*, 2002; Holsen *et al.*, 1997; Ondov *et al.*, 1997; Scudlark *et al.*, 1994), the coastal marine atmosphere adjacent to large urban and industrial centers is strongly impacted by pollution emissions, resulting in high pollutant loading in ambient air. Another pollution problem results from markedly high emissions from gasoline-powered vehicles and fixed sources.

Previous studies have determined that atmospheric aerosols decrease local visibility (Shuetzle, 1983) and have adverse effects on human health (Dockery *et al.*, 1992, 1993; Spengler *et al.*, 1990; Shuetzle, 1983; Pedersen *et al.*, 1999). Furthermore, Epidemiological studies conducted in several countries show consistent associations of exposure to ambient particulates with adverse health effects including increased mortality, hospitalization for respiratory or cardiovascular disease, and respiratory symptoms and decreased lung function. Based on epidemiological time series studies, dose-response functions were identified between an increase in PM and adverse health effects (El-Fadel and Massoud, 2000).

Atmospheric pollution has become a serious problem in Taiwan, especially that of O_3 and PM_{10}. In the past decade, emissions of anthropogenic air pollutants have increased drastically in Asia (Lee *et al.*, 2001). In Taiwan, the number of vehicles on main roads has increased from 5.03 to 6.4 million in recent decades (National Statistics, R.O.C.). Additionally, previous study (Nolte *et al.*, 2002) has also revealed that motor vehicle exhaust is an important source of fine particulates. Thus, understanding emissions form traffic includes identification of the sources, which is also crucial for designing control measures. Road traffic involves numerous potential sources of metals, e.g., combustion products from fuel and oil, wear products form tires, brake linings, bearings, coach and road construction materials, and re-suspension of soil and road dust (El-Fadel and hashish, 2001). Various types and different operating modes such idling, stopping, accelerating, and decelerating of vehicles along with a higher density of vehicles at traffic junctions lead to a pollution source problem at traffic junctions (Kumar *et al.*, 2001).

Larger particulates are markedly affected by gravity; fine particulates are relatively more affected by diffusion (Chan and Kwok, 2000). Heavy metals such as lead, cadmium, chromium, arsenic, and mercury, have been made with their formation, enrichment in different size particulates, removal by absorption and adsorption, reaction and so on (Yoo *et al.*, 2002). Generally, anthropogenic emissions of heavy metals was carried by atmospheric particulates, may settle on the superficial air and their depositions might change the ratios among different chemical forms of heavy metals in the air. The analysis of some meteorological conditions, which may influence the fate of atmospheric particulates, is useful for an improvement characterization of heavy metal comportment (Querol, *et al.*, 1998; Colombo *et al.*, 1999). From a toxicological perspective also, airborne PM has important health implications through the inhalation of PM10, which can be deposited in the tracheobronchial and alveolar regions of the lung (Hileman, 1981). It is well established that these inhalable particles have higher concentrations of many potentially toxic trace elements, such as Pb, Cd, V, Fe, Zn, Cr, Ni, Mn and Cu (Hlavay *et al.*, 1992).

The sources of the size-differentiated suspended particulates in aerosols and potential health effects differ. Fang *et al.* (2002) suggested that fine particulates rapidly penetrate the lungs and likely increase the incidence of respiratory and mutagenic diseases. The health-related findings in these studies were associated with either total mass concentration of suspended particulates (TSP) or mass concentration of particulates with aerodynamic diameters <10 μm (PM_{10}) or <2.5 μm ($PM_{2.5}$). Koliadima *et al.* (1998) identified that the concentration, composition, and size of suspended PM at a given site were determined by factors such as meteorological features of the atmosphere, topographical influences, emission sources, and by particle parameters such as density, shape, and hygroscopicity.

Furthermore, dust storms in China have an important impact on ambient air quality in Taiwan. Several studies (Lin, 2001; Chan, 2002; Liu *et al.*, 2006) observed that dust storms in China triggered by cold air masses passing through northern China and Mongolia increase PM concentrations over Taiwan in spring and winter.

Finally, Metallic element profiles for sources of relevance in other studies were listed in Table 1. The metallic element profiles for stationary sources were included that the soil, suspended dust, vehicular emission, industrial process, incinerator, sea salt, coal combustion, metal industry, oil combustion and construction sites. The following metallic elements have been identified as markers for various sources in atmosphere: soil and re-suspended dust—Ca, Mg, Al, Si, Fe and Mn; vehicular emission—Cr, Pb, Cu, Zn, Cd, Sb, Br, Fe and Ba; industrial process—Mn, Zn and K; incinerator—K, Zn and Pb; sea salt—Na, Ca, Mg and K; coal combustion—Cr; metal industry—Fe, Mn and Pb; oil combustion—Ni and V; construction—Al, Fe, Si, Mn and Ti.

Hence, the purpose of this study was to discuss the characteristic concentrations of ambient air particulates and those metallic elements which attached them at three sampling sites in Taichung country, Taiwan. Furthermore, the corrections between atmospheric particulates and meteorological parameters were described by using various statistical methods. The possible sources for metallic element concentrations in atmospheric particulates at this region were also determined in this study.

Table 1. Various major sources for metallic elements

Predominant species	Source	Reference
Ca, Mg, Al, Si, Fe and Mn	Soil and re-suspended dust	Allen et al., 2001; Kumar et al., 2001; Marcazzan et al., 2001; Chao et al., 2002; Sternbeck et al., 2002; Salvador et al., 2004; Lee et al., 2003; Akhlaghi and Kompany-Zareh, 2005; Astorga-Espana et al., 1999; Barona and Romero, 1996; Danielssona et al., 1999; De Bartolomeo et al., 2004; Faus-Kessler et al., 2001; Kyotani and Iwassuki, 2002
Cr, Pb, Cu, Zn, Cd, Sb, Br, Fe and Ba	Vehicular emission	Kumar et al., 2001; Marcazzan et al., 2001; Chao et al., 2002; Sternbeck et al., 2002; Funasaka et al., 2003., Salvador et al., 2004; Shu et al., 2001; Astorga-Espana et al., 1999; Barona and Romero, 1996; Danielssona et al., 1999; De Bartolomeo et al., 2004; Kyotani and Iwassuki, 2002; Loska and Wiechuła, 2003; Machado Mesa et al., 1999
Mn, Zn and K	Industrial process	Allen et al., 2001; Marcazzan et al., 2001; Akhlaghi and Kompany-Zareh, 2005; Barona and Romero, 1996
K, Zn and Pb	Incinerator	De Miranda et al., 2002; Funasaka et al., 2003; Machado Mesa et al., 1999; Manoli et al., 2004
Na, Ca, Mg, and K	Sea salt	Conner et al., 2001; Salvador et al., 2004; Kim et al., 2006;
Cr	Coal combustion	Gao et al., 2002; Pyle et al., 2001
Fe, Mn and Pb	Metal industry	De Miranda et al., 2002; Funasaka et al., 2003; Machado Mesa et al., 1999; Manoli et al., 2004
Ni and V	Oil combustion	Espinosa et al.,2001; Chao and Wong, 2002; Danielssona et al., 1999; Singh et al., 2005
Al, Fe, Si, Mn and Ti	Construction site	Chao and Wong, 2002; Danielssona et al., 1999

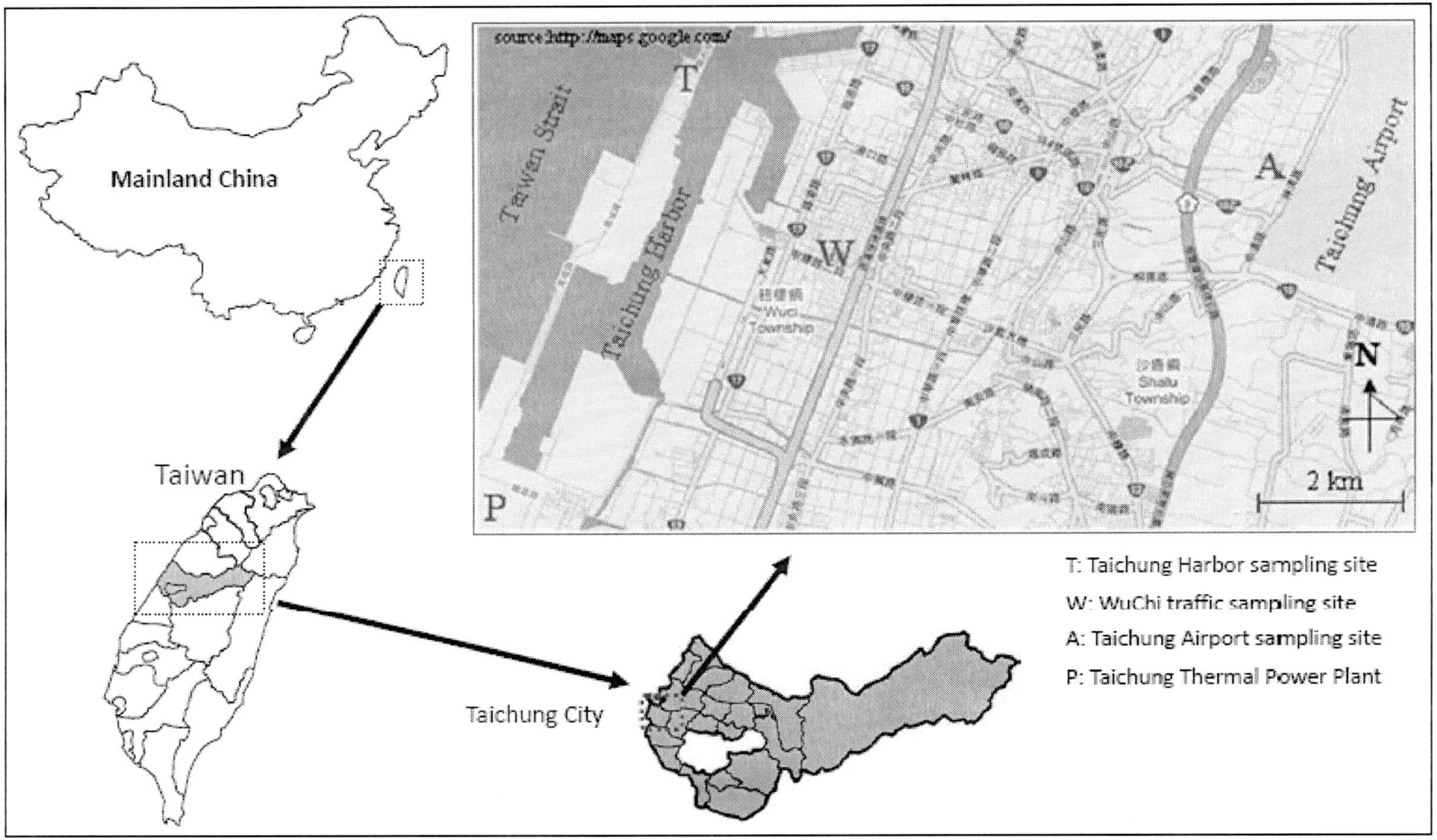

Figure 1. The location of sampling sites for this study near the Taiwan Strait in central Taiwan.

2. Sampling and Analysis Program

2.1. Sampling Sites

Taichung Harbor (TH, 24.282083N and 120.509543E), WuChi Traffic (WT, 24.252066N and 120.531269E) and Taichung Airport (TA, 24.267056N and 120.602567E) sampling sites were located on the western side of central Taiwan. Figure 1 presents the geographic locations of TH, WT and TA, respectively.

Taichung Harbor is an artificial harbor with a maximum of 83 berths located on the west coast of central Taiwan. Additionally, the TH sampling site was set on the roof of a four-story building (HUNG SUN building) which is an office building at the chemical port in Taichung Harbor. This site was typically a harbor sampling site. The TH occupies approximately 3760 acres, 960 of which is water and 2800 acres are of land. In this study, the sampling site was located at the area that deals with chemicals, which is located roughly 100 m on the east side of the Taiwan Strait.

The WT sampling site was located on the east side of Taiwan Strait, roughly 5 km from the WuChi Elementary School building and the WT sampling site was above this building (Height about 10 m). Diesel trucks and cars accounted for most traffic flow in this area. The traffic flow in WT was approximately 3,000 vehicles hr^{-1} during sampling periods. The WT sampling site was located on the southeast side of the TH sampling site roughly 4 km. Additionally, the Taichung Thermal Power Plant (TTPP, 24.214894N and 120.489764E) is located on the southern side of TH approximately 6 km away. The TTPP, which covers 281 hectares, is located along the coast. The plant uses coal combustion to generate power. Each day, the TTPP generates about 4,400 MW of power for central Taiwan. The TTPP is located at the southwest (about 6 km) of the WT sampling site.

TA sampling site was approximately 10 m from ground level. The TA has one runway 3658 m long and 61 m wide. The aircraft parking area is 36280 m^2. The TA sampling site is located on the east side of Taiwan Strait, roughly 10 km. Moreover, the WT sampling site is located on the southwest side of TA, and these two sampling sites was approximately 7 km away from each other.

2.2. Sampling

2.2.1. Samplers

PS-1 Sampler

The PS-1 (GPS1 PUF Sampler, General Metal Work, country) sampler is a complete air sampling system designed to simultaneously collect total suspended airborne particulates (TSP) at flow rate up to 280 L/min. In this study, flow rate was adjusted to 200 L/min. The filter utilized for collecting total suspended particulates was a quartz fiber filter (2500QAT-UP, Pall) with a diameter of 10.2 cm (Pall Corporation, New York, USA). The filter had a collection efficiency of 99.999% for particulate sized >0.5 μm. For particulate sized <0.3 μm, collection efficiency was 99.99%. The filters were first conditioned for 24h under an electric chamber at a humidity of 50±5 % and temperature of 20±5°C prior on and off weighing. Filters were placed in a sealed CD box during transport and for storage.

Universal Air Sampler

The Model 310 Universal Air Sampler ™ (UAS™) is a general-purpose air sampler for atmospheric aerosol sampling and mass concentration determinations, as well as organic or inorganic sampling. The sampler has a default inlet sampling flow rate of 300 L/min. The sampler has an omni-directional inlet, a PM_{10} (10 μm cut) virtual impact classifier, either a $PM_{2.5}$ or PM_{10} virtual impact classifier, and a fine particulate filter. This allows the sampler to be utilized as a high-volume dichotomous sampler for size fractionation of airborne particles in the 0–2.5 μm and 2.5–10 μm aerodynamic size ranges. Air is aspirated at 300 L/min from ambient atmosphere. Particulates with aerodynamic equivalent diameters >10 μm are removed from the air stream by the PM_{10} classifier and discarded. Particulates < 10 μm in diameter flow to the $PM_{2.5}$ classifier located downstream. The Universal Air sampler collected $PM_{2.5}$ and $PM_{2.5\text{-}10}$ using the same filter as the PS-1. Particulates in the 2.5–10 μm range (coarse fraction) were collected on a 62 mm × 165 mm filter, and those < 2.5 μm (fine fraction) were collected on a 200 mm × 250 mm final filter.

2.2.2. Mass Measurement

The particulate mass measurement method weighed and recorded surface weight after sampling in atmosphere that allowed them to equilibrate. This procedure can be divided into the following steps:

1. Weigh after obtaining moisture equilibrium (W_0).
2. Collect atmospheric particulate in the field, record sampling time (T) and the sampler's flow rate (Q).
3. Reweigh after achieving moisture equilibrium (W_1).

The particulate concentration was calculated as follows:

$$concentration = \frac{W_1 - W_0(mg)}{T(\min) \times Q(L/\min)}$$

Furthermore, an Analytical Balance (DENVER INSTRUMENT M-220D) was applied in this weighting process, and the minimum detectable mass for the balance was 0.01 mg and the maximum was 31 g.

2.2.3. Meteorological Analysis

Meteorological analysis utilized the WatchDog weather station Model 525 (Spectrum Technologies, Inc., USA). The weather station provided data for wind speed (WS), wind direction (WD), temperature (Temp.) and relative humidity (RH) during the sampling period.

2.2.4. Chemical Analysis

Following final weighing, all quartz filters used by the PS-1 and UAS samplers were cut into one-eighths and placed in 200 mL bottles for each sampling group. Subsequently, 10 mL ultra-pure HNO_3 was added to digest these particulates at 200–250 °C for 2 h. Thereafter, this solution was added to 30 mL distilled-deionized water. A Hitachi Z-5000 series polarized Zeeman flame atomic absorption spectrophotometer was utilized to measure trace metal concentrations following digestion. This process was the same as that used by Fang *el al.* (2007).

3. Results and Discussion

3.1. Sampling Information

TA Sampling Site

Figure 2 presents the sampling information of TSP, coarse and fine particulate concentrations and meteorological parameters such as temperature, relative humidity, wind speed and prevalent wind direction at TA sampling site during September to December of 2005. Observation results determined that TSP concentrations ranged from 78.41 to 172.4 $\mu g/m^3$, and coarse particulate was 15.35 - 41.42 $\mu g/m^3$. Additionally, fine particulate concentrations ranged from 21.93 to 48.73 $\mu g/m^3$. Furthermore, average TSP concentration was 120.6±30.68 $\mu g/m^3$, and coarse particulate 24.45±7.83 $\mu g/m^3$. Moreover, average fine particulate concentration was 38.13±8.44 $\mu g/m^3$ during sampling period. Besides, the average temperature, relative humidity were 25.55±4.80 □ and 72.15±4.55 %, respectively. Prevailing wind direction was NNE and average wind speed was 4.32±1.92 m/sec during the sampling period at TA sampling site.

WT Sampling Site

Figure 3 presents meteorological parameters and different size particulate concentrations such as TSP, fine, coarse particulate monthly variations during March 2004 and January 2005 at WT sampling site. Measurement results showed that TSP concentrations ranged from 71.13 to 162.5 $\mu g/m^3$, and coarse particulate was 13.68 - 39.12 $\mu g/m^3$. Furthermore, fine particulate concentrations ranged from 28.16 from 68.41 $\mu g/m^3$. Additionally, average TSP concentration was 113.72±34.46 $\mu g/m^3$, and coarse particulate 25.99±8.18 $\mu g/m^3$. Moreover, average fine particulate concentration was 47.38±13.59 $\mu g/m^3$ during sampling period. Besides, the average temperature, relative humidity were 23.14±4.77 □ and 73.82±2.89 %, respectively. Prevailing wind direction was mainly blown from north and average wind speed was 6.85±1.66 m/sec during the sampling period at WT sampling site.

Moreover, average particulate concentrations of TSP in spring (March, April and May), summer (June, July and August), autumn (September, October and November) winter (December and January) were 128.91±30.57, 76.70±7.03, 93.95±18.10 and 143.35±5.85 $\mu g/m^3$, respectively; average particulate concentrations of $PM_{2.5}$ in spring, summer, autumn and winter were 41.84±5.43, 36.52±3.82, 31.65±4.67 and 35.94±1.14 $\mu g/m^3$, respectively. Average particulate concentrations of $PM_{2.5-10}$ in spring, summer, autumn and winter were 26.36±3.36, 22.65±5.62, 15.99±2.71, and 24.08±0.89 $\mu g/m^3$, respectively. Based on comparisons of average TSP concentrations at WT sampling site, the concentration order during sampling periods was as follows: winter > spring > autumn > summer. Based on a comparison of average $PM_{2.5-10}$ and $PM_{2.5}$ concentrations, the concentration order during the sampling periods were as follows: spring > summer > winter > autumn.

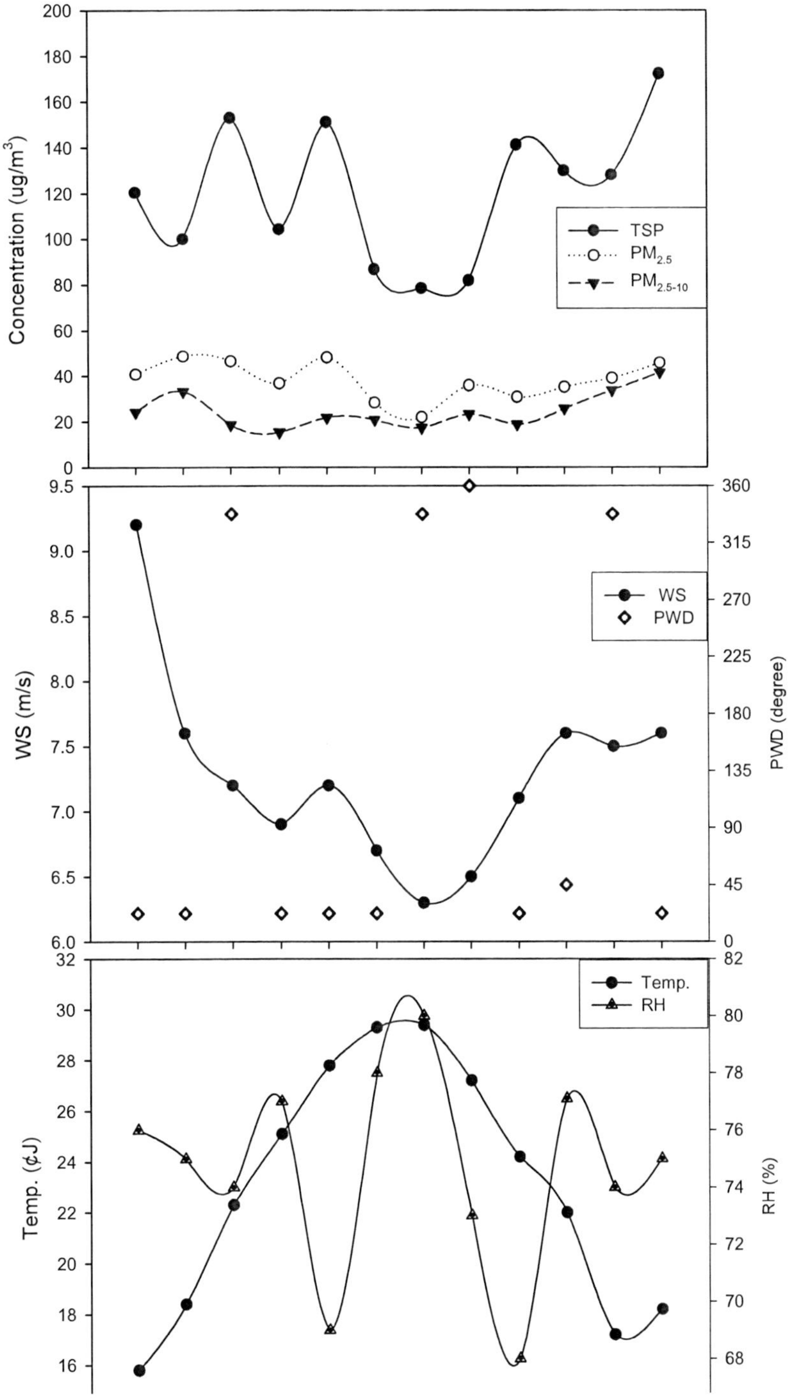

Figure 2. Variations of meteorological parameters and atmospheric particulates at TA sampling site.

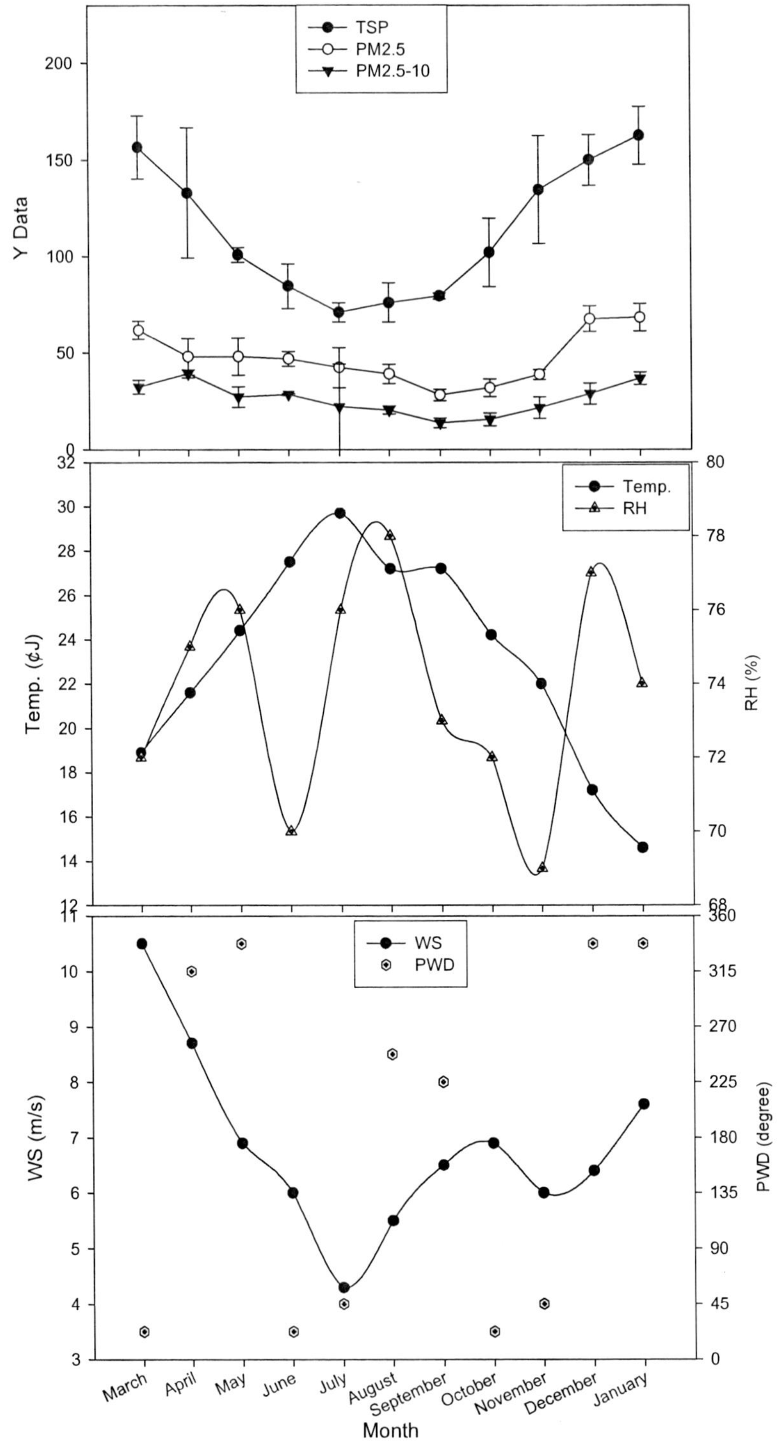

Figure 3. Monthly variations of different meteorological parameters and atmospheric particulates at WT sampling site.

TH Sampling Site

Figure 4 displays TSP, fine, coarse particulate concentrations and meteorological parameters such as temperature, relative humidity, and prevalent wind direction variations between March 2004 and January 2005 at TH sampling site.

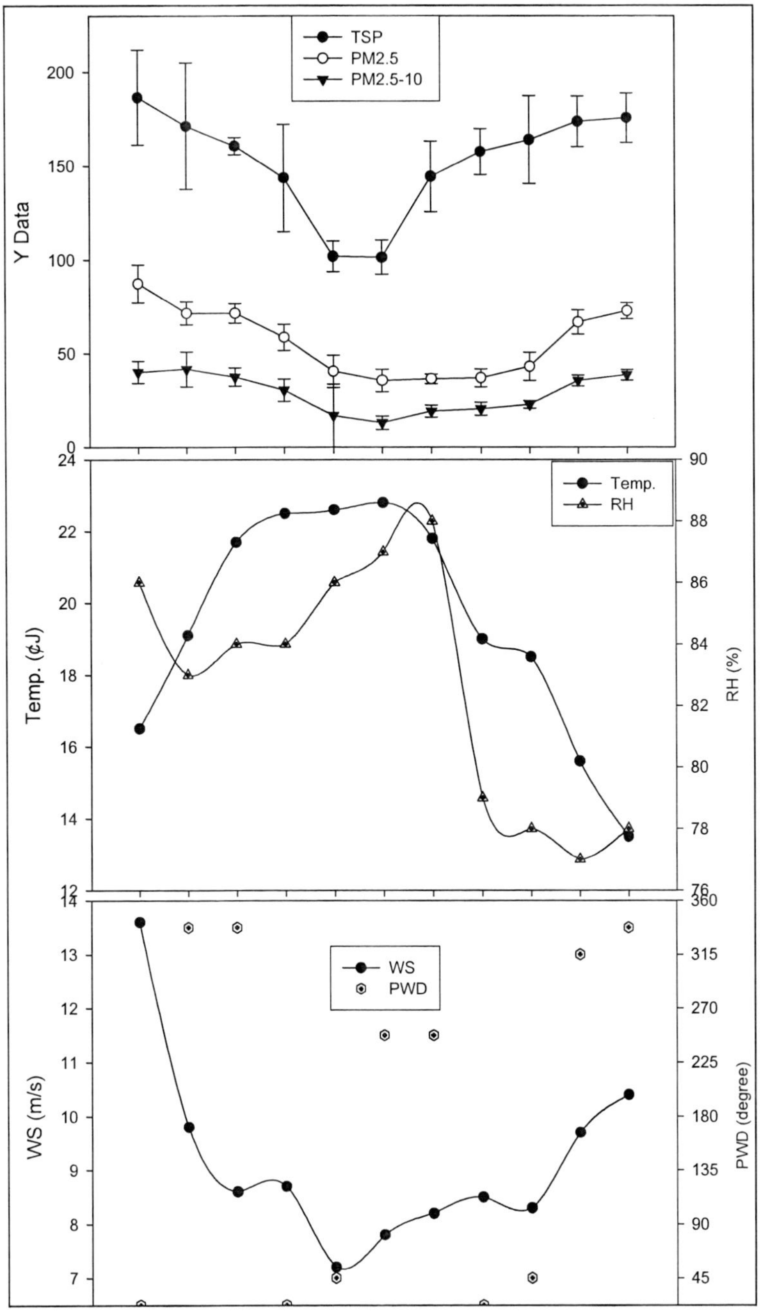

Figure 4. Monthly variations of meteorological parameters and atmospheric particulates at TH sampling site.

Measurement results obtained that TSP concentrations ranged from 101.32 to 186.55 $\mu g/m^3$, and coarse particulate was 16.81 - 41.59 $\mu g/m^3$. Furthermore, fine particulate concentrations ranged from 35.52 to 87.29 $\mu g/m^3$. Additionally, average TSP concentration was 152.76±28.37 $\mu g/m^3$, and coarse particulate 28.69±10.49 $\mu g/m^3$. Moreover, average fine particulate concentration was 56.38±18.53 $\mu g/m^3$ during sampling period. Besides, the average temperature, relative humidity were 19.42±3.18 □ and 82.73±4.03 %, respectively. Prevailing wind direction was principal blown from north and average wind speed was 9.16±1.74 m/sec during the sampling period at TH sampling site.

Additionally, average concentrations of TSP in spring, summer, autumn and winter were 170.65±16.79, 113.47±23.13, 152.56±9.11 and 174.15±2.67 $\mu g/m^3$, respectively. Average particulate concentrations of $PM_{2.5}$ in spring, summer, autumn and winter were 51.50±7.93, 39.54±9.32, 39.65±6.53 and 50.53±3.88 $\mu g/m^3$, respectively. Moreover, average particulate concentrations of $PM_{2.5-10}$ in spring, summer, autumn and winter were 36.60±4.29, 21.19±5.79, 22.51±4.64, and 36.38±1.02 $\mu g/m^3$, respectively. Based on comparisons of average TSP concentrations, the concentration order of sampling periods was as follows: winter > spring > autumn > summer. Additionally, based on a comparison of average coarse and fine particulate concentrations, the concentration order during sampling periods were as follows: spring > winter > autumn > summer.

According to these analyses (TH and WT sampling sites), particulate concentrations (TSP, $PM_{2.5-10}$ and $PM_{2.5}$) in spring and winter were higher than those for other seasons for the WT and TH sampling sites. The primary reason was that the dust storms in China frequently occur in spring and winter, thereby increasing PM concentration in Taiwan. This experimental result was published by Fang *et al.* (2002). Moreover, the strong northeast monsoon has influenced Taiwan including Chinese dust storms in spring and winter and mean wind speeds in spring and winter are higher than those in other seasons.

3.1.1. Comparison of Average Atmospheric Particulates at This Three Sampling Sites

Figure 5 schematizes average concentrations for atmospheric particulates at TA, WT, TH sampling sites during September to December sampling periods. Based on comparisons of average TSP and fine particulate concentrations at these three sampling sites, the concentration order during sampling periods was as follows: TH > WT > TA, and the average coarse particulate concentration order during sampling periods was as follows: TH > TA > WT. Therefore, TSP, fine and coarse particulate concentrations at TH sampling site were highest than those at other two sampling sites. Mean wind speed at TH sampling site higher than those at other sampling site was the main reason for this finding.

3.2. Metallic Elements in Atmospheric Particulates

TA Sampling Site

Figure 6 is a box-plot for those metallic elements concentrations in atmospheric particulates at TA sampling site during sampling period.

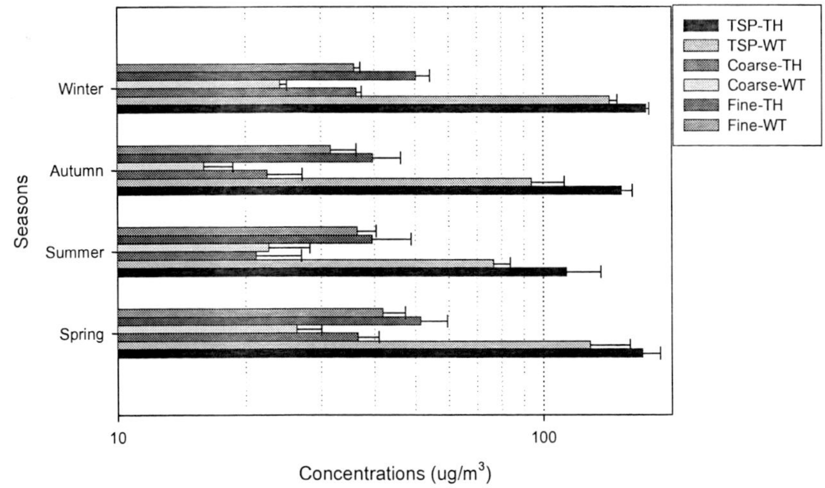

Figure 5. Average seasonal concentrations of TSP, fine and coarse particulates at three sampling sites.

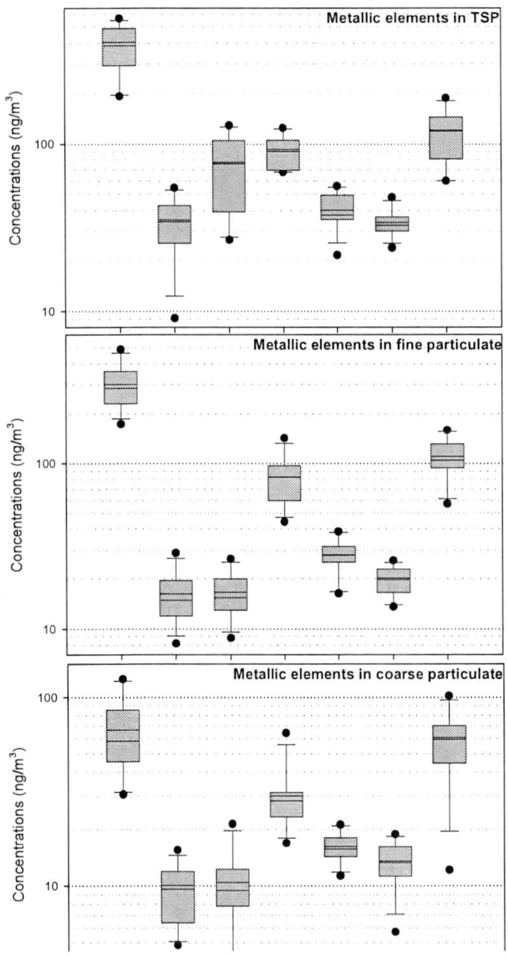

Figure 6. Box-plot of metallic elements for atmospheric particulates at TA sampling site.

Regarding the concentrations of Fe, Mn, Cu, , Zn, Pb, Cr and Mg in TSP were ranged 193.06-561.67 ng/m^3, 9.09-54.66 ng/m^3, 26.73-129.1 ng/m^3, 67.51-125 ng/m^3, 21.68-55.84 ng/m^3, 24-47.9 ng/m^3 and 59.96-187.91 ng/m^3, respectively, and the concentrations in fine particulates were ranged as follows: Fe, 173.59-488.89 ng/m^3; Mn, 8.18-28.8 ng/m^3; Cu, 8.83-26.6 ng/m^3; Zn, 44.41-142.09 ng/m^3; Pb, 16.36-38.82 ng/m^3; Cr, 13.62-26 ng/m^3 and Mg, 57.02-158.72 ng/m^3. Moreover, Fe, Mn, Cu, Zn, Pb, Cr and Mg concentrations in coarse particulates were ranged 30.56-125.03 ng/m^3, 4.87-15.43 ng/m^3, 3.2-21.3 ng/m^3, 20.54-64.4 ng/m^3, 11.3-21.14 ng/m^3, 5.7-18.8 ng/m^3 and 12.1-102.1 ng/m^3, respectively.

The analytical results demonstrate that average metallic elements concentrations in TSP were as follows: Fe, 385.78±117.84 ng/m^3; Mn, 34.28±12.77 ng/m^3; Cu, 77.69±35.46 ng/m^3; Zn, 90.4±19.72 ng/m^3; Pb, 40.18±9.58 ng/m^3; Cr, 34.03±6.12 ng/m^3 and Mg, 119.21±39.81 ng/m^3, and Fe, Mn, Cu, , Zn, Pb, Cr and Mg concentrations in fine particulates were 301.67±89.68 ng/m^3, 16.32±5.65 ng/m^3, 16.69±5.09 ng/m^3, 82.99±26.71 ng/m^3, 28.04±6.57 ng/m^3, 19.86±3.8 ng/m^3 and 110.22±29.95 ng/m^3, respectively. Additionally, Fe, Mn, Cu, , Zn, Pb, Cr and Mg concentrations in coarse particulates were 67.07±29.74 ng/m^3, 9.62±3.3 ng/m^3, 10.45±4.59 ng/m^3, 29.99±12.11 ng/m^3, 16.15±2.88 ng/m^3, 13.25±3.54 ng/m^3 and 59.67±22.92 ng/m^3, respectively. Base on above observations, the results reveal that three primary components of metallic elements in atmospheric particulates were Fe, Mg and Zn at TA sampling site during sampling period.

WT Sampling Site

Figure 7 displays the distribution for Fe, Mg, Pb, Zn, Cr, Mn, and Cu concentrations at WT sampling site during sampling period. The concentrations in TSP were ranged as follows: Fe, 667-1755 ng/m^3; Mg, 350-598 ng/m^3; Pb, 19-102 ng/m^3; Zn, 463-1297 ng/m^3; Cr, 106-546 ng/m^3; Mn, 11-82 ng/m^3; Cu, 217-685 ng/m^3 , and Fe, Mg, Pb, Zn, Cr, Mn, and Cu concentrations in fine particulates were ranged 168-517 ng/m^3, 61-178 ng/m^3, 16-56 ng/m^3, 217-622 ng/m^3, 51-255 ng/m^3, 11-41 ng/m^3 and 94-285 ng/m^3, respectively. Furthermore, Fe, Mg, Pb, Zn, Cr, Mn, and Cu concentrations in coarse particulates were ranged 529-1110 ng/m^3, 103-394 ng/m^3, 3-14 ng/m^3, 113-244 ng/m^3, 27-113 ng/m^3, 11-31 ng/m^3 and 60-201 ng/m^3, respectively.

The examination results obtain that average concentrations of Fe, Mg, Cr, Cu, Zn, Mn, and Pb were 1303±298 ng/m^3, 478±98 ng/m^3, 355±132 ng/m^3, 522±143 ng/m^3, 1081±277 ng/m^3, 54±19 ng/m^3 and 66±26 ng/m^3 in TSP, respectively, and 839±188 ng/m^3, 294±96 ng/m^3, 94±29 ng/m^3, 142±47 ng/m^3, 198±43 ng/m^3, 22±6 ng/m^3 and 10±4 ng/m^3 in coarse particulates, respectively. Additionally, average metallic element concentrations in fine particulates were as follows: Fe, 382±103 ng/m^3; Mg, 144±32 ng/m^3; Cr, 156±57 ng/m^3; Cu, 225±58 ng/m^3; Zn, 491±133 ng/m^3; Mn, 29±9 ng/m^3; and Pb, 39±12 ng/m^3. According above analytical results determine that metallic elements Fe and Zn were the primary components in TSP at WT sampling sites, and Fe and Mg were the predominant components in coarse and fine particulates. However, Zn and Fe were the predominant components in fine particulates at the WT sampling site during sampling period.

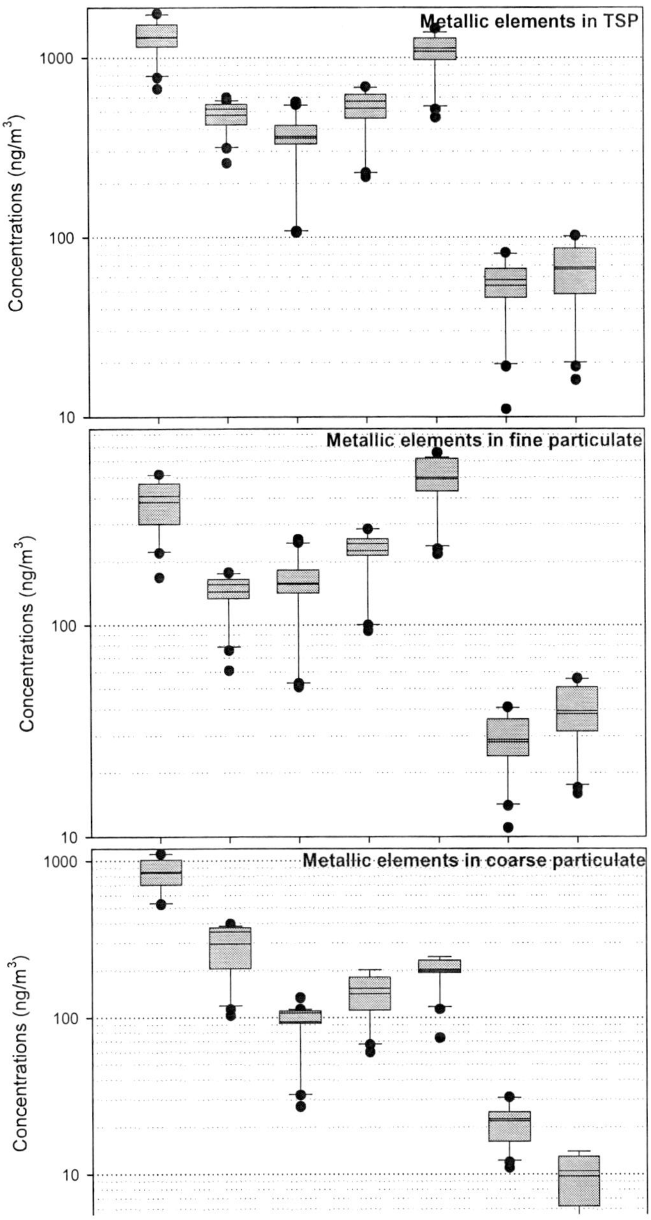

Figure 7. Box-plot of metallic elements for atmospheric particulates at WT sampling site.

TH Sampling Site

Metallic elements concentrations in atmospheric particulates at TH sampling site during sampling period was shown in figure 8 by a box-plot.

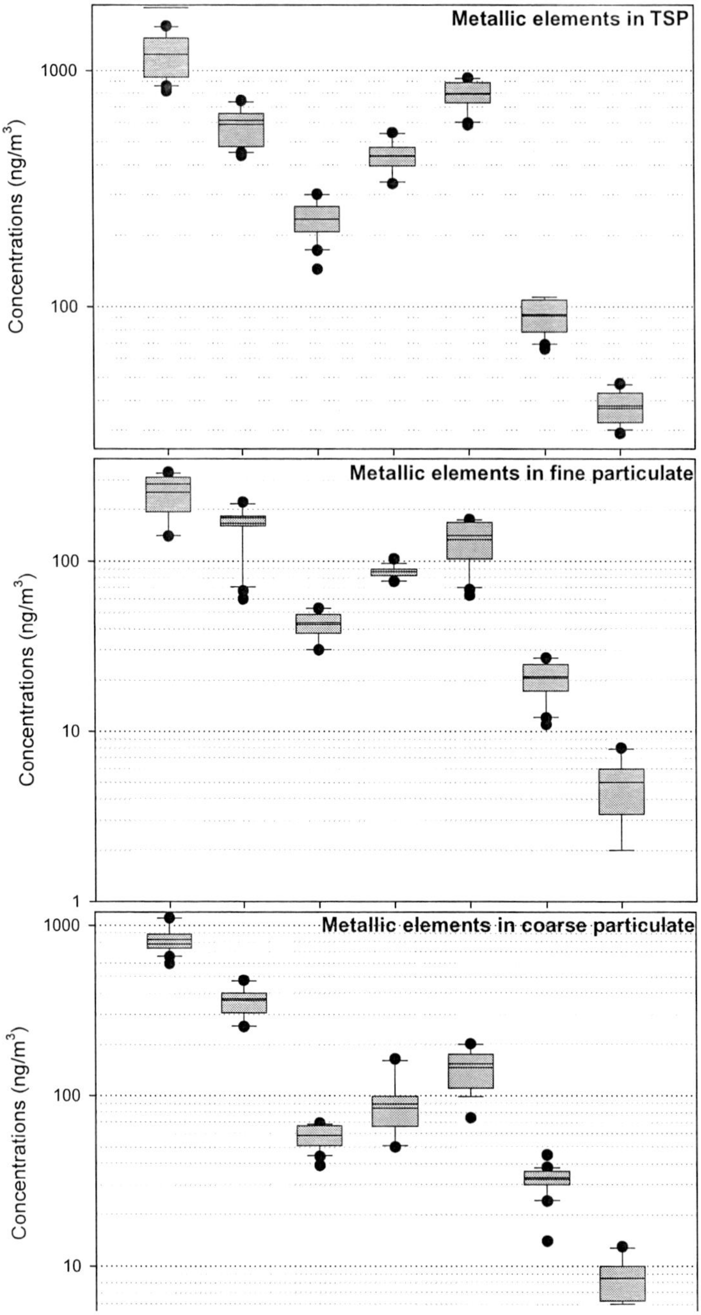

Figure 8. Box-plot of metallic elements for atmospheric particulates at TH sampling site.

The ranges of metallic elements concentrations were as follows: Fe, 817-1543 ng/m^3; Mg, 449-745 ng/m^3; Pb, 29-47 ng/m^3; Zn, 589-927 ng/m^3; Cr, 144-300 ng/m^3; Mn, 66-110 ng/m^3; Cu, 332-546 ng/m^3 in TSP, and Fe, 140-330 ng/m^3; Mg, 60-217 ng/m^3; Pb, 2-8 ng/m^3; Zn, 63-175 ng/m^3; Cr, 30-53 ng/m^3; Mn, 11-27 ng/m^3; Cu, 76-103 ng/m^3 in fine particulates. Additionally, the concentrations of Fe, Mg, Cr, Cu, Zn, Mn, and Pb were ranged 596-1103 ng/m^3, 255-476 ng/m^3, 39-68 ng/m^3, 50-164 ng/m^3, 74-201 ng/m^3, 14-45 ng/m^3 and 4-13 ng/m^3, respectively, in coarse particulate.

The measurement results demonstrate that average metallic element concentrations were as follows: Fe, 1151±236 ng/m^3; Mg, 590±102 ng/m^3; Cr, 235±43 ng/m^3; Cu, 0.436±62 ng/m^3; Zn, 793±104 ng/m^3; Mn, 93±16 ng/m^3; and Pb, 38±6 ng/m^3 in TSP, and Fe, 253±67 ng/m^3; Mg, 167±44 ng/m^3; Cr, 43±7 ng/m^3; Cu, 87±7 ng/m^3; Zn, 134±37 ng/m^3; Mn, 21±5 ng/m^3; and Pb, 5±2 ng/m^3 in fine particulates. Furthermore, average concentrations for Fe, Mg, Cr, Cu, Zn, Mn, and Pb were 825±155 ng/m^3, 363±66 ng/m^3, 58±8 ng/m^3, 89±32 ng/m^3, 146±39 ng/m^3, 32±6 ng/m^3 and 8±2 ng/m^3 in coarse particulates, respectively. Judging from the above results, Fe and Zn were the main components in TSP, while Fe and Mg were the primary components in coarse and fine particulates at the TH sampling sites.

Metallic elements concentrations in atmospheric particulates at three sampling site have been described in detail. These results lead to the conclusion that metallic elements of Fe, Mg, Zn were the predominant components in atmospheric particulates at these three sampling sites. According to previous studies (Allen *et al.*, 2001; Kumar *et al.*, 2001), Fe is a metallic element indicative of crustal, re-suspended dust and metal industry. In this study, the largest source of Fe may come from soil and re-suspended dust. In previous studies (Kumar *et al.*, 2001; Shu *et al.*, 2001), Pb was an effective index in traffic pollutant. However, Zheng *et al.* (2004) indicated that Pb cannot be considered as a traffic emission index as leaded gasoline has been phased out by comparing with lead isotope ratios. Notably, Zn has been considered a good maker for unleaded isotope vehicular emissions (Salvador *et al.*, 2004). Hence, gasoline-powered vehicles were responsible for these results for fine particulates.

3.3. Statistical Analysis

3.4.1. The Correlation between Atmospheric Particulates and Meteorological Parameters

The relationship between atmospheric particulates and meteorological parameters at three sampling sites were estimated by using statically method of Pearson correlation. Furthermore, the simple and multiple regression equations were established to describe the relationship between atmospheric particulates and meteorological parameters.

Table 2 presents the correlation coefficients between atmospheric particulates and meteorological parameters at TA sampling site. The correlation shows a significant negative relationship between coarse particulates and Temp. during sampling period. Furthermore, the relationships between coarse particulates and Temp. are also illustrated in Figure 9, and establish a regression equation for coarse particulates and Temp.. The regression equation for coarse particulates and Temp. during sampling period at TA sampling site is

$$\text{Coarse} = 49.7525 - 1.0966 \times \text{Temp.}$$

The analytical results of Pearson correlation at WT sampling site are presented in Table 3. The correlations show a clear and strong relationship between atmospheric particulates and WS. Additionally, TSP and WS have been shown to be negative correlated with one another, and this similarity result was also obtained between WS and Temp.

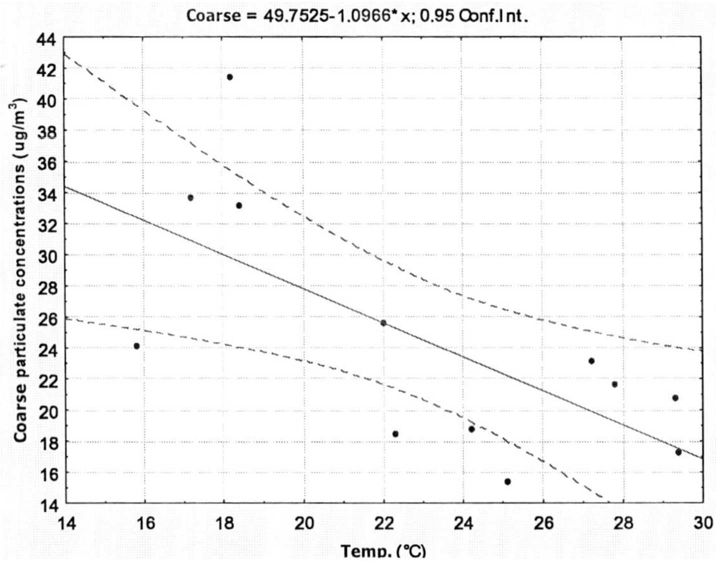

Figure 9. The scatter-plot of coarse particulates and Temp. at TA sampling site.

Table 2. The person correlations of relationship between atmospheric particulates and meteorological parameters at TA sampling sites

TA	TSP	Fine	Coarse	Temp.	RH	WS
Fine	*0.60*					
Coarse	0.38	0.49				
Temp.	-0.45	-0.56	*-0.68*			
RH	-0.52	-0.40	-0.02	0.04		
WS	0.41	0.49	0.39	*-0.83*	-0.06	
PWD	-0.26	-0.21	-0.12	0.15	0.12	-0.41

Note: italic and marked correlations are significant at $p < 0.05$.

Table 3. The person correlations of relationship between atmospheric particulates and meteorological parameters at WT sampling sites

WT	TSP	Fine	Coarse	Temp.	RH	WS
Fine	*0.77*					
Coarse	*0.67*	*0.80*				
Temp.	*-0.97*	*-0.80*	*-0.63*			
RH	-0.15	0.22	0.11	0.02		
WS	*0.68*	0.44	0.57	*-0.61*	-0.19	
PWD	0.25	0.37	0.38	-0.40	*0.67*	0.09

Note: italic and marked correlations are significant at $p < 0.05$.

Generally, the monsoon blows from north were cool and strong in Taiwan, and this phenomena description has been illustrated in Figure 2, previously. Thus, the negative relationship between WS and Temp. has been established and confirmed. Moreover, Figure 10 illustrates the corrections between TSP, Temp. and WS, a multiple regression equation for these three items has also established in this step and described as followed:

$$TSP = 240.6308 + 2.9777 \times WS - 6.3666 \times Temp.$$

Additionally, the corrections between coarse particulate and Temp.; fine particulate and Temp. are also illustrated in Figures 11 and 12. The analytical results concluded that regression equation for Temp. between coarse and fine particulate during sampling period can be presented as followed:

$$Fine = 99.779 - 2.2649 \times Temp.$$

$$Coarse = 51.1091 - 1.0856 \times Temp.$$

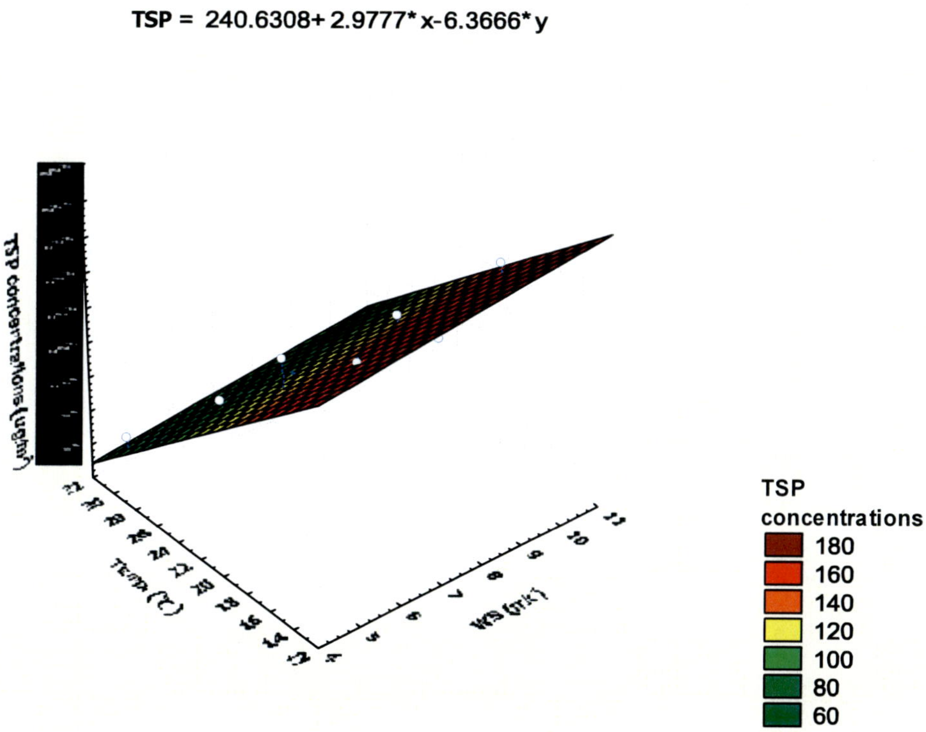

Figure 10. The relationship between TSP, WS and Temp. at WT sampling site.

Table 4 presents the correlation coefficients between atmospheric particulates and meteorological parameters at TH sampling site. The observation results demonstrated that atmospheric particulates seemed to be closely connected to WS. Additionally, TSP and Temp. were also shown to be significantly related, and the relationship between WS and Temp. was the same as that of WT sampling site.

Table 4. The person correlations of relationship between atmospheric particulates and meteorological parameters at TH sampling sites

TH	TSP	Fine	Coarse	Temp.	RH	WS
Fine	*0.73*					
Coarse	*0.82*	*0.96*				
Temp.	*-0.77*	-0.57	-0.60			
RH	-0.54	-0.16	-0.30	*0.71*		
WS	*0.74*	*0.83*	*0.73*	*-0.67*	-0.09	
PWD	0.18	0.29	0.38	-0.22	-0.10	-0.01

Note: italic and marked correlations are significant at $p < 0.05$.

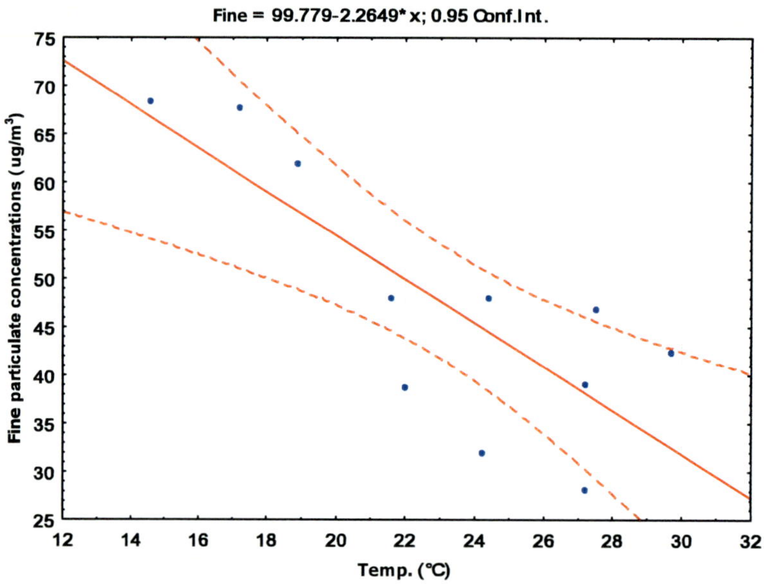

Figure 11. The scatter-plot of fine particulates and Temp. at WT sampling site.

Furthermore, Figure 13 illustrated the relationship between TSP, Temp. and WS, and a multiple regression equation was, thus, established as follow:

$$TSP = 179.2496 + 6.5905 \times WS - 4.4743 \times Temp.$$

Besides, Figures 14 and 15 illustrate the correlations between coarse particulate and WS; fine particulate and WS, respectively. The analytical results obtain a regression equation for each correlation which can be presented as followed:

$$Fine = -24.3011 + 8.8044 \times WS$$

$$Coarse = -11.6222 + 4.3997 \times WS$$

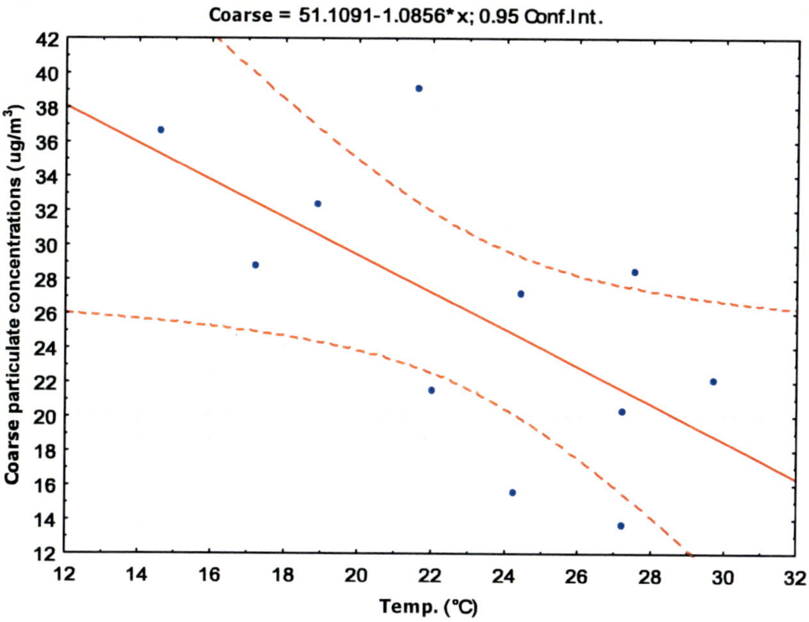

Figure 12. The scatter-plot of coarse particulates and Temp. at WT sampling site.

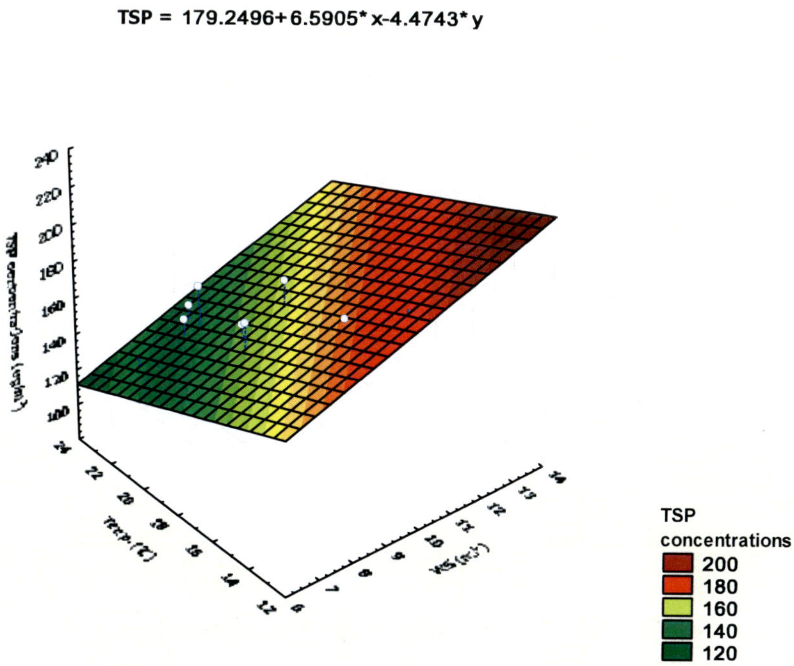

Figure 13. The 3D surface plot of the corrections between TSP, Temp. and WS at TH sampling site.

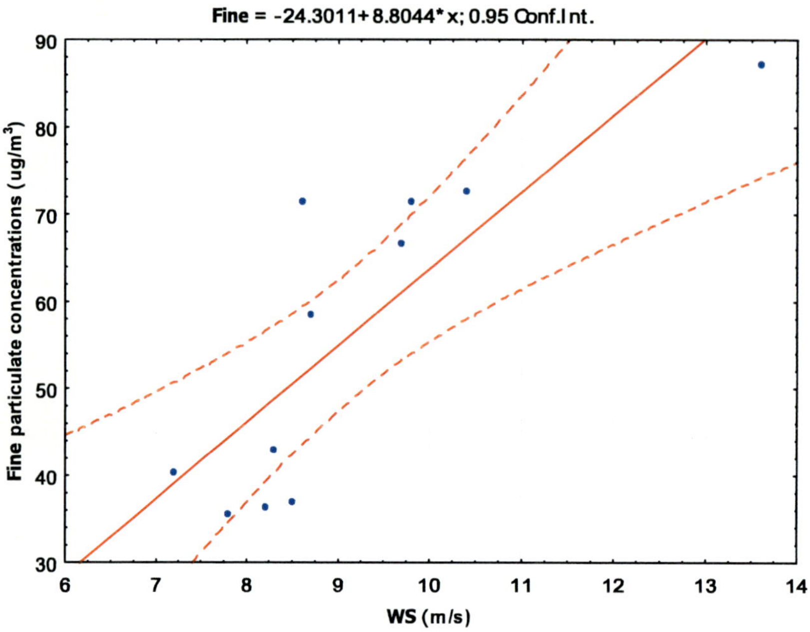

Figure 14. The scatter-plot of coarse particulates and WS at TH sampling site.

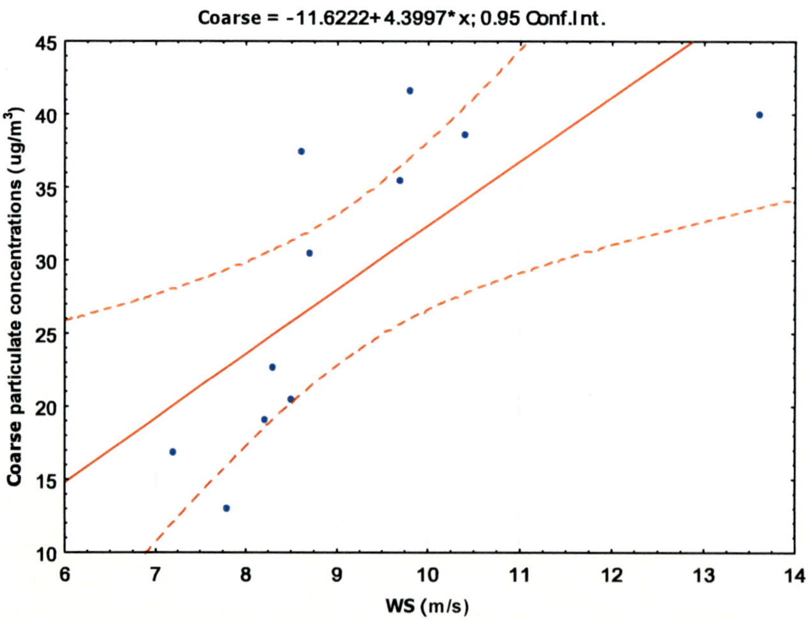

Figure 15. The scatter-plot of fine particulates and WS at TH sampling site.

According to above observations and analysis, the results reveal that WS and Temp. were the main factors which influenced atmospheric particulate concentrations, especially at TH sampling site. The possible proposed reason was that TH sampling site was close to the Taiwan western coast (about 100 m) compared to the rest of the other sampling sites.

Furthermore, TH sampling site has the smallest surface roughness. Thus, wind speed has the least factors, if there is any, to influence the above result.

3.4.2. Factor Analysis of Metallic Element Concentrations in Atmospheric Particulates

The method of factor analysis was utilized to identify the possible sources for metallic elements. The eigenvalues extract by applied principal components with Varimax rotation to obtains greater factors. Once a factor has a great eigenvalue, it can explain more total variance, Kaiser (1974) ever suggested that we can obtained more total variance as long as we have eigenvalue > 1. Additionally, factors extracted also observed from eigenvalue's screeslope. Cattlell (1966) proposed that the steep slope which obtained from the turning point is the place that we need to extract.

At TA sampling site, figure 16 illustrates the results for factor analysis of metallic element concentrations in TSP. The analytical results present that two factors eigenvalues > 1 and explained about 71.53 % of total variance. Factor 1 explained 52.13 % of total variance and had high factor loading for Pb (0.91), Fe (0.87, Cr (0.72 and Zn (0.71), which identified the possible source for vehicle emission and incinerator, and factor 2 accounted for 18.4 % of total variance and identified the possible source for soil. Moreover, metallic element concentrations in coarse particulate were extracted two factors and the results are displayed in figure 17. The analytical results indicated that factor a explained 50.82 % of total variance and had high factor loading for Cr (0.88), Cu (0.82) and Fe (0.7), which identified the possible source for vehicle emission, and factor 2 explained 17.01 % of total variance and identified the possible source for soil and re-suspended dust. Additionally, figure 18 illustrates the results for factor analysis of metallic element concentrations in fine particulate. Three factors were extracted (eigenvalue > 1) by factor analysis, and explained 85 % of total variance. Factor 1 accounted for 47.19 % of total variance and had high factor loading for Mn (0.93), Zn (0.87), Fe (0.85) and Pb (0.84) which identified the possible source for vehicular emission. Factor 2 explained 21.84 % of total variance and had high factor loading for Cr (0.91) which identified the possible source for coal combustion. Factor 3 explained for 16.48 % of total variance and had high factor loading for Mg (0.89) which identified the possible source for sea salt.

At WT sampling site, figures 19, 20 and 21 show the results of factor analysis of metallic element concentrations in atmospheric particulates. As for TSP (Figure 19), two factors were extracted and explained 81.87 % of total variance. Factor 1 accounted for 71.87 of total variance, high factor loading of Zn (0.87), Pb (0.83), Cu (0.82) and Fe (0.79) were contributed by traffic emission. Factor 2 is relative to coal combustion and industrial process. Additionally, factor analysis of metallic element concentrations in coarse particulate were extracted two factors and accounted 82.36 % of total variance (Figure 20). High factor loading of Pb (0.93), Fe (0.92), Cu (0.78) and Mn (0.73) were contributed by traffic emission. High factor loading of Cr (0.94) and Zn (0.81) were observed on factor 2 and the possible source were industrial process and coal combustion. As for fine particulate, the results for factor analysis of metallic element concentrations are illustrated in figure 21. The analytical results show that high factor loadings of Zn (0.89), Pb (0.87), Cu (0.86), Mg (0.85) and Fe (0.7) on factor 1 and explained 69.55 % of total variance. This suggested that traffic vehicle emissions were the main contributors. Factor 2 explained 13.1 % of total variance and had factor loading of Mn (0.94) which identified the possible source for metal industry.

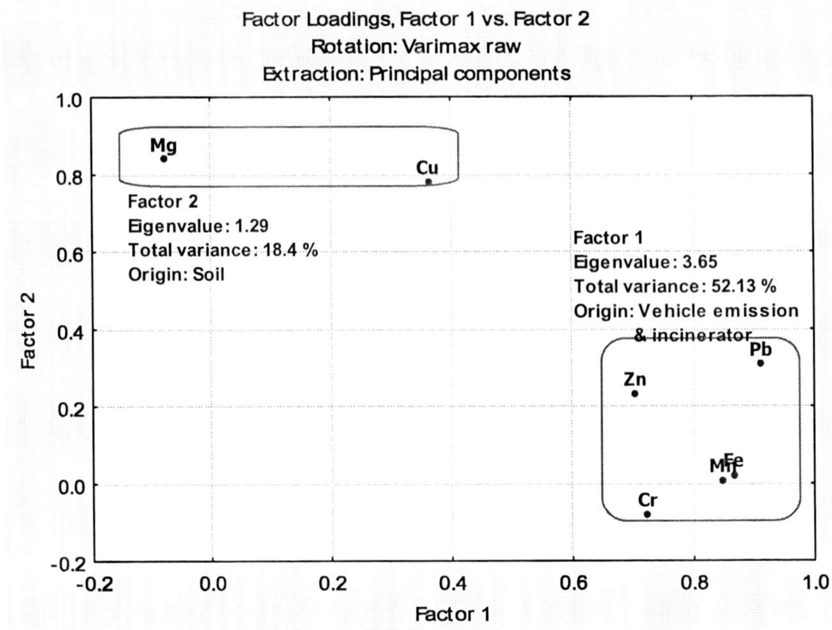

Figure 16. Factor analysis of metallic element concentrations in TSP at TA sampling site.

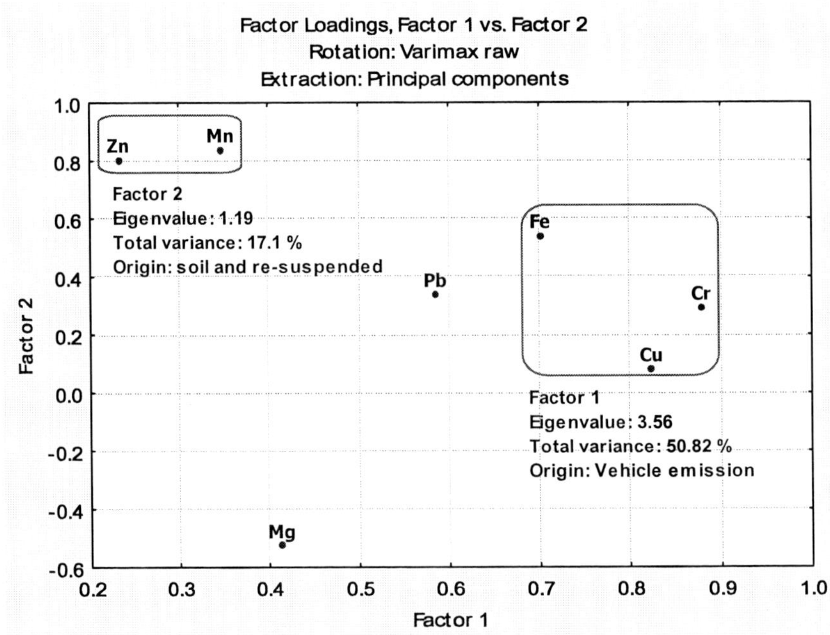

Figure 17. Factor analysis of metallic element concentrations in coarse particulate at TA sampling site.

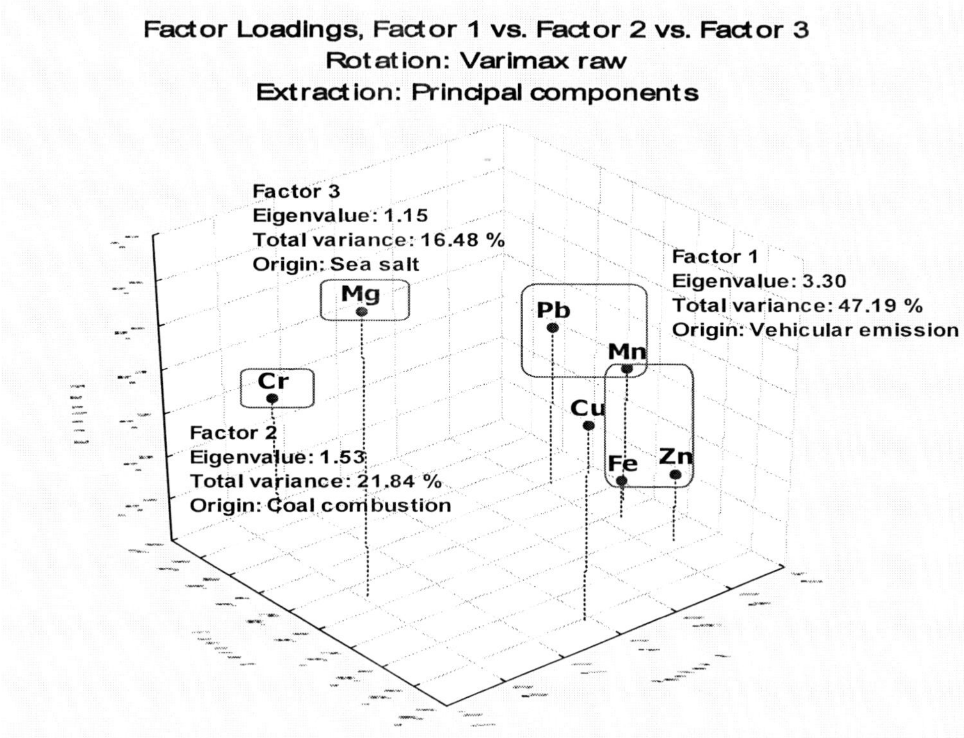

Figure 18. Factor analysis of metallic element concentrations in fine particulate at TA sampling site.

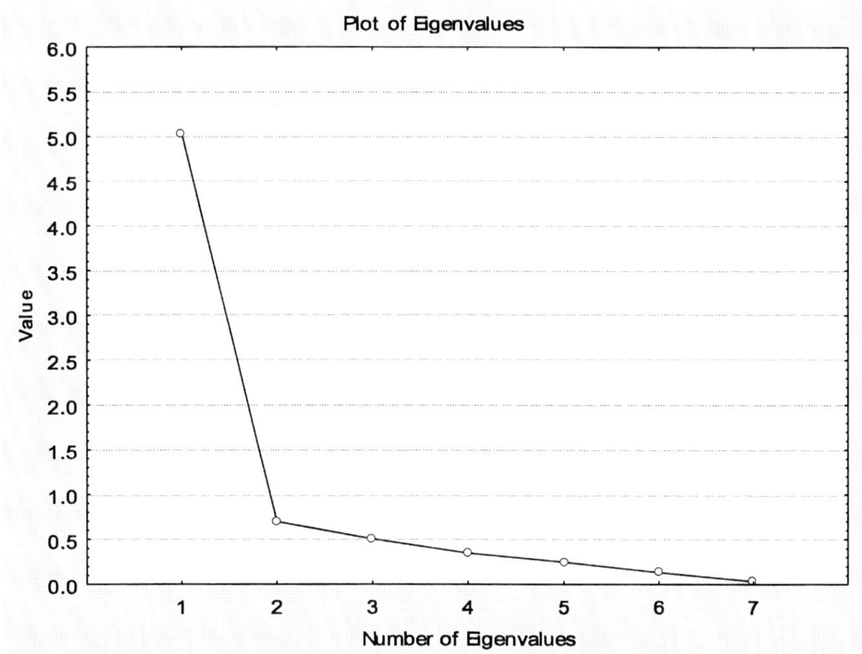

Figure 19. (Continued on next page).

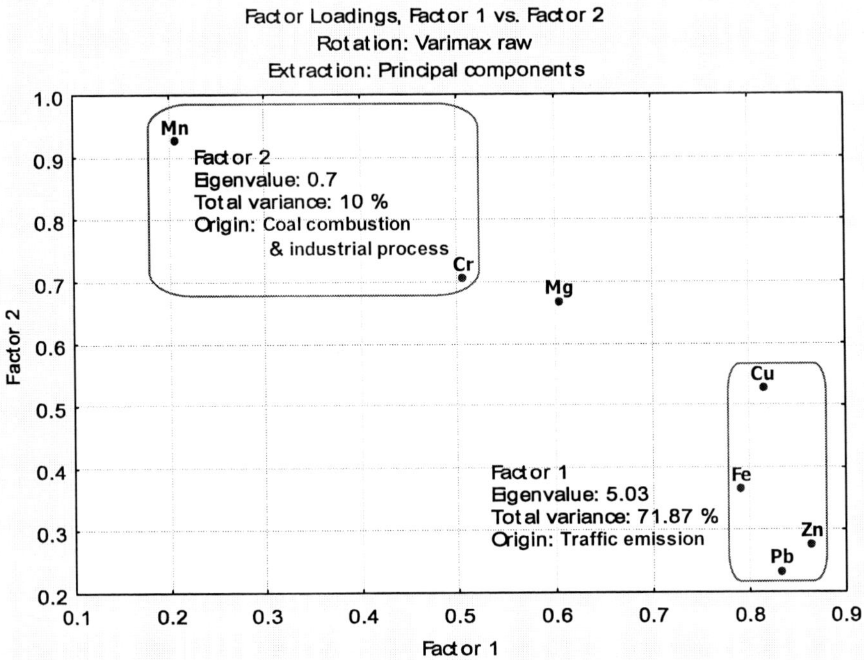

Figure 19. The eigenvalue's screeplot and factor analytical results for metallic element concentrations in TSP at WT sampling site.

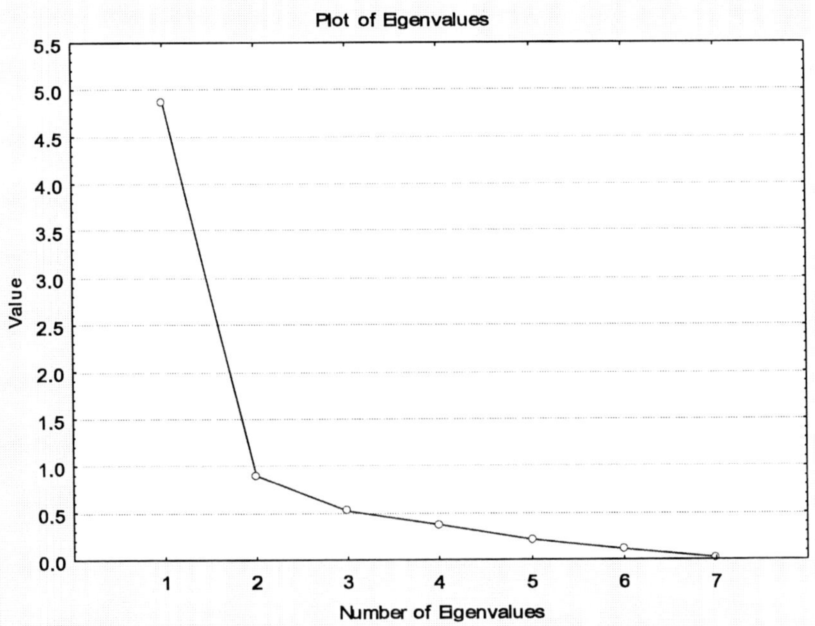

Figure 20. (Continued on next page).

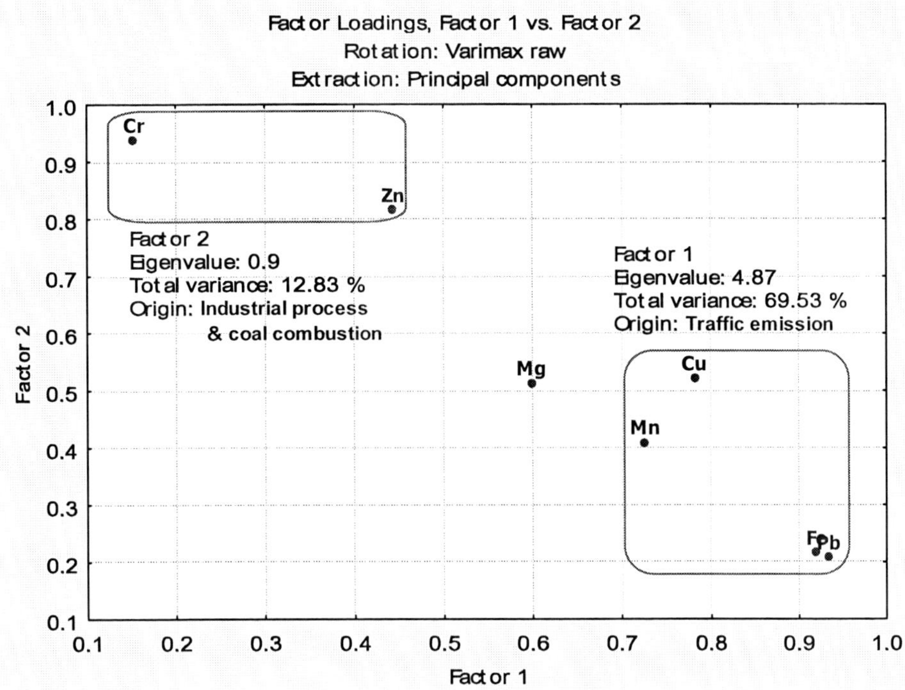

Figure 20. The eigenvalue's screeplot and factor analytical results for metallic element concentrations in coarse particulate at WT sampling site.

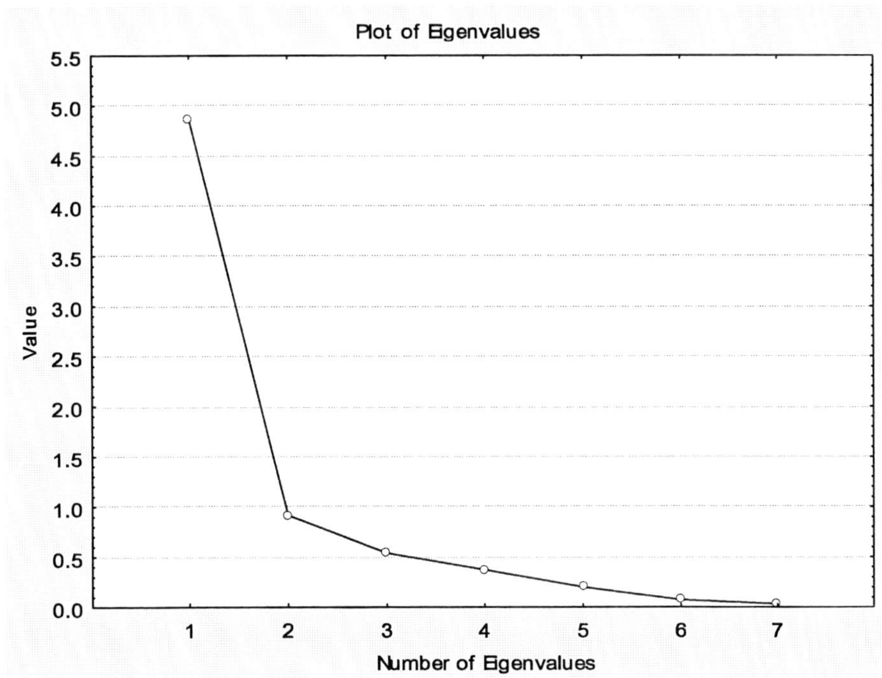

Figure 21. (Continued on next page).

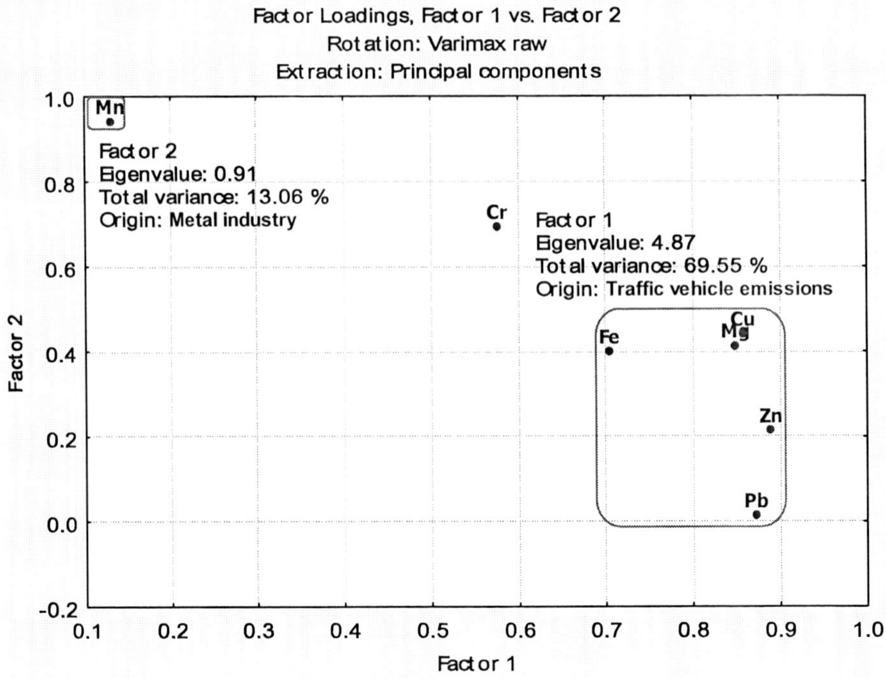

Figure 21. The eigenvalue's screeplot and factor analytical results for metallic element concentrations in fine particulate at WT sampling site.

At TH sampling site, figure 22 illustrates the results for factor analysis of metallic element concentration in TSP. The analytical results show that two factors were extracted and explained 62.06 % of total variance. High factor loadings of Cu (0.88) and Cr (0.71) on factor 1 and explained 46.37 % of total variance. The sea salt was responsible for this result. Factor explained 15% of total variance and had high factor loading of Pb (0.92) and Zn (0.73) which the possible source for vehicular emission. As for coarse particulate, the results for factor analysis of metallic element concentrations are illustrated in figure 23. Clearly, three factors had been extracted by factor analysis. High factor loadings of Zn (0.9) and Mg (0.89) were on factor 1 and interpreted 48.35 % of total variance. The possible source for this finding was sea salt. Factor 2 accounted for 18.27 % of total variance and had high factor loading of Mn (0.94) which the possible source for crustal. Moreover, high factor loadings of Cr (0.81), Cu (0.74) and Pb (0.71) were on factor 3 and interpreted 14.35 % of total variance. The main source for this result was vehicular emission. In fine particulate, figure 24 shows the results for factor analysis of metallic element concentrations. Three factors eigenvalues > 1 and explained about 76.25 % of total variance. High factor loadings of Mg (0.92), Zn (0.78) and Fe (0.72) were appeared on factor 1, and interpreted 39.17 % of total variance. This result suggested that sea salt was the main contributors. Factor 2 interpreted 20 % of total variance and had high factor loading of Cu (0.83) and Pb (0.76) which represented the major contributor was vehicular emission. Additionally, factor 3 explained 17.1 of total variance and had high factor loading of Mn (0.86) and Cr (0.72) which the possible source for coal combustion.

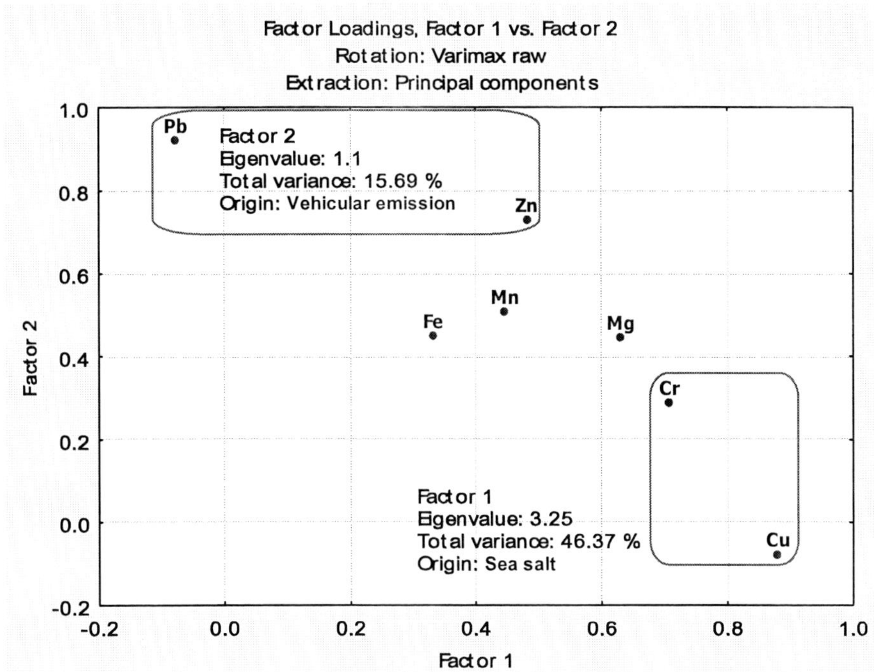

Figure 22. The results of factor analysis for metallic element concentrations in TSP particulate at TH sampling site.

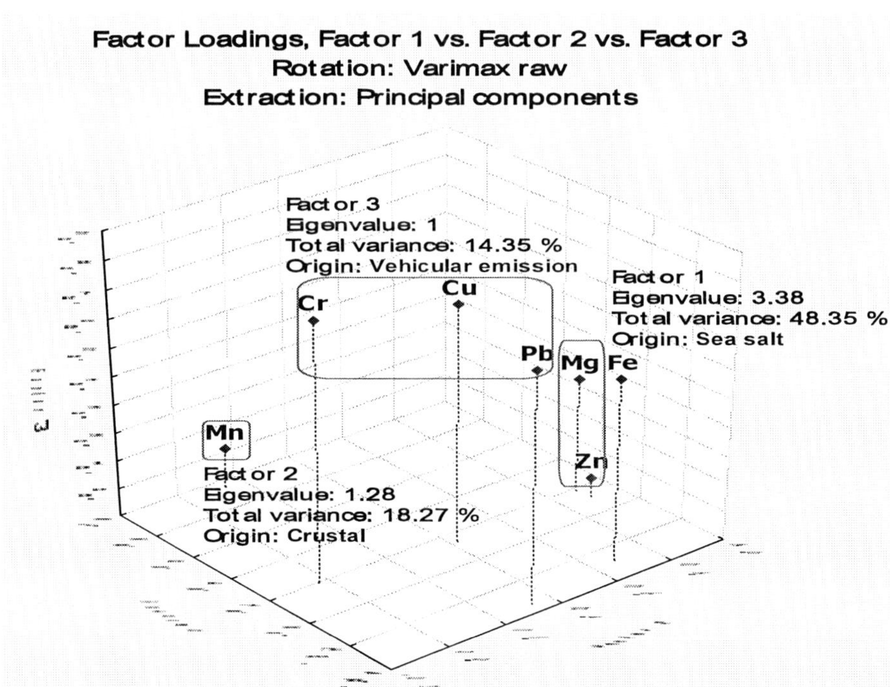

Figure 23. The results of factor analysis for metallic element concentrations in coarse particulate at TH sampling site.

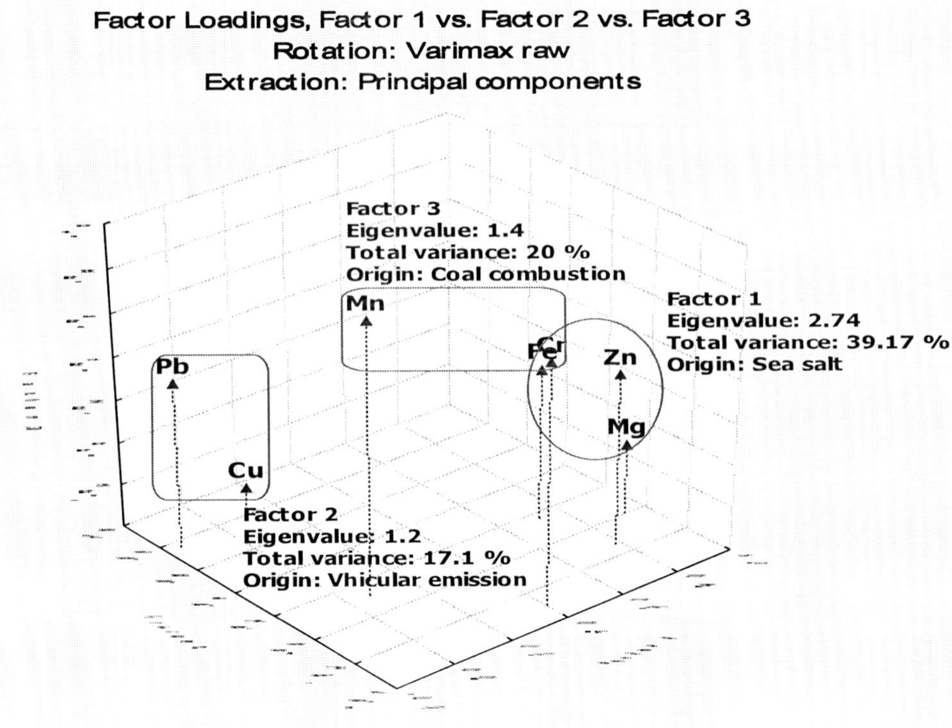

Figure 24. The results of factor analysis for metallic element concentrations in fine particulate at TH sampling site.

Overall, vehicular emission was the main contributor for metallic element concentrations in atmospheric particulates, especially at TA and WT sampling sites. Generally, the high frequency of airplanes took off, landed plus expressed highway which surround this airport were the main pollution sources for TA sampling site. as for WT sampling site, high loading and major express highways which surround this region were the main pollution sources. Moreover , the main contributor for metallic element concentrations in atmospheric particulates at TH sampling site was sea salt. This is because that TH sampling sit is nearly Taiwan Strait (about 10 m). High wind speed which strengthen the effect of the deliver of sea salt. To sump up, the major pollution sources for metallic elements in atmospheric particulates were traffic and sea salts. Then followed by crustal elements and other sources around there regions in central Taiwan. Thus, how to improve the traffic problems has become the most important issue in this area.

CONCLUSION

This study obtained the following conclusions.

1. Average TSP concentrations were 120.6±30.68 μg/m^3, and coarse particulate concentrations were 24.45±7.83 μg/m^3. Moreover, average fine particulate concentrations were 38.13±8.44 μg/m^3 during sampling period (September to

December of 2005) at TA sampling site. Regarding atmospheric particulate concentrations at WT sampling site, the average TSP concentrations were 113.72±34.46 μg/m^3, and coarse particulate concentrations were 25.99±8.18 μg/m^3, respectively. Additionally, average fine particulate concentrations were 47.38±13.59 μg/m^3 during March 2004 and January 2005 at WT sampling site. As for TH sampling site, average TSP concentrations were 152.76±28.37 μg/m^3, and coarse particulate concentrations were 28.69±10.49 μg/m^3. Moreover, average fine particulate concentrations were 56.38±18.53 μg/m^3 between March 2004 and January 2005 at TH sampling site.

2. Generally, atmospheric particulate concentrations (TSP, $PM_{2.5\text{-}10}$ and $PM_{2.5}$) in spring and winter were higher than those for other seasons at the WT and TH sampling sites. The primary reason was that the dust storms in China frequently occurred in spring and winter, thereby increasing PM concentrations in Taiwan. Moreover, the strong northeast monsoon which influenced Taiwan including Chinese dust storms in spring and winter and mean wind speeds in spring and winter are higher than those in other seasons.
3. Comparison average atmospheric particulate concentrations during the same sampling months at these three sampling sites. The comparison results presented that average TSP and fine particulate concentrations at these three sampling sites, the concentration order during sampling periods was as follows: TH > WT > TA, and the average coarse particulate concentration order during sampling periods was as follows: TH > TA > WT. Therefore, TSP, fine and coarse particulate concentrations at TH sampling site were highest than those at other two sampling sites. Mean wind speed at TH sampling site higher than those at other sampling sites was the main reason for this finding.
4. The investigation results revealed that three primary components of metallic elements in atmospheric particulates were Fe, Mg and Zn at TA sampling site. Furthermore, metallic elements Fe and Zn were the primary components on TSP at the TH and WT sampling sites, and Fe and Mg were the predominant components on coarse and fine particulates at the TH, and WT sampling sites. However, Zn and Fe were the predominant components on fine particulates at the WT sampling site.
5. Statistical analytical results revealed that WS and Temp. were the main factors which influenced atmospheric particulate concentrations, especially at TH sampling site. The possible proposed reason was that TH sampling site was close to the Taiwan western coast (about 100 m) compared to the rest of the other sampling sites. Furthermore, TH sampling site has the smallest surface roughness. Thus, wind speed was the least factor, if there is any, to influence the above result.
6. The major pollution sources for metallic elements in atmospheric particulates were traffic and sea salts. Then followed by crustal elements and other sources around there regions in central Taiwan. Thus, how to improve the traffic problems has become the most important issue in this area.

ACKNOWLEDGMENTS

The authors would like to thank the National Science Council of the Republic of China, Taiwan, for financially supporting this research under Contract No. NSC 96-2628-E-241-001-MY3.

REFERENCE

Allen, A. G., Nemitz, E., Shi, J. P., Harrison, R. M., Greenwood, J. C. 2001. Size distributions of trace metals in atmospheric aerosols in the United Kingdom. *Atmospheric Environment*, 35, 4581-4591.

Akhlaghi, Y., Kompany-Zareh, M. 2005. Comparing radial basis function and feed-forward neural networks assisted by linear discriminant or principal component analysis for simultaneous spectrophotometric quantification of mercury and copper. *Anal. Chim. Acta*, 537, 331–338.

Astorga-Espana, M. S., Pena-Mendez, E. M., Garcıa-Montelongo, F. J. 1999. Application of principal component analysis to the study of major cations and trace metals in fish from Tenerife Canary Islands. *Chemometr. Intel. Lab. Syst.*, 49, 173–178.

Barona, A., Romero, F. 1996. Distribution of metals in soils and relationships among fractions by principal component analysis. *Soil Technol.*, 8, 303–319.

Baker, J. E., Poster, D. L., Clark, C. A., Church, T. M., Scudlark, J. R., Ondov, J. M., Dickhut, R. M., Cutter, G. 1997. Loading of atmospheric trace elements and organic contaminants to the Chesapeake Bay. In: Baker, J.E. (Ed.), *Atmospheric Deposition of Contaminants to the Great Lakes and Coastal Waters*. SETAC Press, Pensacola, Florida, 171–194.

Chan, L. Y., Kwok, W. S. 2000. Vertical dispersion of suspended particulates in urban area of Hong Kong. *Atmospheric Environment*, 34, 4403-4412.

Chester, R., Bradshaw, G. F., Corcoran, P. A. 1994. Trace metal chemistry of the North Sea particulate aerosol: concentrations, sources and sea water fates. *Atmospheric Environment*, 28, 2873–2883.

Chan, C. C. 2002. *Adverse health effects of China sandstorms in Taipei*, 2002. EPA-91-FA11-03-D036, Environmental Protection Administration, Taipei.

Cattell, R. B. 1966. The scree test for the number of factors. *Multivariate Behavioral Research,* 1, 245-276.

Colombo, J. C., Landoni, P., Bilos, C. 1999. Sources, distribution and variability of airborne particles and hydrocarbons in La Plata area, Argentina. *Environ. Pollut.*, 104, 305–314.

Chao, C. Y., Wong, K. K. 2002. Residential indoor PM10 and PM2.5 in Hong Kong and the elemental composition. *Atmos. Environ.*, 36, 265–277.

Conner, T. L., Norris, G. A., Landis, M. S., Williams, R. O. 2001. Individual particle analysis of indoor, outdoor, and community samples from the 1998 Baltimore particulate matter study. *Atmos. Environ.*, 35, 3935–3946.

Dockery, D. W., Pope, X., Xu, J. D., Spengler, J. H., Ware, M. E., Fay, B. G., Speizer, F. E. 1993. An association between air pollution and mortality in six US cities. *New England Journal of Medicine*, 329, 1753-1759.

Dockery, D. W., Schwarz, J., Spengler, J. H. 1992. Air pollution and daily mortality; associations with particles and acid aerosols. *Environmental Reswarch*, 59, 362-373.

Danielssona, A., Catob, I., Carmanc, R., Rahm, L. 1999. Spatial clustering of metals in the sediments of the Skagerrak/Kattegat. *Appl. Geochem.*, 14, 689–706.

De Bartolomeo, A., Poletti, L., Sanchini, G., Sebastiani, B., Morozzi, G. 2004. Relationship among parameters of lake polluted sediments evaluated by multivariate statistical analysis. *Chemosphere*, 55, 1323–1329.

De Miranda, R. M., de F´atima Andrade, M., Worobiec, A., Grieken, R.V. 2002. Characterisation of aerosol particles in the S˜ao Paulo Metropolitan Area. *Atmos. Environ.*, 36, 345–352.

Eisenreich, S. J., Hornbuckle, K. C., Achman, D. R. 1997. Air–water exchange of semivolatile organic chemicals in the Great Lakes. In: Baker, J.E. (Ed.), *Atmospheric Deposition of Contaminants to the Great Lakes and Coastal Waters*. SETAC Press, Pensacola, Florida, 109–136.

El-Fadel, M., Massoud, M. 2000. Particulate matter in urban areas: health-based economic assessment. *J. Sci. Total Environ.*, 257, 133–146.

El-Fadel, M., Hashisho, Z. 2001. Vehicular emissions in roadway tunnels: a critical review. *Critical Reviews in Environmental Science and Technology*, 31, 125-174.

Espinosa, A. J. F., Rodr´ıguez, M. T., Rosa, F. J. B. D. L., S´anchez, J. C. J. 2001. Size distribution of metals in urban aerosols in Seville (Spain). *Atmos. Environ.*, 35, 2595–2601.

Fang, G. C., Chang, C. N., Wu, Y. S., Fu, P. P. C., Yang, C. J., Chen, C. D., Chang, S. C. 2002. Ambient suspended particulate matters and related chemical species study in central Taiwan, Taichung during 1998–2001. *Atmospheric Environment*, 36, 1921-1928.

Fang, G. C., Chang, C. N., Wu, Y. S., Lu, S. C., Fu, Peter P. C., Chang, S. C., Cheng, C. D., Yuen, W. H. 2002. Concentration of atmospheric particulates during a dust storm period in central Taiwan, Taichung. *The Science of the Total Environment*, 287, 141-145.

Fang, G. C., Wu, Y. S., Lee, W. J., Chou, T. Y., Lin, I. C. 2007. Ambient air particulates, metallic elements, dry deposition and concentrations at Taichung Airport, Taiwan. *Atmospheric Research*, 84, 280-289.

Faus-Kessler, T., Dietl, C., Tritschler, J., Peichl, L. 2001. Correlation patterns of metals in the epiphytic moss *Hypnum cupressiforme* in Bavaria. *Atmos. Environ.*, 35, 427–439.

Funasaka, K., Sakai, M., Shinya, M., Miyazaki, T., Kamiura, T., Kaneco, S., Ohta, K., Fujita, T. 2003. Size distribution and characteristics of atmospheric inorganic particles by regional comparative study in Urban Osaka, Japan. *Atmos. Environ.*, 37, 4597–4605.

Gao, Y., Nelson, E. D., Field, M. P., Ding, Q., Li, H., Sherrell, R. M., Gigliotti, C. L., Van Ry, D. A., Glenn, T. R., Eisenreich, S. J. 2002. Characterization of atmospheric trace elements on PM2.5 particulate matter over the New York-New Jersey harbor estuary. *Atmospheric Environment*, 36, 1077-1086.

Holsen, T. M., Zhu, X., Khalili, N. R., Lin, J. J., Lestari, P., Lu, C. S., Noll, K. E. 1997. Atmospheric particle size distributions and dry deposition measured around Lake Michigan. In: Baker, J.E. (Ed.), *Atmospheric Deposition of Contaminants to the Great Lakes and Coastal Waters*. SETAC Press, Pensacola, Florida, 35–50.

Hileman, B. 1981. Particulate matter: the inevitable variety. *Environ. Sci.Technol.*, 15, 983–986.

Hlavay, J., Polyak, K., Wesemann, G. 1992. Particle size distribution of minerals phases and metals in dusts collected at different workplaces. *Fresenius J. Anal. Chem.*, 344, 319–321.

Koliadima, A., Athanasopoulou, A., Karaiskakis, G.. 1998. Particulate matter in air of the cities of Athens and Patras (Greece): particle-size distribution and elemental concentrations. *Aerosol Science and Technology*, 28, 292-300.

Kumar, A. V., Patil, R. S., Nambi, K. S. V. 2001. Source apportionment of suspended particulate matter at two traffic junctions in Mumbai, India. *Atmospheric Environment*, 35, 4245-4251.

Kim, K. H., Mishra, V. K., Kang, C. H. Choi, K. C., Kim, Y. J., Kim, D. S. 2006. The ionic compositions of fine and coarse particle fractions in the two urban areas of Korea. *J. Environ. Manage.*, 78, 170–182.

Kyotani, T., Iwassuki, M. 2002. Characterization of soluble and insoluble components in PM2.5 and PM10 fractions of airborne particulate matter in Kofu city, Japan. *Atmos. Environ.*, 36, 639–649.

Kaiser, H. F. 1974. An index of factorial simplicity. *Psychometrika*, 39, 31-36.

Lee, J. H., Kim, Y. P., Moon, K. C., Kim, H. K., Lee, C. B. 2001. Fine particle measurements at two background sites in Korea between 1996 and 1997. *Atmospheric Environment*, 35, 635-643.

Lin T. H. 2001. Long-range transport of yellow sand to Taiwan in Spring 2000: observed evidence and simulation. *Atmospheric Environment*, 35, 5873–82.

Liu, C. M., Young, C. Y., Lee, Y. C. 2006. Influence of Asian dust storms on air quality in Taiwan. *Science of the Total Environment*, 368, 884–897.

Loska, K., Wiechuła, D. 2003. Application of principal component analysis for the estimation of source of heavy metal contamination in surface sediments from the Rybnik Reservoir. *Chemosphere*, 51, 723–733.

Lee, Y. C., Hills, P. R. 2003. Cool season pollution episodes in Hong Kong, 1996–2002. *Atmos Environ.*, 37, 2927–2939.

National Statistics, ROC (Taiwan), http://www.stat.gov.tw/mp.asp?mp=4.

Nolte, C. G., Schauer, J. J., Cass, G. R., Simoneit, B. R. T. 2002. Trimethylsilyl derivatives of organic compounds in source samples and in atmospheric fine particulate matter. *Environment Science and Technology*, 36, 4273-4281.

Machado Mesa, L., Pena Mendez, E., Sanchez Sanchez, M., Garcia Montelongo, F. 1999. Interpretation of heavy metal data from mussel by use of multivariate classification techniques. *Chemosphere,* 38, 1103–1111.

Manoli, E., Kouras, A., Samara, C. 2004. Profile analysis of ambient and source emitted particle-bound polycyclic aromatic hydrocarbons from three sites in northern Greece. *Chemosphere*, 56, 867–878.

Marcazzan, G. M., Vaccaro, S., Valli, G., Vecchi, R. 2001. Characterisation of PM10 and PM2.5 particulate matter in the ambient air of Milan (Italy). *Atmos. Environ.*, 35, 4639–4650.

Ondov, J. M., Quinn, T. L., Battel, G. F. 1997. Influence of temporal changes in relative humidity on size and dry depositional fluxes of aerosol particles bearing trace elements. In: Baker, J.E. (Ed.), *Atmospheric Deposition of Contaminants to the Great Lakes and Coastal Waters*. SETAC Press, Pensacola, Florida, 17-34.

Pyle, G. G., Swanson, S. M., Lehmkuhl, D. M. 2001. Toxicity of uranium minereceiving waters to caged fathead minnows, *Pimephales promelas*. *Ecotoxicol. Environ.* Safety, 48, 202–214.

Pedersen, D. U., Durant, J. L., Penman, B. W., Crespi, C. L., Hemond, H. F., Laflerur, A. L., Cass, G. R.. 1999. Seasonal and spatial in human cell mutagenicity of respirable airborne particles in the northeastern United States. *Environmental Science and Technology*, 33, 4407–4415.

Querol, X., Alastuey, A., Puicercus, J. A., Mantilla, E. Ruiz, C. R. Lopez-Soler, A., Plana, F., Juan, R. 1998. Seasonal evolution of suspended particles around a large coal-fired power station: chemical characterization. *Atmos. Environ.*, 32, 719–731.

Scudlark, J. R., Conko, K. M., Church, T. M. 1994. Atmospheric wet deposition of trace elements to Chesapeake Bay: Y. Gao et al. / Atmospheric Environment 36 (2002) 1077–1086 1085 CBAD study year 1 results. *Atmospheric Environment* 28 (8), 1487–1498.

Shuetzle, D. 1983. Sampling of vehicle emissions for chemical analysis and biological testing. *Environmental Health Perspective*, 47, 65-80.

Spengler, J. D., Brauer, M., Koitratis, P. 1990. Acid air and health. *Environmental Science and Technology*, 24, 946-955.

Singh, K. P., Malik, A. S. 2005. Sinha,Water quality assessment and apportionment of pollution sources of Gomti river (India) using multivariate statistical techniques-a case study. *Anal. Chim. Acta*, 538, 355–374.

Shu, J., Dearing, J. A., Morse, A. P., Yu, L., Yuan, N. 2001. Determining the sources of atmospheric particles in Shanghai, China, from magnetic and geochemical properties. *Atmospheric Environment*, 35, 2615-2625.

Salvador, P., Artíñano, B., Alonso, D. G., Querol, X., Alastuey, A. 2004. Identification and characterisation of sources of PM10 in Madrid (Spain) by statistical methods. *Atmospheric Environment*, 38, 435-447.

Sternbeck, J., Sjodin, A., Andreasson, K. 2002. Metal emissions from road traffic and the influence of resuspension-results from two tunnel studies. *Atmos. Environ.*, 30, 4735–4744.

Yoo, J. I., Kim, K. H., Jang, H. N., Seo, Y. C., Seok, K. S., Hong, J. H., Jang M. 2002. Emission characteristics of particulate matter and heavy metals from small incinerators and boilers. *Atmospheric Environment*, 36, 5057-5066.

Zheng, J., Tan, M., Shibata, Y., Tanaka, A., Li, Y., Zheng, G., Zheng, Y., Shan, Z. 2004. Characteristics of lead isotope ratios and elemental concentrations in PM10 fraction of airborne particulate matter in Shanghai after the phase-out of leaded gasoline. *Atmospheric Environment*, 38, 1191-1200.

In: Airborne Particulates
Editor: Ming Cheng and Wen Liu

ISBN: 978-1-60692-907-0

Chapter 8

Toxicological Mechanisms Underlying the Adverse Effects of Ambient Particulate Matter ($PM_{2.5}$) in the Lung: Contribution of an In-Vitro Approach[*]

Guillaume Garçon[1], Sylvain Billet[1], Zeina Dagher[1], Pierre Gosset[2], Anthony Verdin[1], Frédéric Ledoux[3], Dominique Courcot[3] and Pirouz Shirali[1,†]

[1] LCE EA2598, Toxicologie Industrielle et Environnementale, Maison de la Recherche en Environnement Industriel de Dunkerque 2, 189A, Avenue Maurice Schumann, 59140 Dunkerque, France

[2] Laboratoire d'Anatomie et de Cytologie Pathologique du Groupement Hospitalier de l'Institut Catholique de Lille, Faculté Libre de Médecine, 56, rue du Port, 59046 Lille Cedex, France

[3] LCE EA2598, Catalyse et Environnement, Maison de la Recherche en Environnement Industriel de Dunkerque 1, 145, Avenue Maurice Schumann, 59140 Dunkerque, France

Abstract

Among common ambient air pollutants, particulate matter (PM) is currently under intensive epidemiological and toxicological investigation. Epidemiological evidence suggests a strong relationship between elevated urban environmental levels of PM_{10} and increased incidence of morbidity and mortality due to respiratory and/or cardiovascular diseases in the general population. This association seems to be more closely related to fine particles ($PM_{2.5}$). The consistency between the findings of different epidemiological

---

[*] A version of this chapter was also published in *Air Pollution Research Advances,* edited by Corin G. Bodine published by Nova Science Publishers, Inc. It was submitted for appropriate modifications in an effort to encourage wider dissemination of research.

[†] Phone: +33-3-28237610; Fax: +33-3-28237171; E-mail: Pirouz.Shirali@univ-littoral.fr.

studies argues for a causal association, but the attribution of acute health effects to concentration levels appears very difficult to explain in the light of the current knowledge of the toxicological mechanisms. The mechanisms underlying these adverse effects are not well understood and major questions concerning the specific size fraction, chemical composition, and causative mechanisms leading to the observed health effects remain. The objective of this in-vitro approach was to correlate the composition of PM with cytotoxicity in the lung. Under this framework, we determined the physical and chemical characterization of the $PM_{2.5}$ fraction from a sample collected in Dunkerque, a French seaside City heavily industrialized. Accordingly, we found that 92.15% of PM were equal or smaller than 2.5 μm and their specific surface area was 1 m^2/g. Inorganic (i.e. Fe, Al, Ca, Na, K, Mg, Pb, etc.) and organic (i.e. volatile organic compounds, polycyclic aromatic hydrocarbons, etc.) chemicals, usually associated with natural environment, but also so-called anthropogenic elements, were found in PM, suggesting that much of PM derived from wind-borne dust from the industrial complex, and the heavy motor vehicle traffic. Thereafter, we focused our attention on the capacity of ambient air pollution $PM_{2.5}$ to induce cytotoxicity, oxidative stress, inflammatory, and pro-apoptotic responses in human epithelial lung cells (L132) in culture. We notably reported that short-term in-vitro exposure to collected $PM_{2.5}$ induced oxidative stress conditions (i.e malondialdehyde production; superoxide dismutase activity; 8-hydroxy-2'-deoxyguanosine formation; poly(ADP-ribose) polymerase activity), thereby activating the nuclear factor-kappa B/inhibitory kappa B (NF-κB/IκB) complex, as showed by the concentration- and/or time-dependent increases in the nuclear p65 and cytoplasmic inhibitory kappa B phosphorylation, and the nuclear p65 and p50 DNA binding. Moreover, PM-induced-activation of NF-κB/IκB complex resulted in the excessive mRNA expression and/or protein secretion of various inflammatory mediators (i.e *nitric oxide/inducible nitric oxide synthase, tumor necrosis factor-alpha, interleukine-1 beta , interleukine-8, granulocyte macrophage-colony stimulating factor, interleukine-6, transforming growth factor-beta 1*) in L132 cells. Thereafter, we observed that $PM_{2.5}$-derived reactive oxygen intermediates mediated apoptotic events by activating not only the tumor necrosis factor-alpha-induced pathway (i.e. caspase-8 and -3 activation), but also the mitochondrial pathway (i.e. cytochrome c release from mitochondria, caspase-9 and -3 activation), thereby leading to subsequent changes in the transcription rates of *P53*, *BCL-2*, and *BAX* genes, on the one hand, and DNA fragmentation, on the other hand. Taken together, these results showed that the occurrence of oxidative stress conditions, the subsequent activation of NF-κB/IκB complex, and, therefore, the production of various inflammatory mediators and the development of pro-apoptotic events, that aggravate according to PM concentration and/or incubation time, proceeded cytotoxicity in ambient air pollution $PM_{2.5}$-exposed target L132 cells in culture.

Key words: Human lung epithelial cells (L132), $PM_{2.5}$, physical and chemical characteristics, cytotoxicity, oxidative stress, NF-kB activation, inflammatory response, apoptosis.

INTRODUCTION

The effects of air pollution on health have been the focus of intensive investigation since the severe episodes in Meuse Valley, Belgium, in 1930; Donora, Pennsylvania, in 1948; and London, in 1952 [1]. Air pollution is a real problem of public health for the general population since epidemiological studies have shown that air pollution particulate matter (PM) is associated with the increase of respiratory and cardiovascular mortality and morbidity [for review see 2]. A growing body of evidence have also shown consistent correlation between the exposure to airborne PM and acute exacerbation in morbidity and mortality rates, notably for persons with respiratory and/or cardiovascular diseases [3-5]. However, the increase in health risk occurs even when PM concentrations are below established air quality standard levels [7]. Epidemiological studies have also been carried out using current air quality standards that only encompass the PM size and concentration; these studies did never address the issue of PM composition [8, 9]. Attempts to provide linkages between adverse health effects and PM properties have, therefore, significantly increased [9]. However, despite intensive investigation, the underlying mechanisms of PM-related adverse health effects are still incompletely understood, but a hypothesis currently under investigation is that organic compounds-coated onto PM could played a key role in their toxicity [9-11]. Hence, a better knowledge of the physical and chemical characteristics of PM, which depend from the sources, the emission intensity, and the molecular interactions between their inorganic and organic components, constitutes crucial information for the evaluation of their toxicity, but still revealed to be unsatisfactory to predict its toxicological reactivity [12, 13].

Environmental and/or anthropogenic aerosols are generally heterogeneous mixtures that greatly vary in constituent, particle size and chemical composition [7, 14, 15]. Now, it is clearly evident that specific components of the air particulate alone, or as components of mixtures with other pollutants, are capable to produce adverse biological responses consistent with human morbidity findings [5, 15-17]. Epidemiological evidence from several cities around the world indicated that the greatest health risks correlate with smaller PM (i.e. $PM_{2.5}$ and $PM_{0.1}$), which have the capacity to reach the distal regions of the lung after their inhalation [16]. Therefore, attention has turned to the components that are most likely to have toxic potential, and the fine ($PM_{2.5}$) and the ultrafine ($PM_{0.1}$) particles have been identified as potential mediators of some of the toxicity of PM_{10} [11, 17, 18]. In particular, several studies have hypothesized that particles of smaller size (i.e. $PM_{2.5}$) are more potent than larger (i.e. PM_{10}) to induce cytotoxic and/or inflammatory responses in lung models [11, 15, 19]. Currently there is still some question as to whether the PM_{10}, $PM_{2.5}$ or $PM_{2.5-10}$ are more important to human health [8].

Historically, health effect communities have faced formidable barriers to understanding the significance of ambient PM exposure, related to the highly complex nature of atmospheric aerosols and their various sources of origin. Depending upon their natural and/or anthropogenic emission sources, air pollution PM is a complex mixture of chemical and/or biological elements, such as metals, salts, carbonaceous material, volatile organic compounds (VOC), polycyclic aromatic hydrocarbons (PAH), and/or endotoxins, which can in fact interact and adsorb on condensation nuclei [7, 9, 20-23]. Hence, PM mass, PM number, PM specific surface area, qualitative and/or quantitative determinations of bioavailable transition metals, PAH, VOC, and other particle-bound chemicals are suspected to be important in determining the adverse health effects of air pollution [18, 23-27]. Despite their availability at low-dose or ultra-low-dose, some of the above-mentioned constituents (i.e. transition metals, VOC, PAH, etc.) could be greatly involved in the global cytoxicity of PM [11, 13, 19]. However, the majority of the published studies on PM toxicity have been conducted using complex mixtures, yet their physical and chemical characteristics are generally not well established, and the possible influence of the elements coated-onto the particles are often neglected [28].

Much effort has gone into discerning the possible mechanisms whereby air pollution PM cause adverse health effects in humans [7, 11, 13, 17, 28]. The underlying mechanisms of PM-related disorders in the lung are still incompletely understood, but a hypothesis currently under investigation is that many of them may derive from oxidative stress, initiated by the production of reactive oxygen species (ROS) within affected cells [29]. Accordingly, experimental and/or epidemiological findings from several authors indicated that lung diseases are often associated with prooxidant/antioxidant imbalance, and there is increasing evidence that particulate air pollutants induce acute responses as well as exacerbate existing inflammatory diseases in the lung [11, 15, 29-31]. The generation of oxidants by PM results not only in oxidative damage to cell macromolecules (i.e. lipids, proteins and/or DNA) in affected cells, but also in the activation of cell signaling pathways leading thereby to the activation of specific transcription factors (e.g. nuclear factor-kappa B, NF-κB) and to the subsequent secretion of pro-inflammatory mediators (e.g. cytokines, chemokines, etc.) [19, 32, 33].

In affected cells, high levels of ROS also produce a change in the redox status of the cell, thereby triggering a cascade of the physiopathological events associated with the inflammatory response and, at higher concentrations, apoptosis [28, 29, 32, 34, 35]. Indeed, ROS play a critical role in the activation of cells as messengers in the signal transduction pathways leading to gene expression and/or further protein secretion [31, 36, 37]. For example, increased activation of an oxygen-sensitive transcription factor called NF-κB is thought to occur in several inflammatory conditions, and is thereby involved in the regulation of a large number of genes that control various aspects of the inflammatory response [31, 36-38]. Because pro-inflammatory cytokines regulated by or responsive to NF-kB may also activate NF-kB, this provides a positive feedback loop [40]. In non-stimulated cells, inactive NF-κB is maintained into the cytoplasm by a family of inhibitory proteins called inhibitory kappa B (IκB) [38]. Upon stimulation, IκB is quickly phosphorylated on two critical serine residues (Ser32 and Ser36) by the IκB kinase (IKK) complex, which itself needs to be activated by phosphorylation [36, 37, 39]. Active NF-κB, also constituted by phosphorylated Rel proteins, then translocates into the nucleus where it binds to its consensus site present within the promoters of many genes for both the protective antioxidants, as well as several inflammatory cytokines and/or chemokines [37, 40, 41]. The activation of the NF-κB/IκB complex could, therefore, be evaluated through the cytoplasmic phosphorylation of p65 and/or IκBα, in the one hand,, and the DNA binding of p65 and p50, in the other hand [41].

The pathogenesis of lung injury induced by air pollutants is also mediated in large part by ROS and inflammatory cytokines that might play a crucial role in both the occurrence and the development of acute inflammation [11, 15, 19, 30, 31]. In addition to their deleterious oxidative effects, ROS have been firmly involved in the genesis of inflammatory diseases through the activation of the NF-κB, and the subsequent development of an inflammatory response [19, 32, 33, 42, 43]. The gene transcription of *tumor necrosis factor-alpha (TNF-α)* and *interleukin-1β (IL-1β)* are switched on by activated NF-κB as an early event of acute inflammatory response. Both these pro-inflammatory mediators then acts via paracrine and autocrine pathways to stimulate, directly and/or indirectly, cells to release other inflammatory cytokines: interleukin-8 (IL-8), granulocyte/macrophage-colony stimulating factor (GM-CSF), interleukin-6 (IL-6), and transforming growth factor-β1 (TGF-β1) [28, 29, 42, 44, 45]. Stimulated cells are also known to release large quantities of nitric oxide (NO) as a direct consequence of inducible nitric oxide synthase (iNOS) activation by TNF-α [46]. Such cell-derived mediators sometimes restore normal homeostasis. At other times, however, because of their overproduction or their aberrant release, they appear to act as amplification factors in the complex cascade of acute inflammatory events and other associated pathologies [40].

There is now increasing evidence that both the excessive production of ROS and the protein secretion of pro-inflammatory mediators could directly interact by inducing cell death by apoptosis [19, 28, 47]. In addition to morphological features, apoptosis is defined by complex biochemical processes involving mitochondria, activation of a family of cysteine-aspartate proteases (caspases), and endonuclease-induced nuclear fragmentation [48-50]. Aggarwal (2004) has proposed that the differential expression of signaling factors involved in apoptosis in the human lung exposed to ROS is redox-sensitive and is mediated, at least in part, by a negative feedback mechanism transduced by the NF-κB, and, therefore, by the TNF-α [37]. Within the last few years, a series of biochemical steps has been identified that results in apoptosis, and two main pathways have been delineated in human cells [36]. For instance, the extrinsic pathway of apoptosis is represented by TNF-α-induced apoptosis, which involves the activation of the TNF Family Receptors, TNFRI and TNFRII, and the activation of the initiator caspase-8, which initiate a cascade of additional processing and activation of downstream effector caspase-3 [51]. In the mitochondrial pathway, apoptosis is triggered by the release of cytochrome c (Cyt. c), normally confined to the mitochondrial intermembrane space, into the cytosol, which, in the presence of the Apoptosis protease-activating factor-1 (Apaf-1), recruits and converts the latent apoptosis-promoting pro-caspase 9 to its active form, and activates also an overlapping set of effector caspases including caspase 3 [52]. Active effector caspases mediate the cleavage of apoptosis regulator, the cleavage of housekeeping proteins, and DNA fragmentation, resulting in morphological features of apoptosis [51]. The induction of downstream caspase activity can be regulated by the tumor suppressor protein p53 and the Bcl-2 family members [53]. The role of P53 as a nuclear transcription factor is well characterized; it induces apoptosis by upregulating the transcription of many pro-apoptotic genes and by down regulating anti-apoptotic genes [54]. The Bcl-2 family proteins play pivotal roles in regulation of apoptosis and the balance between pro- and anti-apoptotic Bcl-2 family members determines the mitochondrial response to various apoptotic stimuli [35]. After DNA-damaging events, the multidomain pro-apoptotic proteins (e.g., Bax, Bak, and Bid) translocate from the cytosol to the mitochondrial membrane and induce also the release of a variety of apoptogenic factors, including Cyt. c into the cytosol, thereby contributing to the activation of apoptotic events [35].

In view of the above-described evidence, the underlying mechanisms of actions by which air pollution PM induce adverse health effects are, therefore, of intense scientific interest. This includes notably a call for research on the physical and chemical properties that promote PM toxicity [9, 15, 55]. Hence, to improve the knowledge about the underlying mechanisms of action involved in PM-induced adverse health effects, we have undertaken an extensive in-vitro investigation of the adverse health effects of air pollution PM2.5, collected in Dunkerque, a French seaside city located on the southern coast of the North-Sea, characterized by the proximity of industrial activities and heavy motor vehicle traffic. The Figure 1 showed the experimental design of this multi-step transversal approach. PM samples were also collected in Dunkerque, a French seaside City located on the southern coast of the North-Sea, characterized by the proximity of industrial activities (i.e. iron and steel industry, aluminum industry, oil refinery, basic chemistry, pharmaceutical industry, plant health production, food industry, etc.) and heavy motor vehicle traffic.

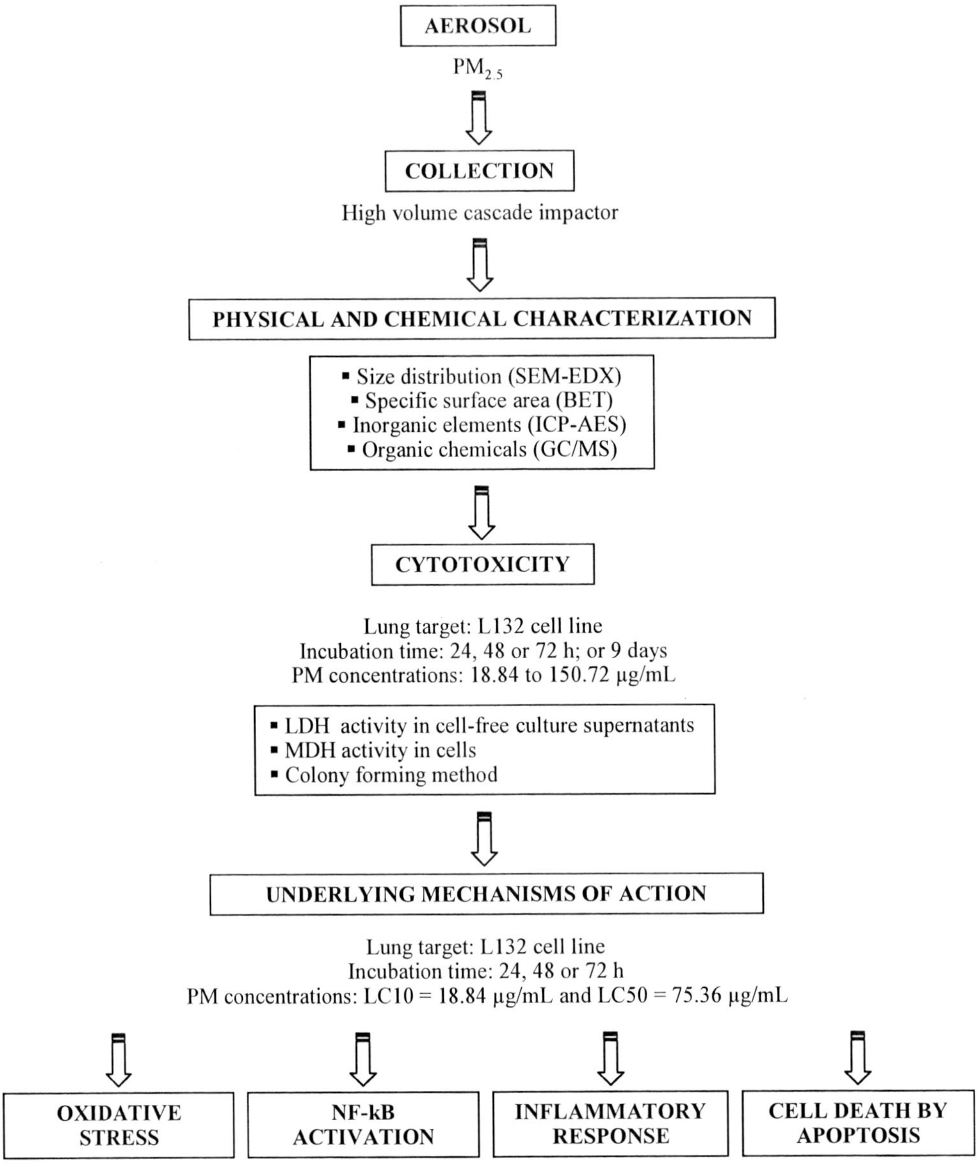

Figure 1.This study consisted in a multi-step in-vitro study including the collection of particle matter (PM), the determination of its physical and chemical characteristics, the evaluation of its cytotoxicity in L132 cells in culture, and the research of the induced-underlying mechanisms of action in these target lung cells: the occurrence of oxidative stress conditions, the activation of .the nuclear factor-kappa B (NF-κB), the development of an inflammatory response, and the induction of cell death by apoptosis. (Scanning electron microscopy coupled with energy dispersive X-ray, SEM-EDX; Brunauer, Emmett, and Teller method, BET method; inductively coupled plasma-atomic emission spectrometry, ICP-AES; gas chromatography-mass spectrometry, GC/MS; mitochondrial dehydrogenase, MDH; lactate dehydrogenase, LDH; lethal concentration at 10%, LC_{10}; lethal concentration at 50%, LC_{50}).

Their physical and chemical characteristics (i.e. size distribution, specific surface area, and inorganic and organic element composition) were carried out. In view of the results arising from this exhaustive physical and chemical characterization, a multi-step in-vitro study was, thereafter, developed [56-60]. There are a number of different approaches that can be taken to assess the toxic effects of inhaled complex mixtures, including air pollution PM. These include epidemiological studies, human clinical studies, animal studies, and in-vitro studies. Each of these approaches has its own strengths and limitations [61]. Risk assessment of complex mixtures is most accurate and defensible when as many of these approaches as possible can be used in an integrated manner to address a specific question. In-vitro toxicology provides a powerful approach to identify underlying mechanisms by which air pollutants damage cells at the cellular, biochemical, and molecular levels. The major limitations of in vitro studies are that cells have been removed from their normal environment, on the one hand, and it can be difficult to expose them in a manner that mimics in-vivo exposure, on the other hand [61]. Nevertheless, many cellular pathways are activated similarly both in vitro and in vivo, and carefully conducted in vitro experiments can be performed using relevant primary cells and exposure scenarios. In agreement with the above-reported statement, in the present study, the model under study was the L132 cell line. L132 cells derive from normal human embryonic lung tissue, present epithelial morphological characteristics, and exhibit typical features of pneumocytes [62]. The use of L132 cells closely rely on their lung origin and their sensitivity to the effects of airborne pollutants [62-64].

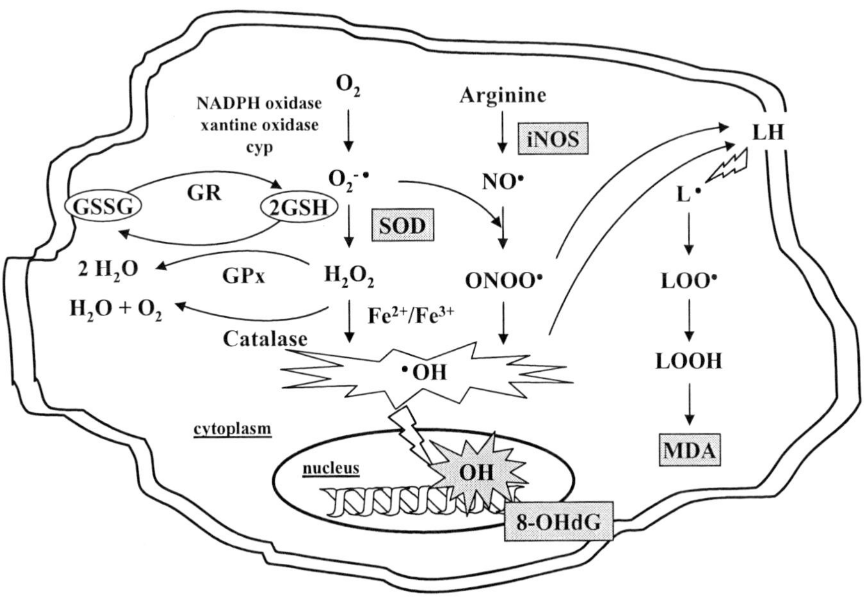

Figure 2. The occurrence of oxidative stress conditions in particulate matter (PM)-exposed L132 cells were carried out through the determination of malondialdehyde (MDA) production, superoxide dismutase (SOD) activity, 8-hydroxy-2'-deoxyguanosine (8-OHdG) formation, and poly(ADP-ribose) polymerase (PARP) activity. (glutathione peroxidase, GPx; glutathione reductase, GR; inducible nitric oxide synthase, iNOS; oxidized glutathione, GSSG; reduced glutathione, GSH; polyunsaturated fatty acid, LH).

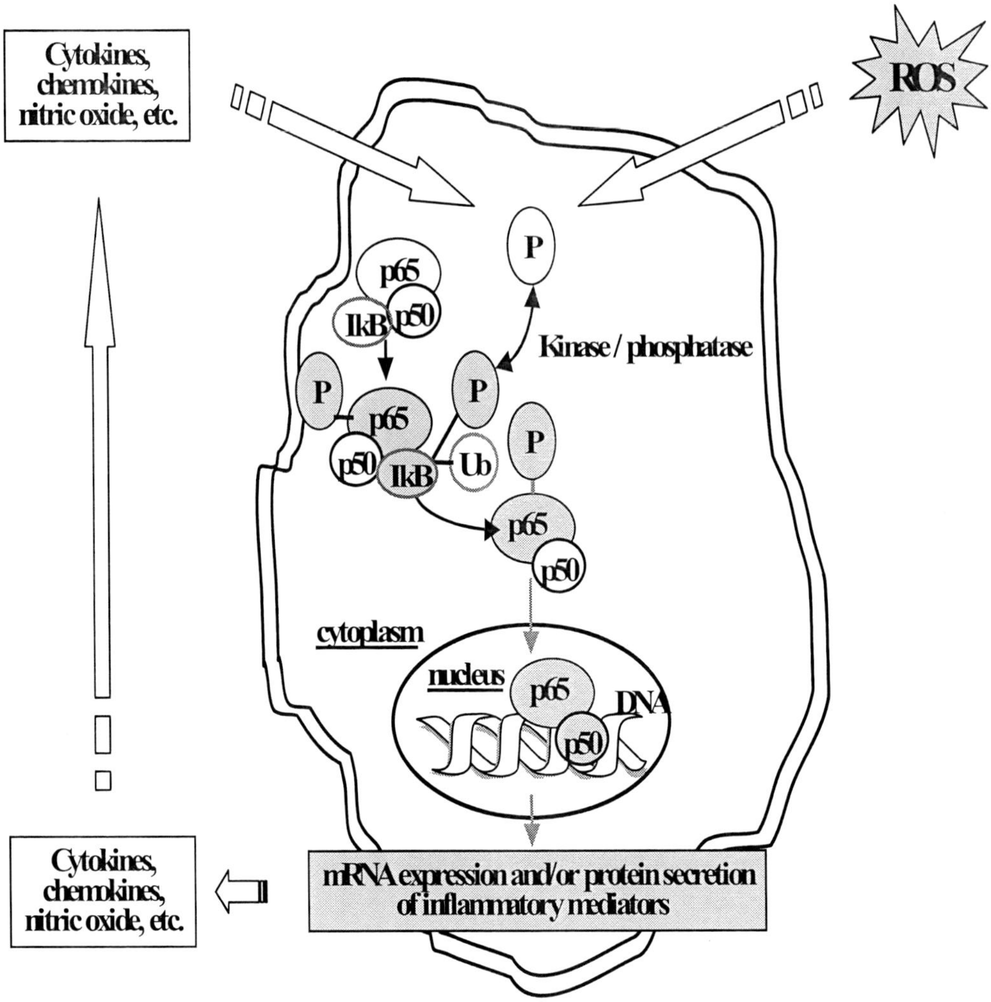

Figure 3. The activation of nuclear factor-kappa B (NF-κB) in particulate matter (PM)-exposed L132 cells were carried out through through the determination of phosphorylated p65 and phosphorylated inhibitory kappa B-alpha (IκBα) protein levels in cytoplasmic extracts, and p65 and p50 DNA bindings in nuclear extracts (phosphorylation, P; reactive oxygen species, ROS; ubiquitination, Ub).

Thereafter, the survival rates of L132 cells exposed to collected PM were studied using the colony forming method, and the cytotoxicity of the airborne PM was evaluated through the determination of the lactate dehydrogenase (LDH) activity in cell-free culture supernatants, and the mitochondrial dehydrogenase (MDH) activity in cultured cells. To contribute to a better knowledge of the underlying adverse mechanisms involved in PM-induced cytotoxicity, we were, thereafter, interested in the key role of PM exposure in the induction of some underlying mechanisms of action in L132 cells: (i) the occurrence of oxidative stress conditions, (ii) the activation of NF-κB, (iii) the development of an inflammatory response, and (iv) the induction of cell death by apoptosis. As shown in Figure 2, PM-induced oxidative damages were investigated through the determination of malondialdehyde (MDA) production, superoxide dismutase (SOD) activity, 8-hydroxy-2'-deoxyguanosine (8-OHdG) formation, and poly(ADP-ribose) polymerase (PARP) activity.

Figure 3 shows that the potential activation of NF-κB/IκB by collected $PM_{2.5}$ in these target respiratory cells was studied through the determination of phosphorylated p65 and phosphorylated IκBα protein levels in cytoplasmic extracts, and p65 and p50 DNA bindings in nuclear extracts.

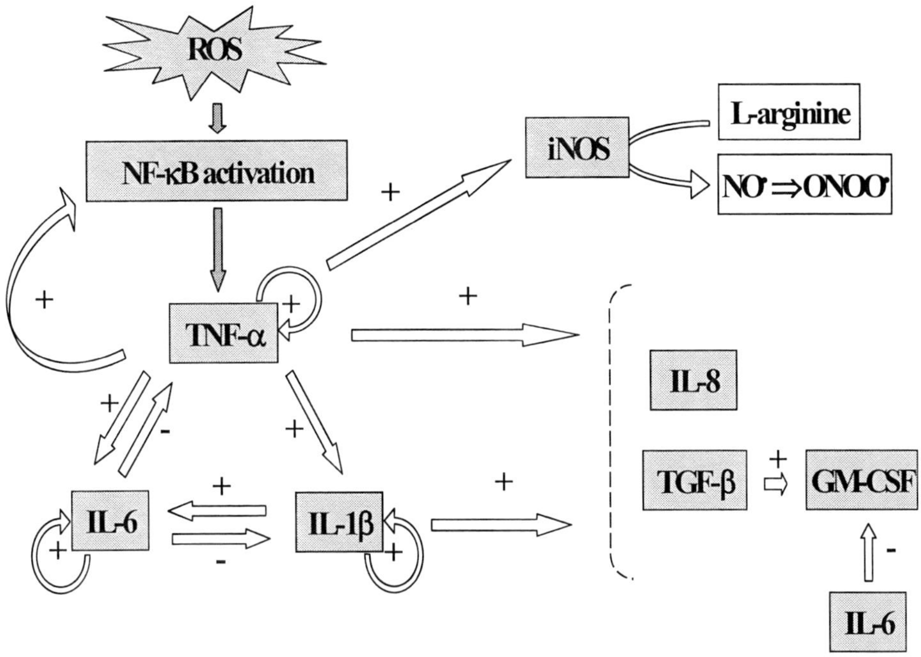

Figure 4. The development of an inflammatory response in particulate matter (PM)-exposed L132 cells were carried out through the determination of protein secretion and/or gene expression of *tumor necrosis factor-alpha (TNF-α), interleukin-1 beta (IL-1β), interleukin-8 (IL-8), granulocyte macrophage-colony stimulating factor (GM-CSF), interleukin-6 (IL-6),* and *tumor growth factor-beta 1 (TGF-β1),* on the one hand, and nitric oxide (NO) and inducible nitric oxide synthase (iNOS), on the other hand. (nuclear factor-kappa B, NF-κB; reactive oxygen species, ROS).

Thereafter, the development of acute inflammatory responses in Dunkerque City's $PM_{2.5}$-exposed L132 cells was evaluated through the determination of protein secretion and/or gene expression of *TNF-α, IL-1β, IL-8, GM-CSF, IL-6,* and *TGF-β1,* on the one hand, and NO and *iNOS*, on the other hand.. (Figure 4). With other respects, the attention was focused on the ability of collected air pollution PM to activate both the TNF-α-induced pathway of apoptosis and the mitochondrial pathway of apoptosis, through the determination of the release of Cyt. c from mitochondria into cytosol, the enzymatic activities of caspases 8, 9 and 3, the gene expressions of *P53, BCL-2 AND BAX*, and the DNA fragmentation, as studied by Terminal deoxynucleotidyl transferase mediated deoxyUridine-5'-triphosphate-biotin Nick-End Labeling (TUNEL), and DNA laddering (Figure 5).

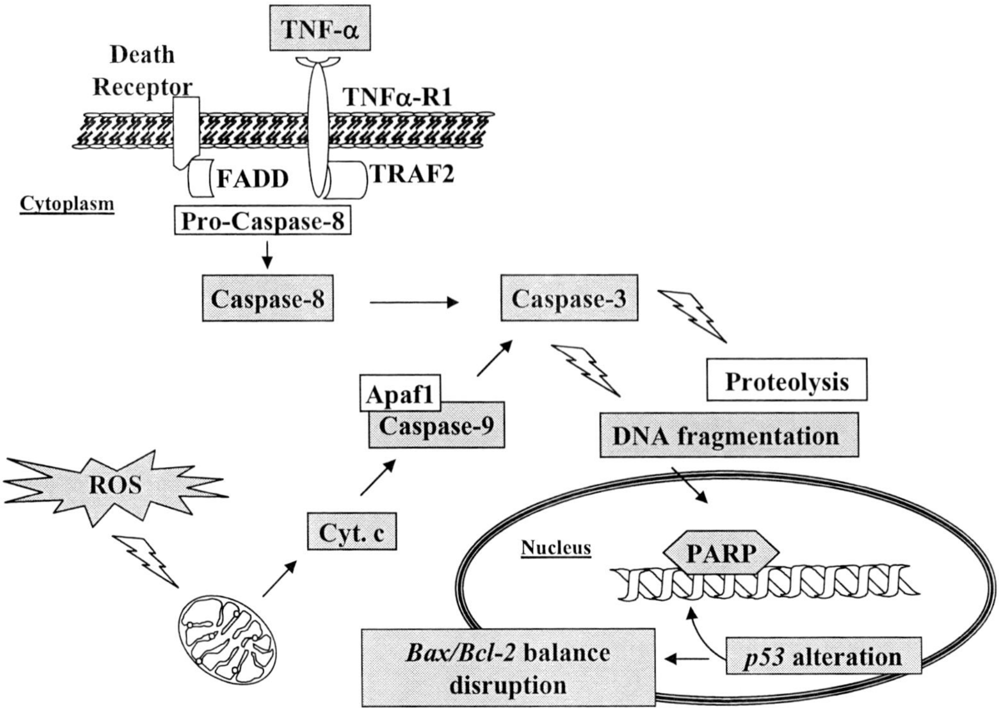

Figure 5. The induction of cell death by apoptosis in particulate matter (PM)-exposed L132 cells were carried out through the determination of the release of the Cytochrome c (Cyt. c) from mitochondria into cytosol, the enzymatic activities of caspases 8, 9 and 3, the gene expressions of *P53, BCL-2 and BAX*, and the DNA fragmentation, as studied by Terminal deoxynucleotidyl transferase mediated deoxyUridine-5'-triphosphate-biotin Nick-End Labeling (TUNEL), and DNA laddering (apoptosis protease-activating factor-1, Apaf-1; cysteine-aspartate proteases, caspase; Fas-associating protein with death domain, FADD; poly(ADP-ribose) polymerase, PARP; tumor necrosis fator family receptor I, TNFRI; tumor necrosis factor receptor-associated factor 2, TRAF2).

Materials and Methods

Chemicals

Minimum essential medium (MEM) with Earle's salts, fetal bovine serum (FBS), L-glutamin, and penicillin/streptomycin solution were provided by In Vitrogen-Life Technologies (Cergy Pontoise, France). Cytotoxicity Detection Kit LDH, Cell Proliferation Reagent WST-1, Dithiothreitol, dNTP, RNase inhibitor, oligo(dT)$_{15}$, Expand Reverse Transcriptase, Taq DNA Polymerase, and In Situ Cell Death Detection Kits were from Roche Diagnostics (Meylan, France). RANSOD reagent kit was purchased from Randox Laboratories (Montpellier-Frejorgues, France). Qiagen (Courtaboeuf, France) supplied the DNeasy Tissue Kit and the RNeasy Tissue Kit. Sigma-Aldrich (Saint-Quentin Fallavier, France) provided the nuclease P1 and the alkaline phosphatase. Japan Institute for the Control of Aging (Haruoka, Japan) supplied the 8-OHdG ELISA kit. The PARP Inhibition Assay, the Quantikine Human Immunoassays for the determination of TNF-α, IL-1β, IL-8, TGF-β1, IL-6, or GM-CSF, the Quantikine Human iNOS Immunoassays, the Total Nitric Oxide Assay, the Quantikine Human Cytochrome c Immunoassays, and Caspase-8, Caspase-9, and Caspase-3 Fluorometric Activity Kits were from R&D Systems Europe (Abingdon, United Kingdom). Active Motif Europe (Rixensart, Belgium) provided Nuclear Extract Kit, Trans-AM NF-κB p65 and p50 Transcription Factor assays. IκB-α Control Cell Extracts, Phospho-NF-κB p65 (Ser536) antibody, Phospho-IκB-α (Ser32) antibody, and Goat anti-rabbit IgG polyclonal antibody conjugated to horseradish peroxidase were from Cell Signaling Technology, Inc. (Danvers, MA, USA). All other reagents for Western Blotting were from GE Healthcare Europe GmbH (Orsay, France). Human IκBα [pS32] ELISA, MessageScreen™ Human Inflammatory Cytokine Set 1 Multiplex PCR© kit, and Mitochondria/Cytosol Fractionation Kits were from was from Biosources Europe SA (Nivelles, Belgium). Proligo France SAS (Paris, France) synthesized primer pairs for Polymerase Chain Reaction (PCR). Sigma-Aldrich (Saint-Quentin Fallavier, France) provided BCA Protein reagents.

PM Sampling, and PM Physical and Chemical Characteristics

PM Sampling: PM were collected in Dunkerque (51°04'N; 2°38'E), a French sea-side City located on the southern coast of the North-Sea, using high volume Sierra model 235 cascade impactor (Sierra Anderson, Smyna, GA, USA). Plates were mounted without any filters and no back up filter was used to maintain a constant aspiration flow rate (68 m^3/h) for a period of 2 weeks. No back up filter was used and the lowest stage was doubled to increase the efficiency of smallest particle ($PM_{0.33}$) sampling. Collection was done continuously during 9 months, from January to September, and impacting system was changed every 15 days; two identical systems were used alternatively. Meteorological data (i.e. wind speed, wind direction, temperature) were obtained from Météo France. The collection urban zone is characterized by the proximity of industrial activities (i.e. iron and steel industry, aluminum industry, oil refinery, basic chemistry, pharmaceutical industry, plant health production, food industry, etc.), and a heavy motor vehicle traffic. After sampling, impacting plates were dried under laminar flow bench during 48 h, then PM were removed from the collection plates and stocked at 4°C. About 2 g of PM have been collected by this way and carefully homogenized.

PM Size Distribution: Scanning electronic microscopy coupled with energy dispersive X-ray analysis (SEM-EDX; LeoTM 438 VP microscope and IXRF analysis system; Zeiss SAS, Le Pecq, France) was used to perform single particle analysis and also to determine PM size distribution. Prior to the analysis, PM was suspended in a n-hexane solution using ultrasonic treatment, and then filtered through a nucleopore filter to obtain well-distributed and dispersed PM, without agglomerates, as suitable for the SEM-EDX analysis. After the filter carbon-coating, the determination of PM size has been done in the automatic mode with SEM operating conditions as follows: accelerating voltage: 25 keV; intensity: 300 nA, and magnification: x 5000. The minimal observable size was 0.1 μm. The total number of studied particles in SEM-EDX was about 2500.

Specific Surface Area: Adsorption data from the Brunauer Emmett Teller (BET) method using a Sorptomatic apparatus (Thermo Electron Corporation, Courtaboeuf, France) provided the specific surface area of PM.

Inorganic Element Composition: The analytical procedure consisted of the sample acidic total digestion of PM by a $HNO_3/HClO_4/HF$ (4/1/1 v/v/v) mixture during 3 h at 130°C, the acid evaporation during 4 h at 170°C, and the subsequent analysis by inductively coupled plasma-atomic emission spectrometry (ICP-AES) using a Perkin-Elmer AAnalyst 600 (Perkin-Elmer Life Sciences France, Courtaboeuf, France).

Organic Compound Composition: After a soxhlet extraction, organic compounds were detected in PM using gas chromatography-mass spectrometry (GC-MS; Brüker EM40, Bruker, Wissembourg, France; Varian Saturn, Varian France, Coutaboeuf, France).

Cells and Culture conditions

The cell line we used originates from the normal lung tissue of a human embryo, and is deposited under the designation L132 in the American Type Culture Collection (ATCC; ATCC number: CCL-5). The morphology of these cells is epithelial, and they exhibit typical features of pneumocytes [62]. L132 cells were cultured in sterile plastic flasks (Corning; Fisher Scientific Labosi SAS, Elancourt, France), in MEM with Earle's salts, containing: 5 % (v/v) FBS, 1% (v/v) L-glutamin (200 mM), 1% (v/v) penicillin (10000 IU/mL), and 1% (v/v) streptomycin (10000 UG/mL) (In Vitrogen-Life Technologies). Exponentially growing cells were maintained at 37°C, in a humidified atmosphere containing 5% CO_2. All the L132 cells we used in all the steps of this study derived from the same initial cell culture.

Sampling of Culture Supernatants and Cells for the Study of the Cytotoxicity

Colony Forming Method. Survival rates of L132 cells exposed to PM were determined using the colony forming method [65]. Briefly, 100 proliferating cells were seeded in culture dishes (60 x 15 mm; Corning; Fisher Scientific Labosi SAS) to determine cell viability after 10 days of incubation at 37°C, in a humidified atmosphere containing 5% CO_2, in the continuous presence of PM without renewing the culture media. Accordingly, 10 culture dishes were chosen at random as control cells (i.e. incubated only with cultured media) and 5 culture dishes per PM concentration were designated at random as exposed cells (i.e. incubated with culture media containing increasing concentrations of PM: 18.84, 37.68, 56.52, 75.36 or 150.72 µg PM/mL). Both the LC_{10} and LC_{50} values were calculated comparing the number of crystal violet-stained cell colonies in exposed culture dishes with the number in non exposed culture dishes.

Released LDH Activity. Released LDH activities were evaluated in cell-free culture supernatants [66]. L132 cells were seeded in 96-well culture plates (Costar; Fisher Scientific Labosi SAS) at a density of 1×10^4 cells/200 µL culture media containing 18.84, 37.68, 56.52 or 75.36 µg PM/mL and incubated at 37°C in a humidified atmosphere containing 5 % CO_2. Non-exposed cells were used as negative controls. Accordingly, 16 replicates were chosen at random as control cells (i.e. incubated only with cultured media) and 8 replicates per PM concentration were designated at random as exposed cells (i.e. incubated with culture media containing 18.84, 37.68, 56.52 or 75.36 µg PM/mL). After 24, 48 or 72 h of incubation, 100 µL-aliquots of culture supernatants were collected cell-free and transferred in new 96-well culture plates to allow the measurement of extracellular LDH activity. The colorimetric assay for the quantification of LDH activity released from the cytosol of damaged cells was performed according to the manufacturer's recommendations (Cytotoxicity Detection Kit LDH, Roche).

MDH Activity. MDH activities were determined in cultured cells [67]. We used the experimental design previously described for the evaluation of released LDH. Non exposed cells were used as negative controls. After 24, 48 or 72 h of incubation, the colorimetric assay for the quantification of cell proliferation and/or cell viability, based on the cleavage of the tetrazolium salt WST-1 by MDH in viable cells, was realized (Cell Proliferation Reagent WST-1, Roche).

Sampling of Culture Supernatants and Cells for the Study of Oxidative Stress, Activation of NF-kB, Inflammatory Response, and Cell Death by Apoptosis

Cell Exposure. Depending on the incubation time, L132 cells were seeded at different density (i.e. 3 x 10^6, 1.5 x 10^6 or 0.75 x 10^6 cells/20 mL of culture media, respectively) in culture plastic flasks (Corning; Fisher Scientific Labosi SAS) and incubated at 37°C, in a humidified atmosphere containing 5 % CO_2, for 24 h. Culture supernatants were removed to eliminate non-adherent cells. Only living L132 cells were also incubated in the continuous presence of PM at their LC_{10} (i.e. 18.84 µg PM/mL) or LC_{50} (i.e. 75.36 µg PM/mL) for 24, 48 or 72 h of incubation, without renewing the culture media. Both the LC_{10} and the LC_{50} values were previously calculated owing to the colony forming method. Non exposed cells were used as negative controls. Accordingly, for each incubation time, 10 culture flasks were chosen at random as control cells (i.e. incubated only with cultured media) and 5 culture flasks per PM concentration were designated at random as exposed cells (i.e. incubated with culture media containing 18.84 or 75.36 µg PM/mL).

Sampling of Culture Supernatants and Cells. After 24, 48 or 72 h of incubation, 1 mL-aliquots of culture supernatants were collected and immediately frozen at - 80°C for the future determination of the biological parameters under study (i.e. protein secretion of TNF-α, IL-1β, IL-8, GM-CSF, IL-6, TGF-β1, and NO concentration), and total protein contents. Adherent cells were removed and centrifuged (500 g, 10 min, 4°C). Cell pellets were washed twice with 10 mL-aliquots of sterile PBS (In Vitrogen-Life Technologies) and cell-aliquots were collected and quickly frozen at - 80°C until the further determination of the biological parameters under study (i.e. MDA production, SOD activity, 8-OHdG formation, PARP activity, DNA binding of p65 and p50, phosphorylation of p65 and IkBα, gene expression of *TNF-α, IL-1β, IL-8, GM-CSF, IL-6,* and *TGF-β1*, iNOS activity, activity of caspase-8, -9, and-3, Cyt. c concentration, gene expression of *P53, BCL-2* and *BAX*, and DNA fragmentation,) and total protein content.

Study of Oxidative Stress

MDA Production. Lipid peroxidation was evaluated in cell lysates using a high performance liquid chromatography (HPLC) MDA assay [68, 69]. Briefly, cell pellets were mixed with 1 mL of 0.1M HCl, extracted by 2 x 3 mL of ethyl acetate, stirred for 5 minutes, and centrifuged at 3,000 g for 10 minutes. After evaporation, the extracts were suspended in 200 µL of methanol. The HPLC system consisted of a Jasco PU-980 pump, equipped with a Nucleosil column (C18, 150 x 4,6 mm, 5 µm size), a Rheodyne 7725 automated injector, a UV Jasco 975 detector and a Shimadzu CR3A integrator (Vasse Industries, Lille, France). The mobile phase was a mixture of 50 mM KH_2PO_4 and methanol 60/40 (v/v) adjusted to pH 6.8 using 1 M KOH. Tetraethoxypropane (Sigma, Saint Quentin Fallavier, France), which hydrolyses to yield one molecule of MDA, was used as the standard, and thiobarbituric acid (TBA) as the reagent. A volume of 50 µL of either standard solutions or methanol extracts was injected in the HPLC system and the MDA-TBA adducts were detected at532 nm.

Total SOD Activity; Total SOD activities were assessed in cell lysates using the RANSOD reagent kit (Randox Laboratories) [69].

8-OHdG Formation. 8-OHdG concentrations were determined using the method originally described by Toyokuni et al. (1997) and modified by Garçon et al. (2001) [69, 70]. Briefly, after DNA extraction (DNeasyTissue Kit, Qiagen) and digestion (nuclease P1 and alkaline phosphatase, Sigma-Aldrich), the formation of the oxidative DNA adduct 8-OHdG was measured using the 8-OHdG ELISA kit (Japan Institute for the Control of Aging).

PARP Activity. PARP activities were studied in cell lysates using the PARP Inhibition Assay (R&D Systems Europe) according to the manufacturer's instructions.

Study of Nuclear-Factor-kappa B Activation

Isolation of Nuclear and Cytoplasmic Protein Fractions. Nuclear and cytoplasmic protein fractions from control and exposed-L132 cells were distinguished using Nuclear Extract Kit (Active Motif, France).

Western Blotting of Phosphorylated p65 and Phosphorylated IκBα Proteins in Cytoplasmic Extracts. Cytoplasmic protein extracts were treated with sodium-dodecyl-sulfate (SDS) loading buffer (60 mM Tris pH 6.8, 0.1 % w/v bromophenol blue, 25 % glycerol, 14.4 mM β-mercaptoethanol, and 2% w/v SDS). Forty microgram of protein extracts and 10 μL of IκBα control cell extracts from TNF-α treated HeLa cells (Cell Signaling Technology) were loaded on a 11 % (w/v) SDS-polyacrylamide gel electrophoresis (SDS-PAGE) at 25 mA for 1h30 min at 4°C in 62.5 mM Tris–HCl, 2 % (w/v) SDS, 10% (w/v) glycerol, 50 mM DDT, 0,01% (w/v) Bromophenol blue, pH 6.8), using an Hoefer SE 260 Small Tanker and an Hoefer Electrophoresis Power Supply EPS 601 (GE Healthcare Europe GmbH). Proteins were transferred onto PVDF membranes at 100 V for 1 h at 10°C in 25 mM Tris–HCl, 192 mM glycine, 10% methanol, pH 8.3, using an Hoefer TE 22 Mighty Small Transfer Apparatus and an Hoefer Electrophoresis Power Supply EPS 2A200 (GE Healthcare Europe GmbH). Membranes were blocked in 2% (w/v) BSA in TBS containing 0.1% (v/v) Tween 20 (TBST) for 1 h at room temperature, and then incubated overnight at 4°C with Phospho-NF-κB p65 (Ser536) antibody or Phospho-IκB-α (Ser32) antibody (Cell Signaling Technology) at 1:1,000 dilution in 2% (w/v) bovine serumalbumine (BSA) in TBST, membranes were washed five times for 5 min in TBST and incubated with goat anti-rabbit IgG polyclonal antibody conjugated to horseradish peroxidase (Cell Signaling Technology) at 1:1,000 dilution in 2% (w/v) BSA in TBST for 1 h at room temperature. After PVDF membrane washing, labeled phosphorylated p65 and IκBα proteins were visualized by chemiluminescence using ECL Advance reagents (GE Healthcare Europe GmbH), according to the manufacturer's recommendations. Chemiluminescence detection was carried out by a G-Box Chemi-16 image station (Syngene Europe, Cambridge, United Kingdom).

DNA Binding of P65 and p50. DNA binding of NF-κB p65 and p50 in nuclear protein extracts were carried out using Trans-AM NF-κB p65 and p50 Transcription Factor assay kits (Active Motif Europe).

Phosphorylated IκBα Protein Concentrations. Phosphorylated IκBα protein concentrations were determined in cytoplasmic fractions using commercially available BioSource Human IκBα [pS32] ELISA (Biosources Europe SA).

Study of Inflammatory Response

Protein Secretion of Pro- and Anti-Inflammatory Mediators. The concentrations of TNF-α, IL-1β, IL-8, GM-CSF, IL-6, and TGF-β1 in collected cell culture supernatants, and the activity of iNOS in cell lysates were determined using commercially available enzyme immunoassays (Quantikine Human Immunoassays, R&D Systems Europe), following the manufacturer's instructions. Total NO concentrations in collected cell culture supernatants were studied using the Griess reaction after the enzymatic conversion of nitrate to nitrite by nitrate reductase (Total Nitric Oxide Assay, R&D Systems Europe).

Gene Expression of Pro- and Anti-Inflammatory Mediators. Total RNA was extracted from collected cells using RNeasy Total RNA System, according to the manufacturer's recommendation. Conditions used for reverse transcription (RT) were as described previously by Garçon et al. (2001) [71]. Primer pairs for *TNF-α, IL-1β, IL-8, GM-CSF, IL-6, TGF-β1, and glyceraldehyde-3-phosphate dehydrogenase (GAPDH)*, were provided by Biosources Europe SA. Using competitive polymerase chain reaction (PCR), the gene expression of each considered gene was evaluated concurrently with the gene expression of *GAPDH*. Five-microliter aliquots of cDNA were added in a final volume of 50 μl of PCR mix, reconstituted following the manufacturer's recommendations.

After the initial denaturing step for 1 min at 95°C, the reaction was initiated by 5 cycles using the following denaturing, annealing, and extension cycling conditions: denaturing for 1 min at 94°C, annealing for 4 min at 60°C, and extension for 2 min at 72°C, using a MJ Research PTC-100 HB (VWR International SA, Fontenay-sous-Bois, France). The reaction was then continued as follows for 35 cycles: denaturing for 1 min at 94°C, annealing for 2.5 min at 68°C, and extension for 2 min at 72°C. After the last cycle, samples were incubated at 72°C for 10 min. The PCR was then stopped by cooling at 4°C. After gel electrophoresis and ethidium bromide intercalation, PCR-amplified products were visualized under UV light and analyzed by a computerized video/densitometry system (Software: GelAnalyst™ v3.1FR; Clara Vision, Orsay, France).

Study of the Cell Death by Apoptosis

Activation of the TNF-α-Induced Apoptotic Pathway and Mitochondrial Apoptotic Pathway. After the isolation of a highly enriched mitochondrial fraction (Mitochondria/Cytosol Fractionation Kits, Biosources Europe SA), Cyt. c concentrations in both the mitochondrial and the cytosolic fractions were determined using the Quantikine Human Cytochrome c Immunoassay (R&D Systems Europe). Caspase-8, 9, and 3 activities in cells were studied using commercially available fluorometric assays (Caspase-8, Caspase-9, and Caspase-3 Fluorometric Activity Kits, respectively, R&D Systems Europe).

Gene Expression of *p53, bcl-2* and *bax*. Total RNA was extracted from cell-aliquots using RNeasy Mini Kits (Qiagen). Conditions for the RT were as described previously by Garçon et al. (2001) [71]. Specific primer pairs and PCR thermocycling conditions were designed as that published by Willey et al. (1998) [72]. After gel electrophoresis and ethidium bromide intercalation, PCR-amplified products were visualized under UV light and analyzed by a computerized video/densitometry system (Software: GelAnalyst™ v3.1FR; Vasse Industries, Lille, France).

DNA Fragmentation. PM-induced DNA fragmentation was carried out using a TUNEL Kit (In Situ Cell Death Detection Kit, Roche Diagnostics, Meylan, France), following the manufacturer's instructions, and measured by an EPICS XLTM Flow Cytometer (Beckman Coulter, Fullerton, CA, USA). The DNA fragmentation in PM-exposed cells was carried out by agarose gel electrophoresis. Briefly, after protein digestion, RNA-free DNA was extracted from cells by DNeasy Tissue Kits (Qiagen), and 5 µg of DNA were suspended in 20 µL DNase-free loading buffer. After gel electrophoresis and ethidium bromide intercalation, DNA fragments were visualized under UV light and analyzed by a computerized video/densitometry system (Software: GelAnalyst™ v3.1FR).

Determination of Total Protein Contents

Total protein contents in the cell lysates and cell culture supernatants were determined with bicinchoninic reagent using the BCA protein reagent (Pierce) [73].

Statistical Analysis

Results were expressed as mean values and ranges (minimum value-maximum value). Data from cell cultures exposed to increasing concentrations of PM were compared with those from non-exposed cell cultures. Thereafter, we looked for correlation between the different parameters under study. Statistical analyses were performed by the Mann-Whitney U test and the non-parametric Spearman test (Software: SPSS for Windows, v10.05, 2000; Paris, France). Statistically significant differences were reported with p values<0.05.

Results

PM Physical and Chemical Characteristics

Figure 6 shows the size distribution results of collected PM as depicted as cumulative frequencies (%) and relative frequencies (%). Accordingly, 64.24 %, 92.15 %, and 98.49 % of PM total number were equal or smaller than 1 µm, 2.5 µm, and 5 µm, respectively. The highest number of collected PM were also detected in size classes included in PM with a geometric diameter inferior or equal to 2.5 µm: 0-0.5 µm (33.63 %), 0.5-1.0 µm (30.61 %), 1.0-1.5 µm (14.33 %), 1.5-2.0 µm (8.69 %), and 2.0-2.5 µm (4.89 %).

Their specific surface area was 1 m^2/g for collected PM, but increased until 50 m^2/g for PM having undergone a thermal desorption at 400°C under secondary vacuum.

Table 1 shows the inorganic elements found in collected Dunkerque city's $PM_{2.5}$. Among them, Fe, Al, Ca, Na, K, Mg, Pb, and Ti were the most abundantly present. In collected PM, there were not only inorganic elements usually associated with natural environment (i.e. Ca, Na, Mg, Ti, Sr, etc.), but also so-called anthropogenic elements (i.e. Fe, Al, Pb, Mn, Zn, Ba, Cu, Cr, Ni, Cd, etc.).

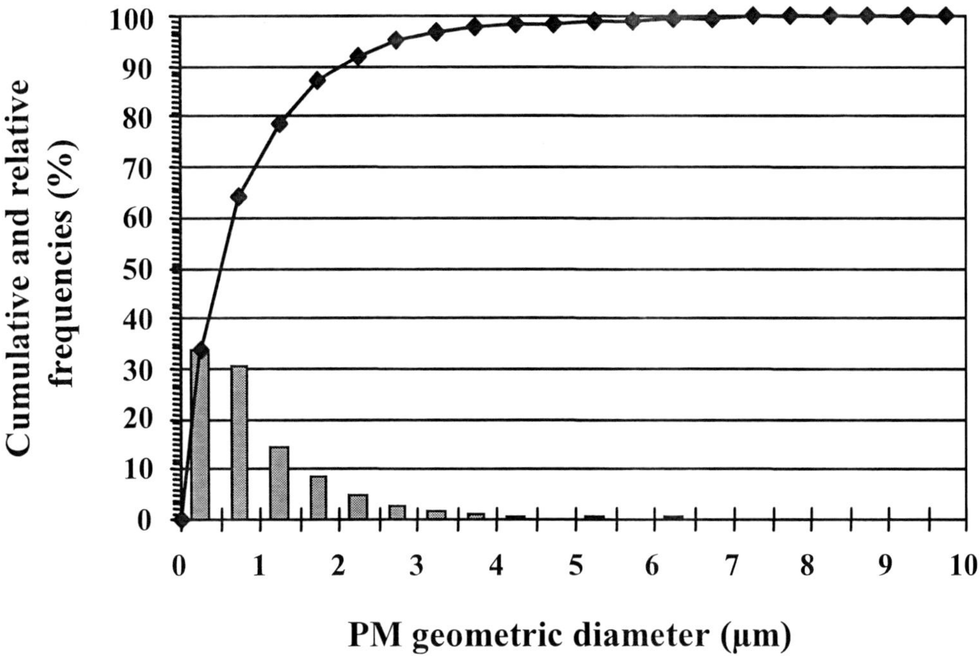

Figure 6. Size distribution of the particle matter (PM) collected in Dunkerque, a French seaside city located on the southern coast of the North-Sea, using high volume cascade impactor, is depicted as cumulative frequency (%; curve) and relative frequency (%; histogram). PM size distribution has been determined using scanning electron microscopy.

Table 1. Inorganic Compounds Detected in PM

Inorganic Compounds	Concentrations (% w/w)	Inorganic Compounds	Concentrations (% w/w)
Fe	7.84	Ba	0.064
Al	5.83	Cu	0.037
Ca	4.95	Cr	0.032
Na	1.88	V	0.030
K	0.97	Ni	0.029
Mg	0.81	Sr	0.022
Pb	0.80	Li	0.011
Ti	0.51	Mo	0.004
Mn	0.352	Cd	< 0.001
Zn	0.184	Co	< 0.0006

Table 1 shows the concentrations (% w/w) of the inorganic compounds detected in the collected particulate matter (PM).

Table 2 shows the list of the organic compounds detected in PM samples. In particular, there were various VOC and PAH coated-onto collected PM.

Table 2. Organic Compounds detected in PM

Organic Compounds	Concentrations (μg/g)	Organic Compounds	Concentrations (μg/g)
Benzene	106.5	Phenanthrene	28.3
Toluene	42.2	Dibutylphtalate	110.1
Nonene	9.3	Eicosane	1.4
Xylene	65.4	Fluoranthene	4.7
Styrene	13.5	Henicosane	5.2
1-Ethenyltrimethylbenzene	15.4	Docosane	17.4
Decane	64.9	Tricosane	8.7
Diethylbenzene	170.3	Tetracosane	39.7
Limonene epoxyde	20.5	Pentacosane	12.2
Tetramethylbenzene	89.5	Bis(Ethyl.Hexyl)phtalate	41.4
Naphtalene	38.1	Acenaphtene	11.1
Methylnaphtalene	141.9	Fluorene	6.3
Dimethylnaphtalene	90.2	Pyrene	4.7
Heptadecane	15.1	Benzo(a)anthracene or Chrysene	4.9
Anthracene	47.1		

Table 2 shows the concentrations (μg/g) of the organic compounds detected in the collected particulate matter (PM).

Cytotoxicity of PM

PM induced a concentration-dependent cytotoxic effect in proliferating L132 cells, as determined using the colony forming method (Figure 7). The LC_{10} and LC_{50} values were 18.84 and 75.36 μg PM/mL, respectively.

Moreover, PM generated concentration and time-dependent increases of released LDH activity (Table 3). Statistically significant increases of extracellular LDH activity were observed in L132 cells 72 h after their exposure to 56.52 or 75.36 μg PM/mL ($p<0.001$, respectively). The control values for released LDH did not differ depending on the incubation time (24 to 72h), indicating the background of cell death.

At least, PM caused decreases of MDH activity in a concentration and a time-dependent manner (Table 3). Maximal decreases of MDH activity were attained in L132 cells exposed to 75.36 μg PM/mL, 72 h after their incubation ($p<0.001$). In contrast, the control values for MDH went up depending on the incubation time (24 to 72h), reflecting the exponentially growth of the negative controls.

Table 3. Cytotoxicity of PM in L132 cells

		Concentration of PM (µg/mL)				
		0	18.84	37.68	56.52	75.36
LDH activity (U)	24 h	0.271	0.267 [a]	0.244 [a]	0.301 [a]	0.319 [a]
		0.196-0.374	0.301-0.437	0.183-0.405	0.199-0.476	0.229-0.449
	48 h	0.300	0.287 [a]	0.307 [a]	0.377 [a]	0.363 [a]
		0.271-0.493	0.309-0.463	0.252-0.404	0.219-0.447	0.292-0.459
	72 h	0.377	0.401 [a]	0.505 [a]	1.438 [d]	1.381 [d]
		0.269-0.481	0.271-0.497	0.300-0.696	1.007-1.696	1.033-1.945
MDH activity (U)	24 h	1.087	0.799 [d]	0.721 [d]	0.505 [d]	0.414 [d]
		0.795-1.138	0.704-0.966	0.644-789	0.481-0.730	0.333-0.490
	48 h	1.794	0.744 [d]	0.699 [d]	0.499 [d]	0.197 [d]
		1.655-1.939	0.671-0.835	0.670-0.725	0.475-0.715	0.140-0.310
	72 h	2.449	0.101 [d]	0.075 [d]	0.021 [d]	0.069 [d]
		1.914-2.679	0.055-0.134	0.045-0.089	0.013-0.049	0.033-0.079

Table 3 shows the extracellular lactate dehydrogenase (LDH) activity (U) in L132 cell-free culture supernatants, and the mitochondrial dehydrogenase (MDH) activity (U) in L132 cells 24, 48 or 72h after their incubation in the continuous presence of increasing concentrations of particle matter (PM; 18.84, 37.68, 56.52 or 75.36 µg PM/mL) without renewing the culture media. Non-exposed cells were used as negative controls. These values are depicted as mean values and ranges (minimum value-maximum value) of 16 replicates for negative controls and 8 replicates for every PM concentrations. ([a] = not significant, and [d] = $p<0.001$; Mann-Whitney U test).

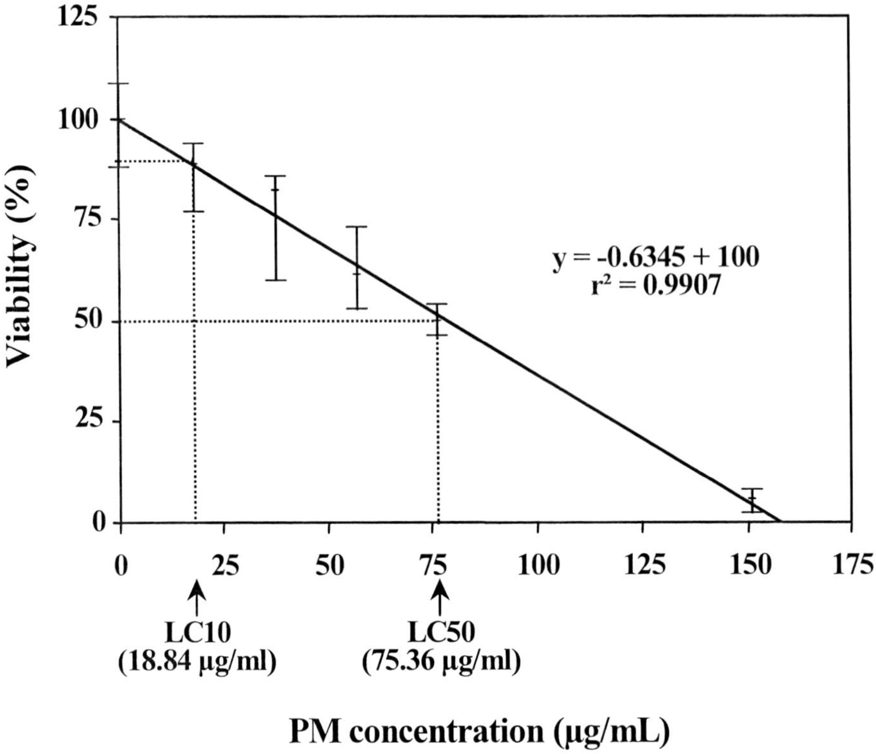

Figure 7. Survival rates (%) of L132 cells 10 days after their incubation in the continuous presence of increasing concentrations of collected particle matter (PM; 18.84, 37.68, 56.52, 75.36 or 150.72 µg PM/mL) without renewing the culture media. Non exposed cells were used as negative controls. These values are depicted as mean values and ranges (minimum value-maximum value) of 10 cell cultures for negative controls and 5 cell cultures for every PM concentrations. Lethal concentrations at 10% and 50% (LC_{10} and LC_{50}, respectively) were calculated comparing the number of crystal violet-stained cell colonies in exposed culture dishes with the number in non exposed culture dishes.

Effects of PM on the Occurrence of Oxidative Stress Conditions

Table 4 showed MDA concentrations in control or PM-exposed proliferating L132 cells. PM exposure induced lipid peroxidation in a concentration and time-dependent manner. Statistically significant increases of MDA concentration were attained in L132 cells 72 h after their exposure to PM at their LC_{10} or LC_{50} (4.17-fold, $p<0.01$, and 11.66-fold, $p<0.01$, respectively), versus control cells. Statistically significant changes in SOD activity were reported in PM-exposed proliferating L132 cells (Table 4).

There were increases of SOD activity in L132 cells 24 h after their exposure to PM at their LC_{10} or LC_{50} (1.18-fold, $p<0.05$, and 1.61-fold, $p<0.01$, respectively), versus control cells. In contrast, decreases in SOD activity were observed in L132 cells 48 h after their exposure to PM at their LC_{50} (1.19-fold, $p<0.01$), and 72 h after their exposure to PM at their LC_{10} or LC_{50} (1.37-fold, $p<0.01$, and 1.41-fold, $p<0.01$, respectively), versus control cells.

Table 4. Effects of PM on the Biological Markers of Oxidative Stress

	Incubation Time and Exposure								
	24 h			48 h			72 h		
	C	LC10	LC50	C	LC10	LC50	C	LC10	LC50
MDA	1.64	1.83 [a]	2.12 [a]	2.64	2.91 [a]	3.16 [a]	3.72	15.50 [c]	43.36 [c]
(µmol/g protein)	1.23-2.23	1.61-2.00	1.76-2.52	1.98-3.23	2.48-3.17	2.94-3.34	2.67-4.94	11.65-18.68	33.25-50.52
SOD	2.93	3.46 [b]	4.72 [c]	3.54	3.37 [a]	2.97 [c]	3.94	2.88 [c]	2.81 [c]
(U/g protein)	1.85-3.89	2.44-5.81	3.95-6.16	3.07-3.98	2.18-3.80	2.56-3.33	3.26-4.70	2.15-3.61	2.34-3.13
8-OHdG	10.32	34.83 [c]	133.4 [c]	9.67	66.89 [c]	162.3 [c]	7.56	89.94 [c]	210.1 [c]
(pg/µg DNA)	7.44-15.22	31.35-46.36	125.8-138.7	7.27-13.51	46.50-78.11	148.5-182.6	5.76-9.18	66.38-103.1	183.9-239.2
PARP	6.79	17.35 [c]	48.82 [c]	6.72	17.29 [c]	47.27 [c]	6.80	21.56 [c]	55.86 [c]
(U/g protein)	5.81-8.08	15.32-19.57	45.99-51.92	5.58-8.67	16.35-19.50	43.70-51.66	6.15-8.29	19.19-23.84	44.29-67.64

Table 4 shows the malondialdehyde (MDA) concentrations (µmol/g protein), superoxide dismutase (SOD) activities (U/g protein), 8-hydroxy-2'-deoxyguanosine (8-OHdG) concentrations (pg/µg DNA), and poly(ADP-ribose) polymerase (PARP) activities (U/g protein) in L132 cells after their incubation in the continuous presence of particle matter (PM) at their LC_{10} (i.e. 18.84 µg PM/mL) or LC_{50} (i.e. 75.36 µg PM/mL) for 24, 48 or 72 h, without renewing the culture media. Non exposed cells were used as negative controls (C). These values are depicted as mean values and ranges (minimum value-maximum value) of 10 replicates for negative controls and 5 replicates for every PM concentrations. ([a] = not significant, [b] = $p<0.05$, and [c] = $p<0.01$; Mann-Whitney U test).

	Incubation Time and Exposure								
	24 h			48 h			72 h		
	C	LC10	LC50	C	LC10	LC50	C	LC10	LC50
p65	0.998	1.230 ^{b}	2.298 ^{c}	1.009	1.514 ^{c}	1.854 ^{c}	0.967	1.700^{c}	1.730 ^{c}
(U/µg protein)	0.689-1.264	1.025-1.441	2.099-2.787	0.608-1.314	1.305-1.979	1.403-1.988	0.708-1.260	1.377-1.879	1.536-1.988
p50	0.533	0.616 ^{b}	0.934 ^{c}	0.496	0.665 ^{c}	0.916 ^{c}	0.581	0.695 ^{c}	0.974 ^{c}
(U/µg protein)	0.438-0.631	0.553-0.702	0.822-1.089	0.417-0.606	0.557-0.784	0.756-1.080	0.463-0.856	0.550-0.781	0.700-1.186
IκBα	20.878	27.436 ^{c}	30.816 ^{c}	23.326	27.354 ^{c}	31.943 ^{c}	25.837	32.175 ^{b}	33.934 ^{b}
(U/µg protein)	14.60-24.35	24.52-31.24	25.82-38.40	18.99-25.85	25.45-28.93	26.50-41.09	16.22-33.74	28.03-40.76	30.21-42.40

Table 5 shows the DNA bindings of p65 and p50 subunits of Nuclear Factor-kappa B (NF-κB; U/µg protein) and cytoplasmic phosphorylated IκBα protein concentrations in L132 cells after their incubation in the continuous presence of particle matter (PM) at their LC_{10} (i.e. 18.84 µg PM/mL) or LC_{50} (i.e. 75.36 µg PM/mL) for 24, 48 or 72 h, without renewing the culture media. Non-exposed cells were used as negative controls (C). These values are depicted as mean values and ranges (minimum value-maximum value) of 10 replicates for negative controls and 5 replicates for every PM concentrations. (^{b} = $p<0.05$ and ^{c} = $p<0.01$; Mann-Whitney U test).

Dunkerque City's PM induced concentration and time-dependent formation of 8-OHdG in PM-exposed proliferating L132 cells (Table 4). Statistically significant increases of 8-OHdG concentration were seen in L132 cells 24 h after their exposure to PM at their LC_{10} or LC_{50} (3.37-fold, $p<0.01$, and 12.91-fold, $p<0.01$, respectively), 48 h after their exposure to PM at their LC_{10} or LC_{50} (6.91-fold, $p<0.01$, and 16.84-fold, $p<0.01$, respectively), and 72 h after their exposure to PM at their LC_{10} or LC_{50} (11.90-fold, $p<0.01$, and 27.79-fold, $p<0.01$, respectively), as compared to non-exposed cells.

Table 4 showed PARP activities in control or PM-exposed proliferating L132 cells. Statistically significant increases of PARP activity were attained in L132 cells 24 h after their exposure to PM at their LC_{10} or LC_{50} (1.41-fold, $p<0.01$, and 1.45-fold, $p<0.01$, respectively), 48 h after their exposure to PM at their LC_{10} or LC_{50} (1.53-fold, $p<0.01$, and 1.68-fold, $p<0.01$, respectively), and 72 h after their exposure to PM at their LC_{10} or LC_{50} (1.28-fold, $p<0.01$, and 1.50-fold, $p<0.01$, respectively), versus control.

Effects of PM on Nuclear-Factor-kappa B Activation

Figure 8 shows the Western Blotting analyses of phosphorylated p65 (75 kDa; Figure 8 A) and phosphorylated IκBα (41 kDa; Figure 8 B) proteins in cytoplasmic extracts from L132 cells. Increases of chemiluminescence signal arising from phosphorylated p65 were observed in PM-exposed L132 cells in a concentration-dependent manner and time-dependent manner, versus controls (Figure 8 A).

Figure 8 B shows that the intensity of the chemiluminescence arising from phosphorylated IκBα increased in PM-exposed L132 cells in a concentration-dependent manner but not in a time-dependent manner, versus non-exposed cells (Figure 8 A).

Table 5 shows increases in nuclear p65 and p50 DNA bindings in PM-exposed proliferating L132 cells. There were also increases in nuclear p65 DNA binding in L132 cells, 24 h after their exposure to PM at their LC_{10} (1.23-fold, $p<0.05$) and LC_{50} (2.30-fold, $p<0.01$), 48 h after their exposure to PM at their LC_{10} (1.50-fold, $p<0.01$) and LC_{50} (1.84-fold, $p<0.01$), and 72 h after their exposure to PM at their LC_{10} (1.76-fold, $p<0.05$) and LC_{50} (1.79-fold, $p<0.01$), versus control cells.

Statistically significant increases in nuclear p50 DNA binding were observed in L132 cells, 24 h after their exposure to PM at their LC_{10} (1.16-fold, $p<0.05$) and LC_{50} (1.75-fold, $p<0.01$), 48 h after their exposure to PM at their LC_{10} (1.34-fold, $p<0.01$) and LC_{50} (1.85-fold, $p<0.01$), and 72 h after their exposure to PM at their LC_{10} (1.20-fold, $p<0.01$) and LC_{50} (1.68-fold, $p<0.01$), as compared to control cells (Table 5). A statistically significant correlation was reported between nuclear p65 DNA binding and nuclear p50 DNA binding (Spearman's rho = 0.686, $p<0.01$).

Statistically significant increases in cytoplasmic phosphorylated IκBα protein concentrations were reported in PM-exposed proliferating L132 cells (Table 5). There were increases in cytoplasmic phosphorylated IκB-α protein concentration in L132 cells 24 h after their exposure to PM at their LC_{10} or LC_{50} (1.31-fold, $p<0.01$, and 1.48-fold, $p<0.01$, respectively), 48 h after their exposure to PM at their LC_{10} or LC_{50} (1.17-fold, $p<0.01$, and 1.37-fold, $p<0.01$, respectively), and 72 h after their exposure to PM at their LC_{10} or LC_{50} (1.25-fold, $p<0.05$, and 1.31-fold, $p<0.05$, respectively), versus control cells. Significant Table 5. Effects of PM on Nuclear-Factor-kappa B/Inhibitory-kappa B Activation correlations were observed between nuclear p65 DNA binding and cytoplasmic phosphorylated IκB-α protein concentration (Spearman's rho = 0.696, p<0.01), on the one hand, and nuclear p50 DNA binding and cytoplasmic phosphorylated IκB-α protein concentration (Spearman's rho = 0.687, p<0.01), on the other hand.

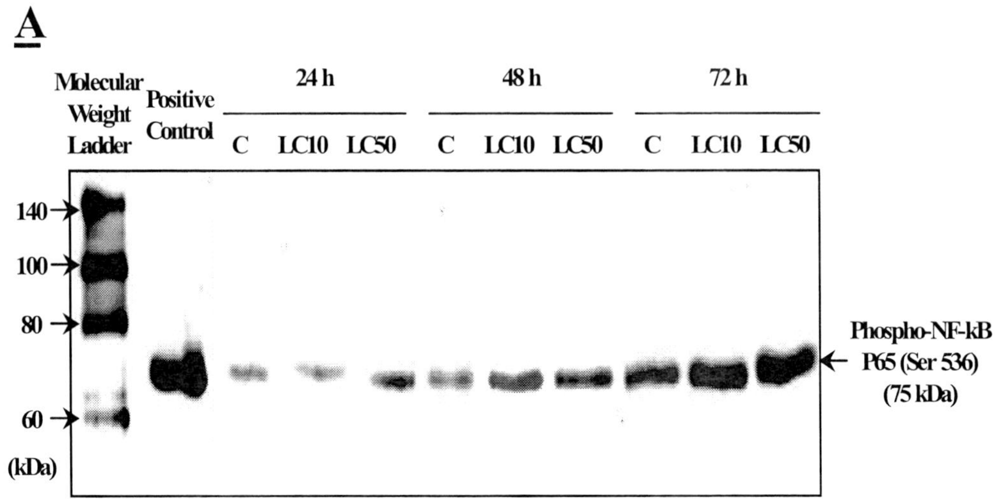

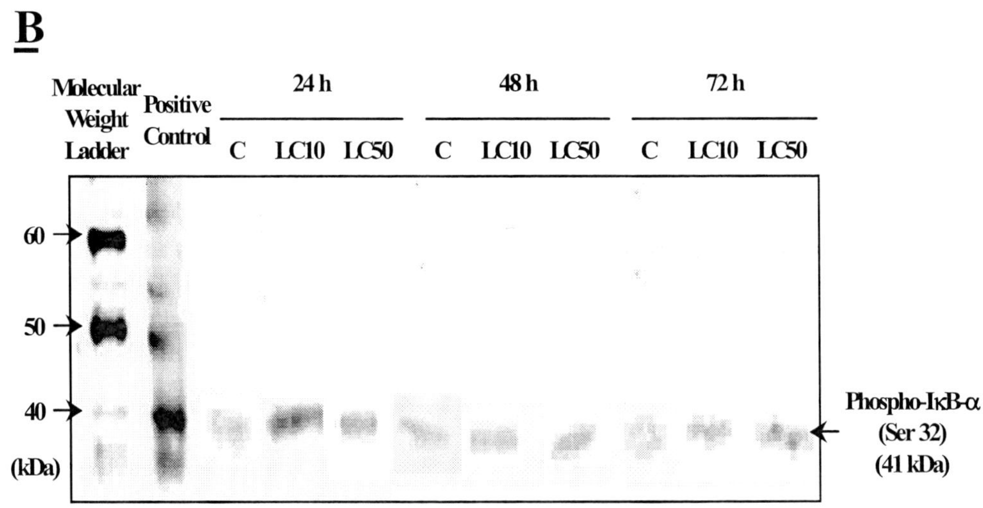

Figure 8. Western blotting analyses of phosphorylated nuclear factor-kappa B (NF-κB) p65 (75 kDa; Figure 8 A) and phosphorylated inhibitory kappa B (IκB-α; 41 kDa; Figure 8 B) in cytoplasmic extracts from L132 cells after their incubation in the continuous presence of particle matter (PM) at their LC_{10} (i.e. 18.84 μg PM/mL) or LC_{50} (i.e. 75.36 μg PM/mL) for 24, 48 or 72 h, without renewing the culture media. Non-exposed cells were used as negative controls. Total cell extracts from HeLa cells prepared with tumor necrosis factor-alpha treatment serve as positive control.

Table 6. Concentrations of pro-inflammatory mediators (TNF-α, IL-1β, IL-8, GM-CSF, iNOS, NO) in the culture supernatants of PM-exposed L132 cells

	Incubation Time and PM Exposure								
	24 h			48 h			72 h		
	C	LC10	LC50	C	LC10	LC50	C	LC10	LC50
TNF-α	2.55	2.49 [a]	15.5 [c]	2.71	5.85 [c]	27.1 [c]	2.40	20.9 [c]	45.8 [c]
(pg/g protein)	2.09-3.09	2.23-2.72	13.3-18.3	2.32-3.28	4.28-7.11	25.2-29.5	2.19-2.65	19.7-23	41.3-49.2
IL-1β	0.81	0.78 [a]	7.31 [c]	0.88	7.31 [c]	21.9 [c]	0.82	11.6 [c]	43.9 [c]
(pg/g protein)	0.65-0.94	0.74-0.85	5.52-8.132	0.78-1.07	6.04-10.12	18.2-25.5	0.73-1.03	10.3-12.5	39.5-50.3
IL-8	3.45	4.59 [a]	22.2 [c]	3.21	17.8 [c]	59.9 [c]	8.6	49.3 [c]	201.4 [c]
(pg/g protein)	2.89-6.42	3.66-6.64	16.6-27.9	2.67-4.05	14.4-20.8	56.3-64.9	6.7-10. 3	41.4-60.6	179-214.5
GM-CSF	1.05	1.07 [a]	1.10 [a]	1.03	0.99 [a]	1.05 [a]	1.57	1.78 [a]	1.80 [a]
(pg/g protein)	0.89-1.24	0.93-1.35	0.98-1.27	0.87-1.45	0.86-1.13	0.97-1.11	1.01-2.01	1.43-2.24	1.55-2.28
iNOS	10.76	25.56 [c]	30.25 [c]	12.54	31.99 [c]	41.43 [c]	14.56	52.46 [c]	129.4 [c]
(U/mg protein)	7.95-16.53	20.04-27.75	29.60-34.36	8.17-19.79	23.58-39.80	30.56-57.46	10.15-26.07	40.84-59.41	93.00-155.8
NO	6.79	17.35 [c]	48.82 [c]	6.72	17.29 [c]	47.28 [c]	6.80	21.57 [c]	55.86 [c]
(µmol/g protein)	5.81-8.08	15.32-19.57	46.00-51.92	5.58-8.67	16.35-19.50	43.71-51.66	6.15-8.29	19.20-23.84	44.29-67.64

Table 6 shows the concentrations of tumor necrosis factor-alpha (TNF-α; pg/mg protein), interleukin-1 beta (IL-1β; pg/mg protein), interleukin-8 (IL-8; pg/mg protein), granulocyte/macrophage-colony stimulating factor (GM-CSF; pg/mg protein), inducible nitric oxide synthase (iNOS) activities (U/g protein), and nitric oxide (NO) concentrations (µmol/g protein) in the culture supernatants of L132 cells 24, 48 or 72 h after their incubation in the continuous presence of particle matter (PM) at their LC_{10} (i.e. 18.84 µg PM/mL) or LC_{50} (i.e. 75.36 µg PM/mL), without renewing the culture media. Non exposed cells were used as negative controls (C). These values are depicted as mean values and ranges (minimum value-maximum value) of 10 replicates for negative controls and 5 replicates for every PM concentrations ([a] = not significant, and [c] = $p<0.01$; Mann-Whitney U test).

Effects of PM on the Development of the Inflammatory Response

Dunkerque City's PM induced TNF-α secretion by L132 cells in a concentration-dependent manner and time-dependent manner (Table 6). Significant increases of TNF-α concentrations were determined in the culture supernatants of L132 cells, 24 h after their exposure to PM at their LC_{50} (6.05-fold, $p<0.01$), 48 h after their exposure to PM at their LC_{10} (2.16-fold, $p<0.01$) and LC_{50} (10.03-fold, $p<0.01$), and 72 h after their exposure to PM at their LC_{10} (8.72-fold, $p<0.01$) and LC_{50} (19.10-fold, $p<0.01$), versus control.

In a similar manner, significant increases of IL-1β concentrations were seen in the culture supernatants of L132 cells, 24 h after their exposure to PM at their LC_{50} (9.01-fold, $p<0.01$), 48 h after their exposure to PM at their LC_{10} (8.30-fold, $p<0.01$) and LC_{50} (24.89-fold, $p<0.01$), and 72 h after their exposure to PM at their LC_{10} (14.05-fold, $p<0.01$) and LC_{50} (53.46-fold, $p<0.01$), compared to control cells (Table 6). There were concentration-dependent and time-dependent increases of IL-8 secretion by PM-exposed L132 cells (Table 6). Statistically significant increases of IL-8 secretion were observed in the culture supernatants of L132 cells, 24 h after their exposure to PM at their LC_{50} (6.41-fold, $p<0.01$), 48 h after their exposure to PM at their LC_{10} (5.55-fold, $p<0.01$) and LC_{50} (18.67-fold, $p<0.01$), and 72 h after their exposure to PM at their LC_{10} (5.72-fold, $p<0.01$) and LC_{50} (23.37-fold, $p<0.01$), versus control cells. There was no significant change in GM-CSF secretion by L132 cells exposed to PM at both their LC_{10} and LC_{50}, compared to control cells (Table 6). Table 6 shows iNOS activities in control or PM-exposed proliferating L132 cells. There were statistically significant increases of iNOS activity in L132 cells 24 h after their exposure to PM at their LC_{10} or LC_{50} (2.37-fold, $p<0.01$, and 2.81-fold, $p<0.01$, respectively), 48 h after their exposure to PM at their LC_{10} or LC_{50} (2.55-fold, $p<0.01$, and 3.30-fold, $p<0.01$, respectively), and 72 h after their exposure to PM at their LC_{10} or LC_{50} (3.60-fold, $p<0.01$, and 8.88-fold, $p<0.01$, respectively), versus control. The releases of NO in the culture supernatants of control or PM-exposed proliferating L132 cells are shown in Table 6. Statistically significant increases of NO concentration were seen in L132 cells 24 h after their exposure to PM at their LC_{10} or LC_{50} (2.55-fold, $p<0.01$, and 7.18-fold, $p<0.01$, respectively), 48 h after their exposure to PM at their LC_{10} or LC_{50} (2.57-fold, $p<0.01$, and 7.04-fold, $p<0.01$, respectively), and 72 h after their exposure to PM at their LC_{10} or LC_{50} (3.53-fold, $p<0.01$, and 8.21-fold, $p<0.01$, respectively), as compared to control.

PM-exposed L132 cells released IL-6 in a concentration-dependent manner and time-dependent manner (Table 7). Increases of IL-6 secretion were reported in the culture supernatants of L132 cells, 24 h after their exposure to PM at their LC_{50} (1.21-fold, $p<0.01$), 48 h after their exposure to PM at their LC_{10} (1.15-fold, $p<0.01$) and LC_{50} (1.53-fold, $p<0.01$), and 72 h after their exposure to PM at their LC_{10} (1.33-fold, $p<0.01$) and LC_{50} (1.73-fold, $p<0.01$), versus control cells. PM caused concentration-dependent and time-dependent releases of TGF-β1 by L132 cells (Table 7). Accordingly, increases of TGF-β1 concentrations in the culture supernatants of L132 cells were seen 48 h after their exposure to PM at their LC_{50} (1.34-fold, $p<0.01$), and 72 h after their exposure to PM at their LC_{10} (1.32-fold, $p<0.01$) and LC_{50} (1.77-fold, $p<0.01$), compared to control cells.

Table 7. Concentrations of anti-inflammatory mediators (IL-6, TGF-β) in the culture supernatants of PM-exposed L132 cells

	Incubation Time and PM Exposure								
	24 h			48 h			72 h		
	C	LC10	LC50	C	LC10	LC50	C	LC10	LC50
IL-6 (pg/g protein)	1375 1140-1495	1415 [a] 1210-1719	1670 [c] 1494-1985	1229 983-1329	1419 [c] 1336-1534	1890 [c] 1588-2270	1727 1441-1947	2299 [c] 1796-2473	2486 [c] 1933-2678
TGF-β (pg/g protein)	52.4 41.5-62.8	51.3 [a] 39.3-63.3	48.3 [a] 38.6-58.4	97.7 82.0-124.3	99.1 [a] 85.5-113.2	131.4 [c] 113 -145. 7	145.4 109.3-171	258.7 [c] 235.5-285	192.3 [c] 156.5-227

Table 7 shows the concentrations of interleukin-6 (IL-6; pg/mg protein) and transforming growth factor-beta (TGF-β; pg/mg protein) in the culture supernatants of L132 cells 24, 48 or 72 h after their incubation in the continuous presence of particle matter (PM) at their LC_{10} (i.e. 18.84 µg PM/mL) or LC_{50} (i.e. 75.36 µg PM/mL), without renewing the culture media. Non exposed cells were used as negative controls (C). These values are depicted as mean values and ranges (minimum value-maximum value) of 10 replicates for negative controls and 5 replicates for every PM concentrations ([a] = not significant, and [c] = $p<0.01$; Mann-Whitney U test).

Table 8. mRNA expressions of inflammatory cytokines in PM-exposed L132 cells

	Incubation Time and PM Exposure								
	24 h			48 h			72 h		
	C	LC10	LC50	C	LC10	LC50	C	LC10	LC50
tnf-α/ gapdh	0.269 0.224-0.331	0.33 [b] 0.246-0.409	0.481 [c] 0.396-0.569	0.226 0.135-0.328	0.295 [b] 0.237-0.337	0.513 [c] 0.433-0.594	0.272 0.189-0.333	0.381 [c] 0.292-0.477	0.655 [c] 0.530-0.784
il-1β/ gapdh	0.419 0.353-0.49	0.477 [b] 0.382-0.523	0.673 [c] 0.478-0.794	0.374 0.295-0.483	0.541 [c] 0.475-0.595	0.925 [c] 0.738-1.133	0.451 0.340-0.570	0.617 [c] 0.592-0.649	1.006 [c] 0.839-1.155
il-8/ gapdh	1.031 0.922-1.191	1.054 [a] 0.769-1.188	1.356 [b] 0.973-1.685	0.979 0.786-1.182	1.305 [c] 1.129-1.374	1.559 [c] 1.329-1.911	0.915 0.632-1.145	1.349 [c] 1.201-1.535	1.542 [c] 1.338-1.898
gm-csf/ gapdh	0.314 0.225-0.386	0.368 [a] 0.312-0.450	0.393 [c] 0.351-0.421	0.338 0.248-0.471	0.347 [a] 0.300-0.389	0.484 [c] 0.397-0.622	0.385 0.262-0.576	0.381 [a] 0.346-0.413	0.564 [c] 0.404-0.756
il-6/ gapdh	1.171 1.069-1.244	1.154 [a] 1.026-1.313	1.348 [c] 1.196-1.431	1.173 1.076-1.334	1.376 [b] 1.133-1.534	1.643 [c] 1.539-1.853	1.235 1.024-1.373	1.455 [c] 1.324-1.578	1.712 [c] 1.537-1.892
tgf-β1/ gapdh	0.597 0.478-0.691	0.635 [a] 0.487-0.725	0.638 [a] 0.431-0.770	0.600 0.482-0.780	0.601 [a] 0.447-0.707	0.975 [c] 0.711-1.175	0.550 0.409-0.655	0.708 [c] 0.616-0.788	1.144 [c] 0.956-1.352

Table 8 shows the mRNA expressions of tumor necrosis factor-alpha (tnf-α), interleukin-1β (il-1β), interleukin-8 (il-8), granulocyte/macrophage-colony stimulating factor (gm-csf), interleukin-6 (il-6), and transforming growth factor-β1 (tgf-β1) in L132 cells 24, 48 or 72 h after their incubation in the continuous presence of particle matter (PM) at their LC_{10} (i.e. 18.84 μg PM/mL) or LC_{50} (i.e. 75.36 μg PM/mL), without renewing the culture media. Non exposed cells were used as negative controls (C). The ratio between cytokine DNA band and glyceraldehyde-3-phosphate dehydrogenase (gapdh) DNA band are shown as mean values and ranges (minimum value-maximum value) of 10 replicates for negative controls and 5 replicates for every PM concentrations (a = not significant, b = $p<0.05$, and c = $p<0.01$; Mann-

The mRNA expression of inflammatory cytokines (i.e. *TNF-α, IL-1β, IL-8, GM-CSF, IL-6,* and *TGF-β1)* in Dunkerque City's PM-exposed L132 cells, are shown in Table 8. Statistically significant induction of *TNF-α* transcript were observed in PM-exposed L132 cells in a concentration-dependent manner and time-dependent manner. Accordingly, increases in *TNF-α* mRNA expressions were observed in L132 cells, 24 h after their exposure to PM at their LC_{10} (1.23-fold, $p<0.05$) and LC_{50} (1.79-fold, $p<0.01$), 48 h after their exposure to PM at their LC_{10} (1.30-fold, $p<0.05$) and LC_{50} (2.27-fold, $p<0.01$), and 72 h after their exposure to PM at their LC_{10} (1.40-fold, $p<0.01$) and LC_{50} (2.40-fold, $p<0.01$), compared to control cells. Significant increases in *IL-1β* mRNA expressions were seen in L132 cells, 24 h after their exposure to PM at their LC_{10} (1.14-fold, $p<0.05$) and LC_{50} (1.65-fold, $p<0.01$), 48 h after their exposure to PM at their LC_{10} (1.44-fold, $p<0.01$) and LC_{50} (2.47-fold, $p<0.01$), and 72 h after their exposure to PM at their LC_{10} (1.37-fold, $p<0.01$) and LC_{50} (2.23-fold, $p<0.01$), versus control cells. There were concentration-dependent and time-dependent increases of *IL-8* mRNA expressions in PM-exposed L132 cells. Increases of *IL-8* transcript were determined in L132 cells, 24 h after their exposure to PM at their LC_{50} (1.31-fold, $p<0.05$), 48 h after their exposure to PM at their LC_{10} (1.33-fold, $p<0.01$) and LC_{50} (1.59-fold, $p<0.01$), and 72 h after their exposure to PM at their LC_{10} (1.47-fold, $p<0.01$) and LC_{50} (1.69-fold, $p<0.01$), versus control cells. Statistically significant, 1.25-fold, 1.43-fold, and 1.43-fold increases in *GM-CSF* gene expression were reported in L132 cells, 24 h, 48 h and 72 h, respectively, after their exposure to PM at their LC_{50} ($p<0.01$), compared to control cells. Moreover, IL-6 gene expression was enhanced in PM-exposed L132 cells in a concentration-dependent manner and time-dependent manner. Significant increases of *IL-6* mRNA expression were reported in L132 cells, 24 h after their exposure to PM at their LC_{50} (1.15-fold, $p<0.01$), 48 h after their exposure to PM at their LC_{10} (1.17-fold, $p<0.05$) and LC_{50} (1.40-fold, $p<0.01$), and 72 h after their exposure to PM at their LC_{10} (1.21-fold, $p<0.01$) and LC_{50} (1.39-fold, $p<0.01$), versus control cells. PM induced concentration-dependent and time-dependent transcription of *TGF-β1* in L132 cells. Increases of *TGF-β1* mRNA expression in L132 cells were reported 48 h after their exposure to PM at their LC_{50} (1.63-fold, $p<0.01$), and 72 h after their exposure to PM at their LC_{10} (1.29-fold, $p<0.01$) and LC_{50} (2.09 -fold, $p<0.01$), versus control cells.

Effects of PM on the Induction of Cell Death by Apoptosis

Table 9 shows that Dunquerque City's $PM_{2.5}$ activated the TNF-α-induced pathway of apoptosis. Indeed, collected $PM_{2.5}$ induced caspase-8 and -3 activities in control or PM-exposed proliferating L132 cells. There were also increases of caspase-8 activity in L132 cells, 24 h after their exposure to PM at their LC_{10} (1.3-fold, $p<0.05$) and LC_{50} (1.8-fold, $p<0.01$), 48 h after their exposure to PM at their LC_{10} (1.1-fold, $p<0.05$) and LC_{50} (1.6-fold, $p<0.01$), and 72 h after their exposure to PM at their LC_{10} (1.5-fold, $p<0.05$) and LC_{50} (1.9-fold, $p<0.01$), versus control cells. Statistically significant increases of caspase-3 activity were observed in L132 cells, 24 h after their exposure to PM at their LC_{10} (1.6-fold, $p<0.01$) and LC_{50} (3.0-fold, $p<0.01$), 48 h after their exposure to PM at their LC_{10} (1.6-fold, $p<0.01$) and LC_{50} (3.1-fold, $p<0.01$), and 72 h after their exposure to PM at their LC_{10} (2.1-fold, $p<0.01$) and LC_{50} (3.0-fold, $p<0.01$), as compared to control cells. There were significant correlation between TNF-α concentration and caspase-8 activity, on the one hand, and

Table 9. Effects of PM on the TNFα-Induced Apoptotic Pathway and the Mitochondrial Apoptotic Pathway

	Incubation Time and Exposure								
	24 h			48 h			72 h		
	C	LC10	LC50	C	LC10	LC50	C	LC10	LC50
Caspase 8 (rfu/mg protein)	3.63 2.53-5.53	4.76 [b] 3.96-5.82	6.38 [c] 5.34-7.69	3.63 2.73-4.80	4.17 [b] 3.59-4.88	5.93 [c] 4.15-7.42	2.90 2.06-3.95	4.33 [b] 2.93-5.33	5.43 [c] 4.04-6.78
Caspase 3 (rfu/mg protein)	0.69 0.58-0.75	1.08 [c] 0.91-1.29	2.05 [c] 1.73-2.37	0.71 0.56-0.91	1.15 [c] 0.89-1.31	2.21 [c] 1.72-3.02	0.65 0.42-0.92	1.34 [c] 1.05-1.80	1.95 [c] 1.73-2.09
Caspase 9 (rfu/mg protein)	2.27 1.75-3.26	2.74 [c] 2.23-3.36	3.99 [c] 3.24-4.78	2.44 1.82-4.45	3.32 [c] 2.11-4.71	4.27 [c] 3.37-4.89	2.02 1.55-2.85	3.00 [c] 2.41-4.02	4.30 [c] 3.75-5.03
Cyt.c $_M$ (ng/mg protein)	27.10 20.49-39.03	20.60 [b] 15.82-28.69	16.15 [c] 14.19-19.41	16.60 13.02-19.31	13.72 [b] 10.91-16.90	11.07 [c] 9.21-12.95	16.05 11.99-22.08	7.55 [d] 6.56-8.54	6.82 [c] 5.98-8.16
Cyt.c $_C$ (ng/mg protein)	6.77 4.08-9.42	10.12 [b] 8.77-12.66	12.35 [c] 10.19-13.42	5.64 3.06-7.49	7.38 [b] 5.44-8.77	9.77 [c] 8.45-12.26	6.49 5.09-8.27	7.89 [b] 6.31-9.26	10.47 [c] 9.52-11.76
Cyt.c $_{C/M}$	0.26 0.18-0.45	0.51 [c] 0.35-0.63	0.77 [c] 0.65-0.95	0.34 0.18-0.50	0.56 [c] 0.32-0.80	0.90 [c] 0.72-1.33	0.41 0.34-0.50	1.05 [c] 0.92-1.35	1.54 [c] 1.27-1.69

Table 9 shows the caspase-8, -3 and -9 activities (rfu/mg protein), and the cytochrome c (Cyt.c) concentrations (ng/mg protein) in mitochondrial and cytosolic fractions (Cyt.c $_M$ and Cyt.c $_C$, respectively), in L132 cells after their incubation in the continuous presence of particle matter (PM) at their at their Lethal Concentration at 10% (LC_{10}; 18.84 µg PM/mL) or at 50% (LC_{50}; 75.36 µg PM/mL) for 24, 48 or 72 h of incubation, without renewing the culture media. Non-exposed cells were used as negative controls (C). These values are depicted as mean values and ranges (minimum value-maximum value) of 10 replicates for negative controls and 5 replicates for every PM concentrations. ([a] = not significant, [b] = $p<0.05$, and [c] = $p<0.01$; Mann-Whitney U test).

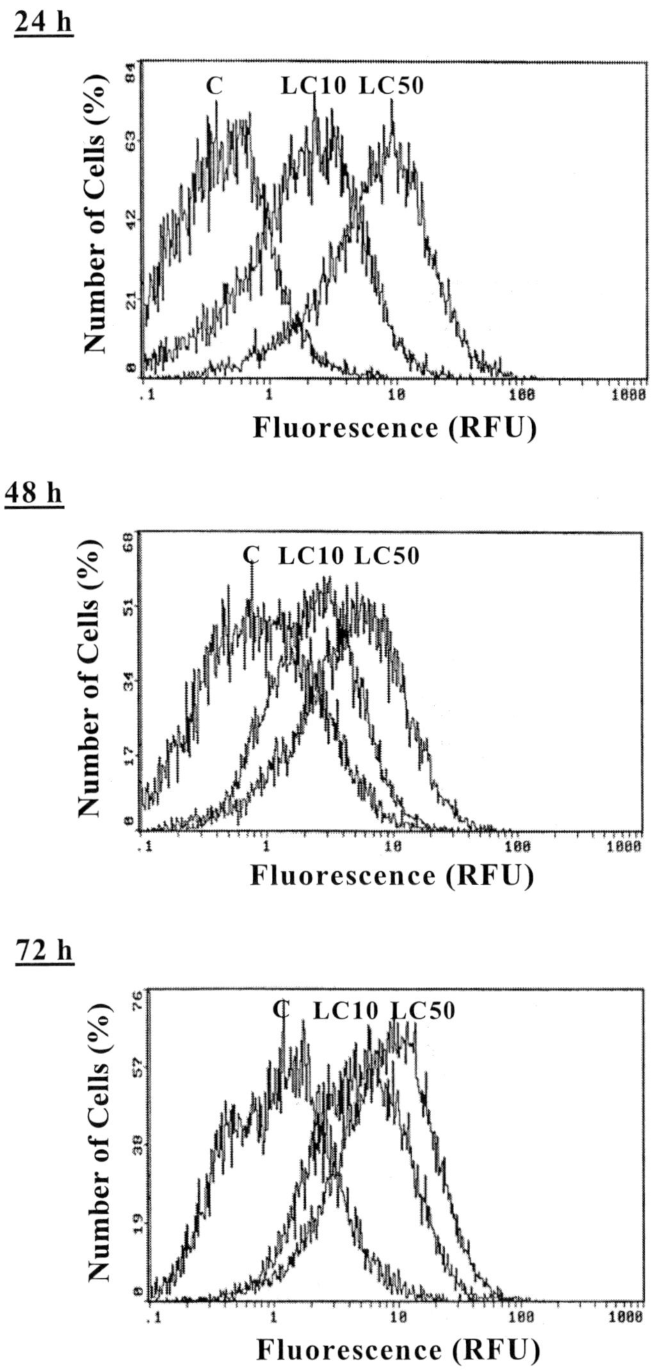

Figure 9. Terminal deoxynucleotidyl transferase mediated deoxyUridine-5'-triphosphate-biotin Nick-End Labeling (TUNEL) of L132 cells after their incubation in the continuous presence of particle matter (PM) at their LC_{10} (i.e. 18.84 μg PM/mL) or LC_{50} (i.e. 75.36 μg PM/mL) for 24, 48 or 72 h, without renewing the culture media. Non-exposed cells were used as negative controls (C).

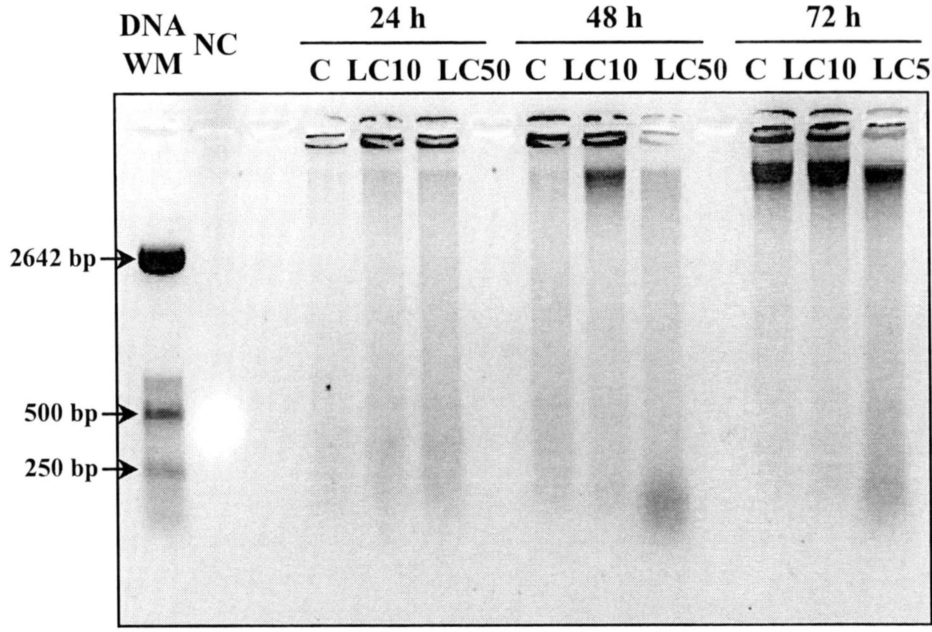

Figure 10. DNA laddering of L132 cells after their incubation in the continuous presence of particle matter (PM) at their LC_{10} (i.e. 18.84 µg PM/mL) or LC_{50} (i.e. 75.36 µg PM/mL) for 24, 48 or 72 h, without renewing the culture media. Non-exposed cells were used as negative controls (C). High molecular weight DNA remained trapped in wells, while DNA that had been digested during apoptosis migrated through the gel and was resolved into multiples of 180-200 base pair fragments (DNA weight marker, DNA WM; base pair, bp).

The mRNA expression of *P53, BCL-2* and *BAX* genes in Dunkerque City's PM-exposed L132 cells are summarized in Table 10. Statistically significant induction of *P53* transcript were observed in L132 cells, 24 h after their exposure to PM at their LC_{10} (1.45-fold, $p<0.05$) and LC_{50} (1.51-fold, $p<0.05$), 48 h after their exposure to PM at their LC_{10} (1.32-fold, $p<0.01$) and LC_{50} (1.29-fold, $p<0.01$), and 72 h after their exposure to PM at their LC_{10} (1.16-fold, $p<0.05$) and LC_{50} (1.23-fold, $p<0.05$), compared to control cells. In contrast, decreases in *BCL-2* mRNA expressions were seen in L132 cells, 24 h after their exposure to PM at their LC_{10} (0.64-fold, $p<0.05$) and LC_{50} (0.64-fold, $p<0.05$), 48 h after their exposure to PM at their LC_{10} (0.68-fold, $p<0.05$) and LC_{50} (0.73-fold, $p<0.05$), and 72 h after their exposure to PM at their LC_{10} (0.82-fold, $p<0.05$) and LC_{50} (0.76-fold, $p<0.05$), versus control cells. In return, increases of *BAX* transcript were determined in L132 cells, 48 h after their exposure to PM at their LC_{10} (1.63-fold, $p<0.05$) and LC_{50} (1.64-fold, $p<0.01$), and 72 h after their exposure to PM at their LC_{10} (1.61-fold, $p<0.01$) and LC_{50} (1.62-fold, $p<0.01$), versus control cells. Statistically significant increases of the ratio between the transcripts of *BAX* and *BCL-2* were reported in L132 cells, 24 h after their exposure to PM at their LC_{10} (1.62-fold, $p<0.01$) and LC_{50} (1.78-fold, $p<0.01$), 48 h after their exposure to PM at their LC_{10} (2.83-fold, $p<0.01$) and LC_{50} (2.58-fold, $p<0.01$), and 72 h after their exposure to PM at their LC_{10} (1.91-fold, $p<0.01$) and LC_{50} (2.19-fold, $p<0.01$), compared to control cells.

Table 10. Effects of PM on mRNA Expressions of P53, Bcl-2, and Bax Genes

	Incubation Time and Exposure								
	24 h			48 h			72 h		
	C	LC10	LC50	C	LC10	LC50	C	LC10	LC50
p53/β-actine	0.585 0.507-0.835	0.850 [b] 0.723-1.197	0.883 [b] 0.698-1.118	0.657 0.502-0.888	0.864 [c] 0.715-1.015	0.845 [c] 0.748-0.932	0.849 0.725-0.956	0.982 [b] 0.963-1.012	1.045 [b] 0.912-1.230
bcl-2/β-actine	1.346 0.972-1.802	0.855 [b] 0.621-1.018	0.858 [b] 0.667-1.012	1.175 0.883-1.536	0.804 [b] 0.525-1.009	0.862 [b] 0.711-1.049	1.190 0.981-1.556	0.977 [b] 0.875-1.071	0.900 [b] 0.645-1.022
bax/β-actine	0.820 0.534-0.971	0.868 [a] 0.607-1.423	0.979 [a] 0.707-1.311	0.815 0.555-0.989	1.329 [b] 0.947-1.673	1.333 [c] 1.054-1.708	0.836 0.628-1.158	1.349 [c] 1.238-1.602	1.354 [c] 1.041-1.546
bax/bcl-2	0.636 0.510-0.954	1.032 [b] 0.655-1.468	1.131 [c] 0.898-1.296	0.602 0.415-0.766	1.706 [c] 1.126-2.185	1.552 [c] 1.322-1.850	0.707 0.576-0.996	1.351 [c] 1.198-1.662	1.546 [c] 1.024-2.006

Table 10 shows the mRNA expressions of *P53, BCL-2*, and *BAX* genes in L132 cells after their incubation in the continuous presence of particle matter (PM) at their at their Lethal Concentration at 10% (LC_{10}; 18.84 µg PM/mL) or at 50% (LC_{50}; 75.36 µg PM/mL) for 24, 48 or 72 h of incubation, without renewing the culture media. Non-exposed cells were used as negative controls (C). The ratio between *P53, BCL-2*, or *BAX* DNA band, on the one hand, and the *β-ACTIN* DNA band are shown as mean values and ranges (minimum value-maximum value) of 10 replicates for negative controls and 5 replicates for every PM concentrations. ([a] = not significant, [b] = $p<0.05$, and [c] = $p<0.01$; Mann-Whitney U test).

Moreover, statistically significant correlations were observed between the mRNA expressions of *P53* and *BCL-2* (Spearman's rho= - 0.438, $p<0.01$),the transcripts of *P53* and *BAX* (Spearman's rho = 0.474, $p<0.01$), and the mRNA expression of *p53* and the ratio between the transcripts of *BAX* and *BCL-2* (Spearman's rho = 0.620, $p<0.01$).

Figure 9 shows Dunkerque city's PM-induced L132 caspase-dependent apoptosis as studied by TUNEL. Accordingly, flow cytometry study indicated increases of fluorescent intensity related to fluorescein labels of genomic DNA strand breaks in L132 cells exposed to PM at their LC_{10} and LC_{50}. DNA laddering showed, in a concentration- and time- dependent manner, the formation of a high molecular weight fragments in L132 cells (Figure 10). The presence of a low molecular weight fragments was reported in L132 cells 48 h and 72 h after their exposure to PM at their LC_{50}.

Discussion

In this work, the main physical and chemical characteristics of Dunkerque City's PM were determined. Size distribution results of collected PM showed that 92.15 % of PM total number were equal or smaller than 2.5 µm, thereby suggesting that they are human respirable. According to epidemiological findings from several cities around the world, the greatest health risks correlate with smaller PM which have the ability to reach the distal regions of the lung after inhalation [16]. Components such as transition metals, VOC, and PAH are known to be associated with PM, either through adherence to the core particle or as an integral component of the particle [34]. A total of 20 inorganic elements were found in Dunkerque City's PM. Among them, Fe, Al, Ca, Na, Pb, Mg, and Mn were the most abundantly present and constitute not only inorganic elements usually associated with natural environment (i.e. Ca, Mg, Na), but also so-called anthropogenic elements (i.e. Al, Mn, Pb, Zn). The specific surface area (i.e. 1 m^2/g for collected PM, but increased until 50 m^2/g for PM having undergone a thermal desorption at 400°C under secondary vacuum) of the PM under study showed the adsorption of various chemicals. Accordingly, VOC and PAH-coated onto PM were notably detected. It is also interesting to report that, according to the considered PAH, PAH concentrations in Dunkerque City's PM were equal or smaller than those measured in PM samples collected in two locations in Paris [13]. Taken together, the results arising from the physical and chemical characterization of Dunkerque City's PM suggest that much of the collected PM derived from wind-borne dust from the industrial complex and the heavy diesel motor vehicle traffic. The chemical composition of Dunkerque City's PM was also closely related to the anthropogenic emission sources located near the sampling site; apart the heavy motor diesel vehicle traffic, petrochemistry, metallurgy, iron and steel industry, chemical synthesis, pharmaceutical formulation, and agroalimentary factory must be cited. Hence, the Dunkerque City's PM consisted in a very complex mixture of chemicals, generally present at relatively low doses, thereby rendering very difficult the share of toxicity ascribable to the different fractions (inorganic vs organic, notably).However, Baulig et al. (2004) reported that the physical and chemical properties of particles revealed to be unsatisfactory to predict their levels of biological reactivity [13]. Probably, the different signaling pathways elicited by each particle component (inorganic elements, organic compounds, etc.) interact and/or converge to give a similar cellular response. Hence, to better evaluate the adverse health effects of PM exposure, it will be of great interest to consider not only the physical and chemical characteristics of PM, but also to determine their toxicological reactivity in target cells.

The adverse effects of ambient PM on human health have been a concern for decades [2]. Attempts to provide linkages between adverse health effects and PM properties have significantly increased [9]. However, despite intensive investigation, the underlying mechanisms of PM-related adverse health effects are still incompletely understood, but a hypothesis currently under investigation is that organic compounds-coated onto PM could played a key role in their toxicity [10, 11, 34]. We showed, therefore, that Dunkerque City's PM lead to concentration-dependent cytotoxicity in proliferating L132 cells, as determined using the colony forming method. PM induced concentration-dependent and time-dependent increases of released LDH activity. Statistically significant increases of extracellular LDH activity were observed in L132 cells 72 h after their exposure to the highest PM concentrations, thereby suggesting late alterations of membrane integrity and/or permeability. In contrast, PM seemed to early affect the MDH activity, which is a measure of both the cell toxicity and the cell proliferation, since the formazan conversion depends on the amount as well as the mitochondrial metabolic competence of the cells. Indeed, PM caused decreases of MDH activity in a concentration and a time-dependent manner, and maximal decrease of MDH activity was attained in L132 cells exposed to LC_{50}, 72 h after their incubation. Hence, these results partly correlate to the observed concentration and time dependent growth alteration assessed through the survival rate analysis. Taken together, these results showed that the PM under study induced a global cytotoxicity in the considered lung in-vitro system.

Experimental and/or epidemiological findings from several authors indicate that lung diseases are often associated with prooxidant/antioxidant imbalance, and there is increasing evidence that particulate air pollutants induce acute responses as well as exacerbate existing oxidative damage in the lung [11, 15, 29-31]. Hence, we investigated the capacity of Dunkerque City's $PM_{2.5}$ to induce oxidative stress conditions in this lung cell target. Firstly, we found that in-vitro exposure to these PM induce lipid peroxidation in a concentration and a time-dependent manner. This result is consistent with the data about LDH activity suggesting the occurrence of significant alteration of membrane fluidity or integrity only 72h after cell exposure to PM. Significant changes in SOD activity (i.e. initially increases then decreases) were reported in PM-exposed L132 cells, versus control cells. Accordingly, it has already been described that the ROS generated in oxidative stress conditions could alter proteins and thereby damage some antioxidant enzymes [73-76]. The literature also indicated that SOD is not particularly protected against ROS and that this process could contribute to its inactivation. Since SOD is principally a protective enzyme, its depletion or lowered activity can lead to adverse effects through the accumulation of ROS in cells and oxidative damages in biological macromolecules [68, 69, 78]. Among them, DNA is oxidized to a number of damaged products in-vitro and one of these products, 8-OHdG, has been suggested for a marker of oxidative damage in DNA in-vitro and in-vivo [79]. We showed that PM induced a concentration and time-dependent formation of 8-OHdG in L132 cells. According to Shi et al. (2003), PM are able to generate hydroxyl radicals and therefore to induce formation of 8-OHdG [80]. Such an excessive production of oxidative DNA adducts seemed to cause a concentration and time-dependent activation of PARP in PM-exposed cells. PARP activation and Poly(ADP-ribose) polymerization are also proposed to be important for the regulation of many cellular processes such as DNA repair, cell death, chromatin functions, and genomic stability [81]. The activation of PARP is an immediate cellular reaction to DNA strand breakage as induced by alkylating agents, ionizing radiation or oxidants [82-84]. Accordingly, recent data in the literature reported that some of the main covalent metals (i.e. Fe, Al, Pb, Mn, Zn) detected in Dunkerque City's PM could be firmly involved in redox systems, which can lead to the initiation of radical reactions [29, 30, 68, 69, 80, 86]. The associations between metals in PM soluble fractions and lung responses have been observed in in-vivo and in-vitro systems [86]. Ghio and Devlin (2001) have suggested that the disparate effects of PM extracts seen in their study resulted from physical or chemical differences other than metal content and catalyzed oxidants cannot be excluded [30]. In addition, it is now well known that the metabolic activation by enzyme-catalyzed reactions of the VOC and/or the PAH-coated onto PM could result in the excessive production of ROS capable of interfering with cell homeostasis [10, 68, 69].

Recent evidence have shown that ROS play a critical role in the activation of cells as messengers in the signal transduction pathways leading to gene expression and /or further protein secretion [31, 37, 38]. Increased activation of an oxygen-sensitive transcription factor called NF-κB is thought to occur in several inflammatory conditions, and is thereby involved in the regulation of a large number of genes that control various aspects of the inflammatory response [31, 37-39]. We showed that L132 exposure to collected PM resulted in a concentration- and/or time-dependent activation of NF-κB/IκB, as evaluated through the phosphorylation on Ser536 of p65 subunit of NF-κB, on the one hand, and the phosphorylation on Ser32 of IκB-α, one the other hand. Further evidence in support of this idea was provided by the observation that DNA binding activity of NF-κB (i.e. p65 and p50 subunits) was increased in PM-exposed-L132 cells, in a concentration-dependent manner and time-dependent manner. According to the literature, ROS-induced activation of NF-kB involves the activation of IKK, which in turn leads to the phosphorylation, ubiqutination, and degradation of IkB-α, leading to NF-kB activation [36, 3è, 47]. Taken together, these results suggested that, in L132 cells, NF-kB/IκB activation was clearly linked to the presence of air pollution PM, and to the generation of oxidative damage. The fact that PM of various types (i.e. $PM_{2.5}$, PM_{10}, organic dust, organic chemical, residual oil fly ash, etc.) activates NF-κB has been addressed by other authors [10, 19, 23, 34, 38, 87, 88]. More accurately, in these experiments, the activation of NF-κB by such types of PM might rely on the presence of inorganic elements (i.e. Fe, Al, Pb, Mn, Zn) and/or organic compounds (VOC, PAH, etc.), which were detected in Dunkerque City's PM. The pathways to NF-κB activation are, however, extremely complex and vary with agent, cell type, and specific conditions [87].

The mechanisms for PM-induced adverse health effects are still not well understood, but inflammation seems to be of importance [10, 28, 29]. The pathogenesis of lung injury induced by air pollutants is also mediated in large part by ROS and inflammatory cytokines that might play a crucial role in both the occurrence and the development of acute inflammation [11, 15, 19]. In addition to their oxidative effects, ROS have been implicated in the regulation of NF-κB, a transcription factor firmly involved in the genesis of inflammatory diseases [19, 32, 33, 42, 43]. While we have reported that Dunkerque city's $PM_{2.5}$ could generate oxidative stress events and the subsequent activation of NF-κB in L132 cells, we investigate their potential role on protein secretion and/or gene transcription of inflammatory cytokine in these relevant target cells. One of the main findings was that *in-vitro* exposure to Dunkerque city's $PM_{2.5}$ lead to significant increases in a concentration-dependent manner and in a time-dependent manner in both the protein secretion and/or the gene expression of members of the coordinated network of pro- and anti-inflammatory cytokines (i.e. *TNF-α, IL-1β, IL-8, GM-CSF, IL-6,* and *TGF-β1*). In addition, it is important to highlight that the mRNA expressions of inflammatory mediators, except for *GM-CSF*, firmly corroborate PM-induced profiles of cytokine secretion. Hence, there seems an apparent discrepancy between the results on the protein secretion and the gene expression of *GM-CSF*. However, cytokine gene expression generally occurs through post-transcriptional mechanisms [89]. Several authors have reported the existence of inhibitory pathways that post-transcriptionally regulate GM-CSF expression [91, 92]. Their data strongly suggested a model for gm-csf regulation involving post-transcriptional and/or translational mechanisms in A549 cells [44, 90, 91]. Taken together, the present findings are consistent with data reported by other authors, that showed TNF-α play a key role in the particle-induced inflammation in the lung and upregulate with IL-1β the production of secondary mediators including IL-6, IL-8, GM-CSF, and TGF-β1 by lung cells [10, 92, 93]. We observed that the studied anti-inflammatory cytokines (i.e. IL-6 and TGF-β1) are constitutively and/or differently expressed, and their excessive secretion in PM-exposed cells suggest a negative feedback in inflammation [92-94]. Extremely high rates of IL-6 illustrate the installation of such a negative feedback [92, 95, 96]. Stimulated cells are also known to release large quantities of NO as a direct consequence of iNOS activation by TNF-α [46]. Accordingly, we showed that PM induced increases of iNOS activity in a concentration and time-dependent manner in PM-exposed cells. Accordingly, many other *in-vitro* studies have shown that different *in-vitro* lung models are capable to induce inflammatory reactions in response to anthropogenic stimuli (VOC, PAH, organic dust, mineral particle, diesel exhaust particles, urban PM_{10}) [7, 10, 11, 18, 28, 32-34, 42, 45, 68-70, 90, 97-103]. In agreement with the literature, in these studies, the induction of excessive cytokine secretion, and, therefore, the development of the inflammatory response might rely on the ability of these air pollutants to generate ROS, which have been firmLy implicated in the stringent regulation of NF-κB, and therefore, of the cytokine network [89]. It has been proposed that effects of PM-exposure should be related to their physical and chemical characteristics, on one hand, and their potency to induce inflammatory effects that aggravate according to time, in the other hand [28, 86]. Recent data in the literature reported also that some of the covalent metals (i.e. Fe, Pb, Cu, Mn, etc.) present in Dunkerque City's $PM_{2.5}$ could be involved in redox systems, which can lead to the initiation of radical reactions [29, 68-70, 86]. The association between metals in soluble fractions of particles and pulmonary responses have been observed *in-vivo* and *in-vitro* systems [86, 73, 101, 102]. In addition, the metabolic activation of some VOC, semi-VOC or PAH by enzyme-catalyzed reactions results in the excessive production of ROS capable of interfering with cell homeostasis [10, 68-70, 103, 104]. Some inherent characteristics of the soluble and/or insoluble fraction of urban air PM are also crucial for the level of cytokine protein secretion and/or gene expression.

In agreement with the literature, high levels of ROS also produce a change in the redox status of the cell, thereby triggering a cascade of events associated with inflammation and, at higher concentrations, apoptosis [29, 34, 35]. Although many pathways of apoptosis may exist, only two main caspase cascades have been elucidated in detail in mammalian cells [51, 105, 106]. Hence, we focused our attention on the activation of these two main caspase cascades by Dunkerque City's $PM_{2.5}$. We showed that in-vitro short-term exposure to air pollution PM activated the TNF-α-induced pathway of apoptosis. Dunkerque City's PM also induced TNF-α secretion by L132 cells, and the subsequent activation of caspase-8 and -3 activities. There were also significant correlation between TNF-α concentration and caspase-8 activity, on the one hand, and caspase-8 activity and caspase-3 activity, on the other hand. The results regarding the TNF-α-induced pathway of apoptosis by air pollution PM are in agreement with other observations previously reported using in-vitro systems by Nel et al. (2006) [26]. In other respects, we found that air pollution PM activates the mitochondrial pathway of apoptosis. Indeed, concentration and time-dependent formation of 8-OHdG was reported in PM-exposed proliferating L132 cells, thereby indicating the occurrence of oxidative stress conditions. Several authors have reported that oxidant-induced DNA damage is a potent stimulus of mitochondrial dysfunction as detected by the release of agents normally confined to the mitochondrial intermembrane space, such as Cyt. c and is critical to caspase activation and, therefore, apoptosis propagation [26, 35, 107]. We also showed that in-vitro short-term exposure to Dunkerque City's PM induced the release of Cyt. c, a necessary component of non-receptor-mediated-caspase activation, from the mitochondria into the cytosol, and the subsequent activation of caspase-9 and -3. Statistically significant correlations were seen between 8-OHdG formation, Cyt. c release, and caspase-9 and -8 activities. One of the novel findings of the present work was, therefore, that Dunkerque City's PM could at least induce apoptotic events in human lung epithelial cells through the activation of both the extrinsic and the intrinsic pathways. In agreement with other results, in-vitro short term exposure of target lung cell to some of the metals (i.e. Fe, Al, Pb, Mn, Zn) or the organic chemicals (i.e. VOC, PAH, etc.) detected in Dunkerque City's $PM_{2.5}$ have already been demonstrated as inducing oxidative stress conditions, inflammatory response, and/or apoptosis in lung target systems [7, 11, 18, 28, 33, 34, 45, 68-70, 90, 97-103, 108, 109].

In other respects, we observed not only a transcriptional induction of pro-apoptotic p53 gene, but also an imbalance between the respective transcription rates of bax and bcl-2 genes in PM-exposed cells. Accordingly, several lines of evidence, supportive of a key role for P53, Bcl-2, and Bax in the control of cell cycle, the regulation of downstream caspase activity, and the subsequent induction of apoptosis, have arisen [35, 53]. Reisman and Loging (1997) have described that TNF-α enhanced p53 mRNA expression [110]. In addition, in-vitro and developing lung ex vivo studies have reported that the reduced transcription of bcl-2 gene, coding for a mitochondrial targeted anti-apoptotic protein, could not yet inhibit Cyt. c release from mitochondria and protect against oxidative stress-induced apoptosis [47].

Evidence presented herein extended these observations by showing that the TNF-α and mitochondria-regulated apoptotic pathways mediated DNA fragmentation in PM-exposed proliferating L132 cells. Results arising from TUNEL, DNA laddering, and PARP activity firmly supported the findings that in-vitro short-term exposure to Dunkerque City's PM induced DNA fragmentation. It is now well known that the activation of downstream effector caspase -3 resulted in the cleavage of DNA [111]. With other respects, DNA laddering showed the presence of both the low and the high molecular weight fragments that indicates the activation of random caspase-dependent and independent DNA fragmentation in PM-exposed human lung epithelial cells. Nevertheless, the relative contributions of caspase-dependent and caspase-independent post-mitochondrial events to mediate cell death by apoptosis remain a matter of debate.

However, before concluding on the above-presented results, caution is necessary when tackling the question of the helpfulness of in-vitro models to improve the extrapolation to the experiments in animals and/or the situations in humans. Further consideration of this key question indicated a general lack of information about the potential uses of data generated from in-vitro systems in the risk assessment process. Scientific statements have notably highlight the potential discrepancies between the mechanisms of action occurring in in-vitro systems and those appearing in animal systems, and whether the in-vitro observations have any real relevance to the mechanisms of PM-induced health effects. Moreover, there are other limitations of such in-vitro approaches in toxicology. For example, it is often very difficult to extrapolate from the dose applied to individual cells in culture to the inhaled doses that reach the epithelial cells in animals and/or humans. Indeed, it is likely that the onetime application of PM to a target cell is remarkably different than short-term or long-term inhalation exposures of animals and/or humans. The concerns of extrapolating from cells also should be kept in mind prior to claiming that these approaches explain the mechanisms of PM toxicity. However, despite these limitations, in-vitro experiments continue to be a prerequisite for a preliminary understanding of mechanism of action that, in general, is currently obtainable only through animal and/or human studies. The observations reported in in-vitro approaches require also further validation through experiments in animals and studies in humans before to be validated.

Conclusion

To contribute to improve the knowledge of the adverse health effects induced by air pollution PM, we have undertaken an extensive investigation of the underlying mechanisms of action nduced by air pollution $PM_{2.5}$ collected in Dunkerque, a French seaside City located on the southern coast of the North-Sea, and characterized by the proximity of industrial activities (i.e. iron and steel industry, aluminum industry, oil refinery, basic chemistry, pharmaceutical industry, plant health production, food industry, etc.) and the heavy diesel motor vehicle traffic a French seaside City strongly industrialized. The physical and chemical properties of Dunkerque City's PM suggest that much of the collected PM derived from wind-borne dust from the industrial complex, and the heavy diesel motor vehicle traffic. The results reported in the present chapter show that the collected Dunkerque City's $PM_{2.5}$ is a complex and heterogeneous mixture of various inorganic and organic chemicals. The high concentrations of transition metals (i.e. Fe, Al, Pb, Mn, Zn) and the presence of organic compounds (i.e. VOC and PAH) found in the collected PM under study might be firmly involved in the induction of different underlying mechanisms of action, which are closely associated with the occurrence of adverse lung effects in humans. Taken together, these results showed that the occurrence of oxidative stress conditions, the subsequent activation of NF-κB/IκB complex, and, therefore, the production of various inflammatory mediators and the subsequent development of pro-apoptotic events, that aggravate according to PM concentration and/or incubation time, proceeded cytotoxicity in ambient air pollution $PM_{2.5}$-exposed target L132 cells in culture. It is greatly possible that there are important interactive effects of the components that form the complex mixture of PM in determining the overall adverse health effects of the PM.

With other respects, the example of the multi-step in-vitro study reported in this chapter shows that in-vitro studies can be a very useful approach to identify the underlying mechanisms of action by which air pollutant PM induce adverse health effects in humans. In-vitro studies are ideally suited for rapid screening of many pollutants, or of components of complex mixtures, as such as air pollutants PM. They are relatively inexpensive, especially compared with in vivo studies, and can be performed rapidly. The major limitation of in vitro studies is that cells have been removed from their normal environment. There are no neighboring cells or tissues to interact with and no blood to supply potentially important factors or nutrients. In addition it can be difficult to expose cells in a manner that mimics in-vivo exposure. Hence, it must be demonstrated that cultured cells respond in a similar manner to air pollution PM as lung target cells in-vivo. Nevertheless, many cellular pathways are activated similarly both in-vitro and in-vivo and carefully conducted in-vitro experiments can be performed using relevant primary cells and exposure scenarios. However, the burden is always on the investigator to demonstrate the relevance of results obtained from in-vitro studies, preferably by showing that similar results can be obtained using selected in vivo experiments. It is also of fundamental importance to keep in mind that no extrapolation of in-vitro exposure data to improve human and animal exposure data will be possible without respecting careful restriction. Accordingly, in the due course of the multi-step in-vitro approach we reported in this chapter, the induction of the underlying mechanisms of action (i.e. oxidative stress, NF-κB activation, inflammatory response, and apoptosis) under study by air pollutant PM in exposed-animals will be one of the major challenges in further studies. In future work, rodent short-term exposure to collected Dunkerque City's $PM_{2.5}$ will be achieve, the sum of which will larger contribute to elucidate the underlying mechanisms of action which are in-vivo involved in the development of the respiratory disorders occurring following PM-exposure., and, thereafter, to put forward biological markers of exposure and/or effects for detecting PM-induced adverse health effects in the lung at early events, and ideally at subclinical reversible changes. A successful prevention of lung diseases induced by occupational or environmental exposure to air pollution PM, therefore, largely relies on the capability to detect adverse lung effects at a stage when they are still reversible or at least not yet compromising lung function. Moreover, knowledge about the predictive value of these biological markers will be particularly useful for preventive purposes.

Clearly, epidemiology studies are the most relevant and in an ideal world would provide all the information needed for accurate risk assessment. However, in reality this is not the case and controlled human and animal exposure studies are needed to complement epidemiology studies by providing biological plausibility and a better understanding of the underlying pathophysiological mechanisms that are responsible for causing an adverse health event reported in epidemiology studies. Finally, well-conducted in-vitro studies have the potential to characterize cellular and molecular mechanisms that underlie the pathophysiological processes, notably caused by air pollution PM exposure.

Acknowledgments

The research described in this article has been supported by grants from the University of Littoral-Opale Coast, the Franco-British Interreg IIIA European Program (French Convention n°24), the Research Institute of Industrial Environment, the Urban Community of Dunkerque, the French Ministry for Higher Education and Research (Convention n°16848-2005), and the Regional Council of Nord/Pas-de-Calais (Convention n°04 06 0092).

REFERENCES

[1] Pope, C.A., Hansen, M.L., Long, R.W., Nielsen, K.R., Eatough, N.L., Wilson, W.E., Eatough, D.J. (2004). Ambient particulate air pollution, heart rate variability, and blood markers of inflammation in a panel of elderly subjects. *Environ. Health Perspect., 112*, 339-345.

[2] Brunekreef, B. & Holgate, S. (2002). Air pollution and health. *Lancet, 360*, 1233-1242.

[3] Levy, J.I., Hammit, J.K., Spengler, J.D. (2000). Estimating the mortality impacts of particulate matter: what can be learned from between-study variability? *Environ. Health Perspect., 108*, 109-117.

[4] Pope, C.A. & Kanner, R.E. (1993). Acute effects of PM_{10} pollution on pulmonary function of smokers with mild to moderate chronic obstructive pulmonary disease. *Am. Rev. Respir. Dis., 147*, 1336-1340.

[5] Schwartz, J., Slater, D., Larson, T.V., Pierson, W.E., Koenig, J.Q. (1993). Particulate air pollution and hospital emergency room visits for asthma in Seattle. *Am. Rev. Respir. Dis., 147*, 826-831.

[6] Schwartz, J., Dockery, D.W., Neas, L.M. (1996). Is daily mortality associated specially with fine particles? *J. Air Waste Manag. Assoc., 46*, 927-939.

[7] Alfaro-Moreno, E., Martinez, L., García-Cuellar, C., Bonner, J.C., Murray, J.C., Rosas, I., Ponce-de-Leon Rosales, S., Miranda, J., Osornio-Vargas, A.R. (2002). Biological effects induced in vitro by PM10 from three different zones of Mexico city. *Environ. Health Perspect., 110*, 715-720.

[8] Brunekreef, B. & Forsberg, B. (2005). Epidemiological evidence of effects of coarseairborne particles on health. *Eur. Respir. J., 26*, 309-318.

[9] Englert, N. (2004). Fine particle and human health - a review of epidemiological studies. *Toxicol. Lett., 149*, 235-242.

[10] Liden, J., Ek, A., Palmberg, L., Okret, S., Larsson, K. (2003). Organic dust activates NF-κB in lung epithelial cells. *Respir. Med., 97*, 882-892.

[11] Osornio-Vargas, A.R., Bonner, J.C., Alfaro-Moreno, E., Martinez, L., Garcia-Cuellar, C., Ponce-de-Leon Rosales, S., Miranda, J., Rosas, I. (2003). Proinflammatory and cytotoxic effects of Mexico city air pollution particulate matter in vitro are dependent on particle size and composition. *Environ. Health Perspect., 111*, 1289-1293.

[12] Buschini, A., Cassoni, F., Anceschi, E., Pasini, L., Poli, P., Rossi, C. (2001). Urban airborne particulate: genotoxicity evaluation of different size fractions by mutagenesis tests on microorganisms and comet assay. *Chemosphere, 44*, 1723-1736.

[13] Baulig, A., Poirault, J.J., Ausset, P., Schins, R., Shi, T., Baralle, D., Dorlhene, P., Meyer, M., Lefevre, R., Baeza-Squiban, A., Marano, F. (2004). Physicochemical characteristics and biological activities of seasonal atmospheric particulate matter sampling in two locations of Paris. *Environ. Sci. Technol., 38*, 5985-5992.

[14] Harrison, R.M. & Yin, J. (2000). Particulate matter in the atmosphere: which particle properties are important for its effects on health? *Sci. Total Environ., 249*, 85-101.

[15] Spurny, K.R. (1998). On the physics, chemistry and toxicology of Ultrafine Anthropogenic Atmospheric Aerosols (UAAA): New advances; *Toxicol. Lett., 96/97*, 253-261.

[16] Levy, J.I., Hammit, J.K., Spengler, J.D. (1999). Estimating the mortality impacts of particulate matter: what can be learned from between-study variability. *Environ. Health Perspect.*, *108*, 109-117.

[17] Diociaiuti, M., Balduzzi, M., De Berardis, B., Cattiani, G., Stacchini, G., Ziemacki, G., Marconi, A., Paoletti, L. (2001). The two PM (2.5) and PM (2.5-10) coarse fractions: evidence of different biological activity. *Environ. Res.*, *86*, 254-262.

[18] Monn, C. & Becker, S. (1999). Cytotoxicity and induction of proinflammatory cytokines from human monocytes exposed to fine (PM2.5) and coarse particles (PM10–2.5) in outdoor and indoor air. *Toxicol. Appl. Pharmacol.*, *155*, 245-252.

[19] Shukla, A., Timblin, C., BeruBe, K., Gordon, T., McKinney, W., Driscoll, K., Vacek, P., Mossman, B.T. (2000). Inhaled particulate matter causes expression of nuclear factor (NF)-κB-related genes and oxidant-dependent NF-κB activation in vitro. *Am. J. Respir. Cell. Mol. Biol.*, *23*, 182-187.

[20] Elder, A.C.P., Gelein, R., Finkelstein, J.N., Cox, C., Oberdörster, G. (2000). Endotoxin priming affects the lung response to ultrafine particles and ozone in young and old rats; *Inhal. Toxicol.*, *12*, 85-98.

[21] Knaapen, A.M., Shi, T., Borm, P.J., Schins, R.P. (2002). Soluble metals as well as the insoluble particle fraction are involved in cellular DNA damage induced by particulate matter. *Mol. Cell. Biochem.*, *234/235*, 317-326.

[22] Soukup, J.M. & Becker, S. (2001). Human alveolar macrophage responses to air pollution particulates are associated with insoluble components of coarse material, including particulate endotoxin. *Toxicol. Appl. Pharmacol.*, *171*, 20-26.

[23] Carter, J.D., Ghio, A.J., Samet, J.M., Devlin, R.B. (1997). Cytokine production by human airway epithelial cells after exposure to an air pollution particle is metal-dependent; *Toxicol.. Appl. Pharmacol.*, *146*, 180-188.

[24] Frampton, M.W., Ghio, A.J., Samet, J.M., Carson, J.L., Carter, J.D., Devlin, R.B. (1999). Effects of aqueous extracts of PM10 filters from the Utah Valley on human airway epithelial cells. *Lung Cell. Mol. Physiol.*, *21*, L960-L967.

[25] Kennedy, T., Ghio, A.J., Reed, W., Samet, J., Zagorski, J., Quay, J., Carter, J., Dailey, L., Hoidal, J.R., Devlin, R.B. (1998). Copper-dependent inflammation and nuclear factor-κb activation by particulate air pollution. *Am. J. Respir. Cell Mol. Biol.*, *19*, 1-14.

[26] Nel, A.E., Xia, T., Madler, L., Li, N. (2006). Toxic potential of materials at the nanolevel. *Science*, *311*, 622-627.

[27] Patterson, E. & Eatough, D.J. (2000). Indoor/outdoor relationships for ambient PM2.5 and associated pollutants: epidemiological implications in Lindon, Utah. *J. Air Waste Manage. Assoc.*, *50*, 103-110.

[28] Hetland, R.B., Cassee, F.R., Refsnes, M., Schwarze, P.E., Lag, M., Boere, A.J.F., Dybing, E. (2004). Release of inflammatory cytokines, cell toxicity and apoptosis in epithelial lung cells after exposure to ambient air particles of different size fractions. *Toxicol. In Vitro*, *18*, 203-212.

[29] Becker, S., Soukup, J.M., Gallagher, J.E. (2002). Differential particulate air pollution induced oxidant stress in human granulocytes monocytes and alveolar macrophages. *Toxicol. In Vitro*, *16*, 209-218.

[30] Ghio, A.J. & Devlin, R.B. (2001). Inflammatory lung injury after bronchial instillation of air pollution particles. *Am. J. Respir. Crit. Care Med.*, *164*, 704-708.

[31] Rahman, I. & MacNee, W. (2000). Oxidative stress and regulation of glutathione in lung inflammation. *Eur. Respir. J.*, *16*, 534-554.

[32] Pozzi, R., De Berardis, B., Paoletti, L., Guastadisegni, C. (2003). Inflammatory mediators induced by coarse (PM2.5–1.0) and fine (PM2.5) urban air particles in RAW 2647 cells. *Toxicology*, *183*, 243-254.

[33] Schins, R.P., Knaapen, A.M., Weishaupt, C., Winzer, A., Borm, P.J.A. (2002). Cytotoxic and inflammatory effects of coarse and fine particulate matter in macrophages and epithelial cells. *Ann. Occup. Hyg.*, *46*, 203-206.

[34] Nel, A.E., Diaz-Sanchez, D., Li, N. (2001). The role of particulate pollutants in pulmonary inflammation and asthma: evidence for the involvement of organic chemicals and oxidative stress. *Curr. Opin. Pulm. Med.*, *7*, 20-26.

[35] Upadhyay, D., Panduri, V., Ghio, A., Kamp, D.W. (2003). Particulate matter induces alveolar epithelial cell DNA damage and apoptosis: role of free radicals and the mitochondria. *Am. J. Respir. Cell Mol. Biol.*, *29*,180-187.

[36] Aggarwal, B.B. (2000). Apoptosis and Nuclear factor-κB: a tale of association and dissociation. *Biochem. Pharmacol.*, *60*, 1033-1039.

[37] Aggarwal, B.B. (2004). Nuclear factor-κB:The enemy within. *Cancer cell.*, *6*, 203-208.

[38] Baldwin, A.S. (2001). The transcription factor NF-kB and human disease. *J. Clin. Invest.*, *107*, 3-6.

[39] Rahman, I., Milier, B., Gilmour, P.S., Watchorn, T., Donaldson, K., Jeffrey, P.K., MacNee, W. (2001). Oxidant-mediated lung epithelial cell tolerance: the role of intracellular glutathione and nuclear factor-kappaB. *Biochem. Pharmacol.*, *62*, 787-794.

[40] Luster, M.I., Simeonova, P.P., Gallucci, R., Matheson, J., 1999. Tumor necrosis factor-α and toxicology. *Crit. Rev. Toxicol.*, *29*, 491-511.

[41] Li, Z.W., Rickert, R.C., Karin, M. (2004). Genetic dissection of antigen receptor induced-NF-κB activation. *Mol. Immunol.*, *41*, 701-714.

[42] Driscoll, K.E. (2000). TNF-alpha and MIP-2: role in particle-induced inflammation and regulation by oxidative stress. *Toxicol. Lett.*, *112/113*,177-183.

[43] Tao, F., Gonzalez-Flecha, B., Kobzik, L. (2003). Reactive oxygen species in pulmonary inflammation by ambient particulates. *Free Radical Biol. Med.*, *35*, 327-340.

[44] Fujii, T., Hayashi, S., Hogg, J.C., Vincent, R., Van Eeden, S.F. (2001). Particulate matter induces cytokine expression in human bronchial epithelial cells. *Am. J. Respir. Cell Mol. Biol.*, *25*, 265-271.

[45] Van Eeden, S.F., Tan, W.C., Suwa, T., Mukae, H., Terashima, T., Fujii, T., Qui, D., Vincent, R., Hogg, J.C. (2001). Cytokines involved in the systemic inflammatory response induced by exposure to particulate matter air pollutants (PM(10)). *Am. J. Respir. Crit. Care Med.*, *164*, 826-830.

[46] Rao, K.M. (2000). Molecular mechanisms regulating iNOS expression in various cell types. *J. Toxicol. Environ. Health.*, *3*, 27-58.

[47] Haddad, J.J. (2004). Redox and oxidant-mediated regulation of apoptosis signaling pathways: immuno-pharmaco-redox conception of oxidative siege versus cell death commitment. *Int. Immunopharmacol.*, *4*, 475-493.

[48] Ganju, N. & Eastman, A. (2003). Zinc inhibits Bax and Bak activation and cytochrome c release induced by chemical inducers of apoptosis but not by death-receptor-initiated pathways. *Cell Death Diff.*, *10*, 652-661.

[49] Kowaltowski, A., Fenton, R., Fiskum, G. (2004). Bcl-2 family proteins regulate mitochondrial reactive oxygen production and protect against oxidative stress. *Free Radical Biol. Med.*, *37*, 1845-1853.

[50] Micheau, O. & Tschopp, J. (2003). Induction of TNF Receptor I-mediated apoptosis via two sequential signaling complexes. *Cell*, *114*, 181-190.

[51] Nunez, G., Benedict, M.A., Hu, Y., Inohara, N. (1998). Caspases: the proteases of the apoptotic pathway. *Oncogene*, *17*, 3237-3245.

[52] Skulachev, V.P. (1998). Cytochrome c in the apoptotic and antioxidant cascades. *FEBS Lett.*, *423*, 275-280.

[53] Geske, F.J. & Gerschenson, L.E. (2001). The biology of apoptosis. *Hum. Pathol.*, *32*, 1029-1038.

[54] Oren, M., Damalas, A., Gottlieb, T., Michael, D., Taplick, Y., Leal, J., Maya, R., Moas, M., Seger, R., Taya, Y., Ben-Ze'ev, A. (2002). Regulation of p53: intricate loops and delicate balances. *Biochem. Pharmacol.*, *64*, 865-871.

[55] Pope, I.I.I.C., Burnett, R.T., Thun, M.J., Calle, E.E., Krewski,D., Ito, K., Thurston, G.D. (2002). Lung cancer, cardiopulmonary mortality, and long-term exposure to fine particulate air pollution. *J. Am. Med. Ass.*, *287*, 1132–1141.

[56] Garçon, G., Dagher, Z., Zerimech, F., Ledoux, F., Courcot, D., Aboukais, A., Puskaric, E., Shirali, P. (2006). Dunkerque city air pollution particulate matter-induced cytotoxicity, oxidative stress and inflammation in human epithelial lung cells (L132) in culture. *Toxicol. In Vitro*, *20*, 519-528.

[57] Dagher Z., Garçon G., Billet S., Verdin A., Gosset P., Ledoux F., Courcot D., Aboukais A., Shirali P. (2007). Role of nuclear factor-kappa B activation in the adverse effects induced by air pollution particulate matter (PM2.5) in human epithelial lung cells (L132) in culture. *J. Appl. Toxicol.*, (DOI: 10.1002/jat.1211)

[58] Dagher, Z., Garçon, G., Gosset, P., Ledoux, F., Surpateanu, G., Courcot, D., Aboukais, A., Puskaric, E., Shirali, P. (2005). Pro-inflammatory effects of Dunkerque city air pollution particulate matter 2.5 in human epithelial lung cells (L132) in culture. *J. Appl. Toxicol.*, *25*, 166-175.

[59] Dagher, Z., Garçon, G., Billet, S., Gosset, P., Ledoux, F., Courcot, D., Aboukais, A., Shirali, P. (2006). Activation of different pathways of apoptosis by air pollution particulate matter (PM2.5) in human epithelial lung cells (L132) in culture. *Toxicology*, *225*, 12-24..

[60] Billet S., Garçon G., Dagher Z., Verdin A., Ledoux F., Cazier F., Courcot D., Aboukais A., Shirali P. (2007). Metabolic activation of organic compounds coated onto PM2.5 in human alveolar lung cells (A549) in culture. *Env. Res.*, (DOI: 10.1016/envres.2007.03.001).

[61] Devlina, R.B., Framptonb, M.L., Ghio, A.J. (2005). In vitro studies: What is their role in toxicology? *Exp.Toxicol. Pathol.*, *57*, 183–188

[62] Erfle, V. & Mellert, W. (1996). Human cell line LC5 and its use. *US Patent*, *5*, 582-967.

[63] Shirali, P., Teissier, E., Marez, T., Hildebrand, F.H., Haguenoer, J.M. (1994). Effect of αNi_3S_2 on arachidonic acid metabolites in cultured human lung cells (L132 cell line). *Carcinogenesis*, *15*, 759-762.

[64] Garçon, G., Shirali, P., Garry, S., Fontaine, M., Zerimech, F., Martin, A., Haguenoer, J.M., Hannothiaux, M.H. (2000). Polycyclic aromatic hydrocarbons-coated onto Fe_2O_3 particles: assessment of cellular membrane damage and antioxidant system disruption in human epithelial lung cells (L132) in culture. *Toxicol. Lett.*, *117*, 25-35.

[65] Shirali, P., Gosset, P., Marez, T., Hachimi, A., Muller, J.F., Haguenoer, J.M. (1995). Cytotoxicity evaluation of hematite (Fe_2O_3), benzo(a)pyrene and pyrene and analysis of particles by Laser Microprobe Mass Analyser (in-vitro). *Cell Biol. Toxicol.*, *11*, 215.

[66] Boland, S., Baeza-Squiban, A., Bonvallot, V., Houcine, O., Pain, C., Meyer, M., Marano, F. (2001). Similar effects induced by diesel exhaust particles from a representative diesel vehicle recovered from filters and standard reference material 1650. *Toxicol. In Vitro*, *15*, 379-385.

[67] Takenouchi, T. & Munekata, E. (1995). Trophic effects of substance P and □-amyloid peptide on dibutyryl cyclic AMP-differentiated human leukemic (HL-60) cells. *Life Sci.*, *56*, 479-484.

[68] Garçon, G., Zerimech, F., Hannothiaux, M.H., Gosset, P., Martin, A., Marez, T., Shirali, P. (2001a). Antioxidant defense disruption by polycyclic aromatic hydrocarbons-coated onto Fe_2O_3 particles in human lung cells (A549). *Toxicology*, *166*, 129-137.

[69] Garçon, G., Garry, S., Gosset, P., Zerimech, F., Martin, A., Hannothiaux, M.H., Shirali, P. (2001b). Benzo(a)pyrene-coated onto Fe_2O_3 particles-induced lung tissue injury: role of free radicals. *Cancer Lett.*, *167*, 7-15.

[70] Garçon, G., Gosset, P., Garry, S., Marez, T., Hannothiaux, M.H., Shirali, P. (2001c). Pulmonary induction of proinflammatory mediators following the rat exposure to benzo(a)pyrene-coated onto Fe_2O_3 particles. *Toxicol. Lett.*, *121*, 107-117.

[71] Toyokuni, S., Tanaka, T., Hattori, Y., Nishiyama, Y., Yoshida, A., Uchida, K., Hiai, H., Ochi, H., Osawa, T. (1997). Quantitative immunohistochemical determination of 8-hydroxy-2'-deoxyguanosine by a monoclonal antibody N451: its application to ferric nitrilotriacetate-induced renal carcinogenesis model. *Lab. Invest.*, *76*, 365-374.

[72] Willey, J.J., Stripp, B.R., Baggs, R.B., Gasiewicz, T.A. (1998). Aryl hydrocarbon receptor activation in genital tubercle, palate, and other embryonic tissues in 2,3,7, 8-tetrachlorodibenzo-p-dioxin-responsive lacZ mice. *Toxicol. Appl. Pharmacol.*, *151*, 33-44.

[73] Smith, P.K., Krohn, R.I., Hermanson, G.T., Mallia, A.K., Gartner, F.H., Provenzano, M.D., Fujimoto, E.K., Goeke, N.M., Olson B.J., Klenk, D.C. (1985). Measurement of protein using bicinchoninic acid. *Anal. Biochem.*, *150*, 76-85.

[74] Dreher, K.L., Jaskot, R.H., Lehmann, J.R., Richards, J.H., McGee, J.K., Ghio, A.J., Costa, D.L. (1997). Soluble transition metals mediate residual oil fly ash induced acute lung injury. *J. Toxicol. Environ. Health.*, *50*, 285-305.

[75] De Zwart, L.L., Meerman, J.H.N., Commandeur, J.N.M., Vermeulen, N.P. (1999). Biomarkers of free radical damage applications in experimental animals and in humans. *Free Radical Biol. Med.*, *26*, 202-226.

[76] Gilks, C.B., Price, K., Wright, J.L., Churg, A. (1998). Antioxidant gene expression in rat lung after exposure to cigarette smoke. *Am. J. Pathol.*, *152*, 269-278.

[77] Mates, J.M. & Sanchez-Jimenez, F. (1999). Antioxidant enzymes and their implications in pathophysiologic processes. *Front Biosci.*, *4*, d339-d345.

[78] Escobar, J.A., Rubio, M.A., Lissi, E.A. (1996). SOD and catalase inactivation by singlet oxygen and peroxyl radicals. *Free Radical. Biol. Med.*, *20*, 285-290.

[79] Tokiwa, H., Sera, N., Nakanishi, Y., Sagai, M. (1999). 8-Hydroxy-deoxyguanosine formed in human lung tissues and the association with diesel exhaust particles. *Free Radic. Biol. Med.*, *27*, 1251-1258.

[80] Shi, T., Knaapen, A.M., Begerow, J., Birmili, W., Borm, P.J.A., Schins, R.P.F. (2003). Temporal variations of hydroxyl radical generation and 8-hydroxy-2'-deoxyguanosine formation by coarse and fine particulate matter. *Occup. Environ. Med.*, *60*, 315-321.

[81] Herceg, Z. & Wang, Z.Q., 2001. Functions of poly(ADP-ribose)polymerase (PARP) in DNA repair, genomic integrity and cell death. *Mutat. Res.*, *477*, 97-110.

[82] Bouchard, V., Rouleau, M., Poirier, G. (2003). PARP-1, a determinant of cell in response to DNA damage. *Exp. Hematol.*, *31*, 446-454.

[83] Burkle, A. (2001). PolyADP-ribosylation, a DNA damage-driven protein modification and regulator of genomic instability. *Cancer Lett.*, *163*, 1-5.

[84] Nicoletti, V.G. & Stella, A.M.G. (2003). Role of PARP under stress conditions: cell death or protection? *Neurochem. Res.*, *28*, 187-194.

[85] Ghio, A.J. & Devlin, R.B. (2001). Inflammatory lung injury after bronchial instillation of air pollution particles. *Am. J. Respir. Crit. Care Med.*, *164*, 704-708.

[86] Hetland, R.B., Myhre, O., Lag, M., Hongve, D., Schawarze, P.E., Refsnes, M. (2001). Importance of soluble metals and reactive oxygen species for cytokine release induced by mineral particles. *Toxicology*, *165*, 133-144.

[87] Churg, A., Xie, C., Wang, X., Vincent R., Wang, R.D. (2005). Air pollution particles activate NF-κB on contact with airway epithelial cell surfaces. *Toxicol. Appl. Pharmacol.*, *208*, 37-45.

[88] Nam, H.Y., Choi, B.H., Lee, J.Y., Lee, S.G., Kim, Y.H., Lee, K.H., Yoon, H.K., Song, J.S., Kim, H.J., Lim, Y. (2004). The role of nitric oxide in the particulate matter (PM2.5)-induced NF-κB activation in lung epithelial cells. *Toxicol. Lett.*, *148*, 95-102.

[89] Winyard, P.G. & Blake, D.R. (1997). Antioxidant redox-regulated transcription factors and inflammation. *Adv. Pharmacol.*, *38*, 403-421.

[90] Bergmann, M., Barnes, P.J., Newton, R. (2000). Molecular regulation of granulocyte macrophage colony-stimulating factor in human lung epithelial cells by interleukin (IL)-1β, IL-4, and IL-13 involves both transcriptional and post-transcriptional mechanisms. *Am. J. Respir. Cell Mol. Biol.*, *22*, 582-589.

[91] Fujii, T., Hayashi, S., Hogg, J.C., Vincent, R., Van Eeden, S.F. (2001). Particulate matter induces cytokine expression in human bronchial epithelial cells. *Am. J. Respir. Cell Mol. Biol.*, *25*, 265-271

[92] Becher, R., Hetland, R.B., Refsnes, M., Dahl, J.E., Dahlman, H.J., Schawarze, P.E. (2001). Rat lung inflammatory responses after in vivo and in vitro exposure to various stone particles. *Inhal. Toxicol.*, *13*, 789-805.

[93] Warshamana, G.S., Corti, M., Brody, A.R. (2001). TNF-alpha, PDGF, and TGF-beta(1) expression by primary mouse bronchiolar-alveolar epithelial and mesenchymal cells: tnf-alpha induces TGF-beta(1). *Exp. Mol. Pathol.*, *71*, 13-33.

[94] Bellocq, A., Azoulay, E., Marulla, S., Flahault, A., Fouqueray, B., Philippe, C., Cadranel, J., Baud, L. (1999). Reactive oxygen and nitrogen intermediates increase

transforming growth factor-beta1 release from human epithelial alveolar cells through two different mechanisms. *Am. J. Respir. Cell Mol. Biol.*, *21*, 128-36.

[95] McRitchie, D.I., Isowa, N., Edelson, J.D., Xavier, A.M., Cai, L., Man, H.Y., Wang, Y.T., Keshavjee, S.H., Slutsky, A.S., Liu, M. (2000). Production of tumour necrosis factor alpha by primary cultured rat alveolar epithelial cells. *Cytokine*, *12*, 644-654.

[96] Yoshida, Y., Maruyama, M., Fujita, T., Arai, N., Hayashi, R., Araya, J., Matsui, S., Yamashita, N., Sugiyama, E., Kobayashi, M. (1999). Reactive oxygen intermediates stimulate interleukin-6 production in human bronchial epithelial cells. *Am. J. Physiol.*, *276*, L900-L908.

[97] Baulig, A., Sourdeval, M., Meyer, M., Marano, F., Baeza-Squiban, A. (2003). Biological effects of atmospheric particles on human bronchial epithelial cells. Comparison with diesel exhaust particles. *Toxicol. In Vitro*, *17*, 567–573.

[98] Becker, S., Soukup, J.M., Gilmour, M.I., Delvin, R.B. (1996). Stimulation of human and rat alveolar macrophages by urban air particulates: effect on oxidant radical generation and cytokine production. *Toxicol. Appl. Pharmacol.*, *141*, 637-648.

[99] Becker, S., Soukup, J.M., Gilmour, I., Devlin, R.B. (1997). Stimulation of human and rat alveolar macrophages by urban air particulates: effects on oxidant radical generation and cytokine production. *Toxicol. Appl. Pharmacol.*, *141*, 637-648.

[100] Bayram, H., Devalia, J.L., Sapsford, R.J., Ohtoshi, T., Miyabara, Y., Sagai, M., Davies, R.J. (1998). The effect of diesel exhaust particles on cell function and release of inflammatory mediators from human bronchial epithelial cells in vitro. *Am. J. Respir. Cell Mol. Biol.*, *18*, 441-448.

[101] Dye, J.A., Adler, K.B., Richards, J.H., Dreher, K.L. (1999). Role of soluble metals in oil fly ash-induced airway epithelial injury and cytokine gene expression. *Am. J. Physiol.*, *277*, L498-L510.

[102] Hetland, R.B., Refsnes, M., Myran, T., Johansen, B.V., Uthus, N., Schwarze, P.E. (2000). Mineral and/or metal content as critical determinants of particle-induced release of IL-6 and IL-8 from A549 cells. *J. Toxicol. Environ. Health*, *60*, 47-65.

[103] Kawasaki, S., Takizawa, H., Takami, K., Desaki, M., Okazaki, H., Kasama, T., Kobayashi, K., Yamamoto, K., Nakahara, K., Tanaka, M., Sagai, M., Ohtoshi, T. (2001). Benzene-extracted components are important for the major activity of diesel exhaust particles Effect on interleukin-8 gene expression in human bronchial epithelial cells. *Am. J. Respir. Cell Mol. Biol.*, *24*, 419-426.

[104] Li, N., Wang, M., Oberley, T.D., Sempf, J.M., Nel, A.E. (2002). Comparison of the pro-oxidative and proinflammatory effects of organic diesel exhaust particle chemical in bronchial epithelial cells and macrophages. *J. Immunol.*, *169*, 4531-4541.

[105] Baker, S.J. & Reddy E.P. (1998). Modulation of life and death by the TNF receptor superfamily. *Oncogene*, *17*, 3261-3271.

[106] Cohen, G.M. (1997). Caspases: the executioners of apoptosis. *Biochem. J.*, *326*, 1-16.

[107] Finkel, T. (2000). Redox-dependent signal transduction. *FEBS Lett.*, *476*, 52-54.

[108] Chin, B.Y., Choi, M.E., Burdick, M.D., Strieter, R.M., Risby, T.H., Choi, A.M. (1998). Induction of apoptosis by particulate matter: role of TNF-alpha and MAPK. *Am. J. Physiol.*, *275*, L942-L949.

[109] Takizawa, H., Ohtoshi, T., Kawasaki, S., Kohyama, T., Desaki, M., Kasama, T., Kobayashi, K., Nakahara, K., Yamamoto, K., Matsushima, K., Kudoh, S. (1999).

Diesel exhaust particles induce NF-κB activation in human bronchial epithelial cells in-vitro: importance in cytokine transcription. *J. Immunol.*, *162*, 4705-4711.

[110] Reisman, D. & Loging, W.T. (1997). Transcriptional regulation of the p53 tumor suppressor gene. Sem. *Cancer Biol.*, *8*, 317-324.

[111] Curtin, J.F., Donovan, M., Cotter, T.G. (2002). Regulation and measurement of oxidative stress in apoptosis. *J. Immunol. Methods*, *265*, 49-72.

In: Airborne Particulates
Editor: Ming Cheng and Wen Liu

ISBN: 978-1-60692-907-0

Chapter 9

TRENDS IN MODELLING OF RADIONUCLIDES UPTAKE BY PARTICULATE MATTER IN THE MARINE ENVIRONMENT USING BOX MODELS*

A. Laissaoui† *and R. El Mrabet*
Centre National de l'Energie, des Sciences et des Techniques Nucléaires B.P. 1382 R.P. 10001 Rabat, Morocco

ABSTRACT

The objective of this paper is to provide a documented discussion on modelling of radionuclides interaction with suspended particulate matter in the marine environment. Aquatic environments occupy a major portion of the earth's surface and, therefore, an understanding of the radionuclide pathways is essential for radioecology purposes. Many radionuclides within such systems exhibit non-conservative behaviour, i.e., they undergo a change in the distribution between the particulate and dissolved phases, due to sorption reactions. The first reactive transport models for non-conservative radionuclides were based upon the equilibrium distribution coefficient, kd. However, recent models are based on kinetic rates of uptake/release of radionuclides between waters and the solid phases when short timescales are involved. Recently, much experimental and modelling effort has been focused on determining those factors which affect the kinetics and the final equilibrium conditions for the uptake of pollutants in aqueous suspensions under dynamic or static situations. Some of the obtained results appear to be either surprising or contradictory and introduce some uncertainty on which parameter values are most appropriate for environmental modelling.

The kinetic box models are widely used in modelling of radionuclides dispersion in aquatic systems, and are based on reversible reactions of constant coefficients. Indeed, several models have been proposed in the literature taking account of parallel, consecutive reactions and some combinations of both depending on the uptake experimental data. Research in this field of radioecology has been stimulated by the high

* A version of this chapter was also published in *Environmental Modelling: New Research,* edited by Paul N. Findley published by Nova Science Publishers, Inc. It was submitted for appropriate modifications in an effort to encourage wider dissemination of research.

† E-mail address: Laissaoui@cnesten.org.ma

level of public concern over health aspects of the presence of contaminant radionuclides in the environment, which has prompted a generally anti-nuclear posture. In this paper we present a review of the most relevant mathematical models developed so far and a brief revision of the historic developments of theories on surface-electrolyte interactions. Laboratory tracing experiments to study the uptake of ^{239}Pu, ^{241}Am, ^{133}Ba and ^{85}Sr in natural aqueous suspensions from several aquatic systems (reservoir, river, estuary and sea) are also presented to illustrate the application of box models covering a large number of environmental situations and different reaction channels.

Keywords: marine radioecology, box models, radionuclides, transfer coefficients.

I. Introduction

Environmental radioactivity studies are of major importance with respect to radioactive waste disposal and future exploitation of nuclear energy. This is due to the high level of public concern over health aspects of the presence of contaminant radionuclides in the environment, which has stimulated a generally anti-nuclear posture. The numerous accidents involving reactors and other nuclear facilities, satellites containing radionuclides, nuclear weapons and large radiation sources (MacKenzie, 2000) have resulted in public distrust regarding the promises of clean energy from the nuclear industry in the mid-20th century.

Aquatic environments occupy a major portion of the earth's surface and, therefore, an understanding of the radionuclide pathways is essential for radioecology purposes. Many radionuclides within such systems exhibit non-conservative behaviour, i.e., they undergo a change in the distribution between the particulate and dissolved phases, due to sorption reactions. Thus, migration of non-conservative radionuclides in aquatic environments is strongly affected by the interaction of the radionuclides with solids present in such systems. Particularly, the uptake and release by particulate suspended mater combined with sedimentation and resuspension of the solids plays a significant role. Indeed, once introduced into the marine environment, particle reactive radionuclides will sorb, more or less strongly, onto solid particles so that they are partitioned between the solid and water. The solid particles can also settle out to the bottom and be removed from the water column. On the other hand, radionuclides in the water can also be adsorbed directly onto the sediment after diffusing across the sediment water interface, always called the boundary layer. Radioactivity can also be released back into the water column by desorption and remobilisation from the sediment and by resuspension of solid material. All of these processes are able to modify the global dynamics of radionuclides and, therefore, must be included in models simulating transport processes. Most of the existing mathematical codes neglect this respect by assuming that the uptake kinetics are so fast that equilibrium is always established.

The first reactive transport models for non-conservative radionuclides were based upon the equilibrium distribution coefficient, kd. This parameter has always been a measure of sorption of contaminants to solids and is defined as the ratio of the quantity of the adsorbate adsorbed per unit mass of solid to the amount of the adsorbate remaining in solution at equilibrium. The widespread use of kd is due to the need for a single parameter describing solid/solution distribution for the purpose of modelling the geochemical partitioning of radionuclides in steady state conditions. However, the use of distribution coefficients is limited by the fact that it can vary by 9 orders of magnitude between different nuclides and 3 orders of magnitude for any particular one (Hamilton-Taylor et al. 1993). This variation depends mainly on the solution composition and the nature of the solid substrate. On the contrary, recent models are based on kinetic rates of uptake/release of radionuclides between waters and the solid phases when short timescales are involved (Perianez, 2004).

Recently, much experimental and modelling effort has been focused on determining those factors which affect the kinetics and the final equilibrium conditions for the uptake of pollutants in aqueous suspensions under dynamic or static situations. Some models include the kinetics as a one-step reversible or irreversible reaction which is often considered as a rough approximation (Laissaoui et al., 1998). However, data reported in the scientific literature show that the interaction of radionuclides with natural solids is often slow and is a two-to three-step process (El Mrabet et al., 2001; Barros & Abril, 2004). Thus, a model taking into account a single reaction is generally unable to properly describe both sorption and release kinetics. It is worth noting that mathematical approaches using box models in the analysis of time-resolved sorption data are capable of providing kinetic parameters which are important in determining the mobility and rate of removal of radionuclides from the water column (Robbins et al., 1992; Smith and Comans, 1996). However, the values of parameters are often different, depending on the number of processes identified in the uptake curves.

In this paper we present a review of the most relevant mathematical models developed so far and a brief revision of the historic developments of theories on surface-electrolyte interactions. Laboratory tracing experiments to study the uptake of ^{239}Pu, ^{241}Am, ^{133}Ba and ^{85}Sr in natural aqueous suspensions from several aquatic systems (reservoir, river, estuary and sea) are also presented to illustrate the application of box models covering a large number of environmental situations and different reaction channels.

II. Theory of Ion Exchange between Water and Suspended Particles

Sorption of radionuclides by particles in aqueous suspensions can be regarded as a physico-chemical process through which the charges on the mineral lattice, generally negative, are compensated by ions of positive charge. This phenomenon can take place on the outer surfaces of the particles as well as on the inner surfaces of the pores, as shown in Figure 1 (Abril ,1998a; Eisma, 1993). Thus, the whole volume of a particle is not really relevant to this process. One of the most common adsorption reactions in aqueous suspensions is ion exchange. In its most general meaning, an ion-exchange reaction involves the replacement of ionic species on a solid phase by another ionic species taken from an aqueous solution in contact with the solid; that is, a previously-adsorbed ion of weaker affinity is exchanged by the particle for a dissolved ion. Most metals in aqueous solution occur as charged ions and thus metal species adsorb primarily in response to electrostatic attraction.

The effect of surface charge on particles is that ions in the aqueous phase distribute themselves so that electrical neutrality is reached. Radionuclide adsorption is a statistical result of ionic collisions over the free surfaces of particles and of the probability of occurrence of different physico-chemical reactions. In most cases, collisions may take place over a non-active site on the surface of the particle and thus the exchangeable ion returns back into solution. Thus, only a fraction of all the collisions will be captured by uncompensated charges (Abril, 1998a).

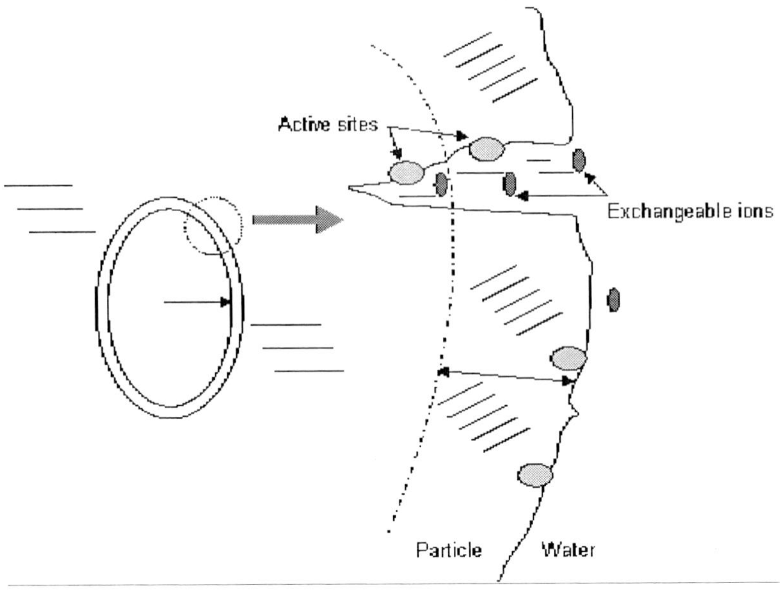

Figure 1. Schematic representation of a particle with its external surface layer.

A revision of the historic development of early theories on surface-electrolyte interactions can be found in Arnold (1978) and Richter (1987). Some of these theories relate only to equilibrium conditions, such as the Gouy-Chapman diffuse double layer and Stern double layer theories or the application of Donan Principles to cation exchange. Non-equilibrium formulations, as presented in Richter (1987), often start from a governing differential equation with constant coefficients, which is stated for each specific adsorption case, such as the Langmuir approach, the Freundlich reaction, for mixed-order reactions, or the competition reactions. Furthermore, of all the complex reactions that take place in heterogeneous systems, the Langmuir adsorption is the most significant. It was originally proposed to describe adsorption of gas molecules onto homogeneous solid surfaces that exhibit one type of adsorption site. Many investigators have tacitly extended the Langmuir adsorption model to describe adsorption of solution species onto solid adsorbents including heterogeneous solids such as suspended particulate matter. It expresses the non-specific adsorption of ions at surfaces; non-specific means that all sorption sites are equal. This feature presents restrictions due to the number of active sites to be occupied which is limited by the available area. For this reason, Langmuir adsorptions show characteristic saturation phenomena. On the other hand, if the adsorption at heterogeneous surfaces is to be described where sites with different adsorption energy exist, then the Freundlich approach will be used to describe the equilibrium state.

III. Development of Kinetic Box Models

Accurate numerical modeling of non-conservative radionuclide behavior in aquatic systems requires the knowledge and quantification of the processes able to modify the global dynamics of such radioactive elements. In this way, some theoretical considerations and experimental studies on the uptake kinetics of dissolved radionuclides have been carried out in recent decades and reported in the scientific literature (Bird & Evenden, 1994; Abril, 1998a and b; Laissaoui et al., 1998; Benes & Cernick, 1992). Therefore, a number of different models have been proposed to qualitatively describe and quantitatively predict the dynamic uptake of radionuclides in aqueous suspensions (Nyffler et al., 1984; Laissaoui & Abril, 1999; El Mrabet et al. 2001; Barros & Abril, 2004). These box models included parallel and/or consecutive as well as reversible uptake kinetic reactions involving one to three steps migration. It may be noted that different models could be fitted to a given experimental data. The goodness of fit of the model should be compared using statistical parameters as the residual sum of squares. This is the simplest criterion for deciding which of the box models is most appropriate (Bunker et al. 2001).

III.1. The One-Step Reversible Reaction

The simplest model (Figure 2) consists of a single reversible reaction quantified by two kinetic transfer coefficients (k_1 for sorption and k_2 for desorption) where the mathematical solution is an exponential decay function for each compartment. The basic assumptions for this case are the following:

- There is no discrimination between reactive sites in the adsorption process.
- There is no interaction between the adsorbed ions.
- In the case of mixed adsorption of different species, they compete for the same sites but competitors for identical sites have different energies.

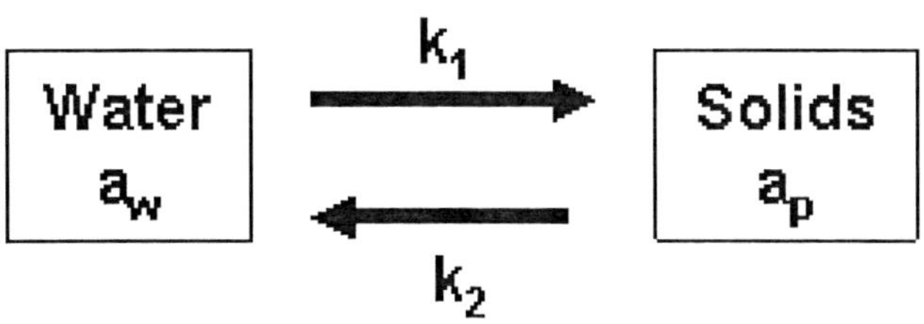

Figure 2. Schematic representation of the one-step model. a_w and a_p are the tracer concentrations in the dissolved phase and in suspended particles respectively. The transfer coefficients k_1 and k_2 are assumed to be constant.

The governing differential equations in the case of a single step model are:

$$\frac{da_w}{dt} = -k_1.a_w + k_2.a_p$$
$$\frac{da_p}{dt} = +k_1.a_w - k_2.a_p \qquad (1)$$

k_1 an k_2 are respectively the forward and backward transfer rates having dimension of $[T^{-1}]$ and can be obtained by fitting the experimentally determined data to the analytical solution of Eqs. 1, which is the following:

$$a_w(t) = \frac{k_1.a_w(0) - k_2.a_p(0)}{k_1 + k_2}\left(e^{-(k_1+k_2).t} - 1\right) + a_w(0) \qquad (2)$$

$a_w(0)$ and $a_p(0)$ are the initial concentrations of the studied radionuclide in the dissolved phase and particles respectively. $a_p(0)$ is always equal to zero since most experimental studies start from uncontaminated particulate matter conditions.

Previous studies devoted to the kinetic aspects of the exchange of radionuclides with particles have been based on mathematical models describing the sorption kinetics using single step reaction model (Jannasch 1988, Laissaoui et al. 1998 among others). This kind of models have shown its applicability in several studies, however, such a mathematical representation fails to describe some experimentally observed phenomena; the sorption of some radionuclides (e.g. Pu and Am) by suspended particulate matter can involve different processes with distinct characteristic time periods: very fast sorption followed by much slower and more extended reaction (Benes et al. 1992). Consequently, to avoid a sub-parameterisation in describing the processes involved in the fixation of radionuclides, it is necessary to consider models of higher orders taking account of the existence of well-defined reaction channels or type of sites on the suspended particles.

III.2. The Two-Step Model

The second model, in order of complexity, is the so called “3-box” or “two-step” model, in which the particle is conceptually divided in two fractions: one associated to the faster reaction channel and another related to the slower uptake. In this case the model, as depicted in Figure 3, can be built in two different ways, considering parallel or consecutive reversible reactions (i.e. four coefficients are needed).

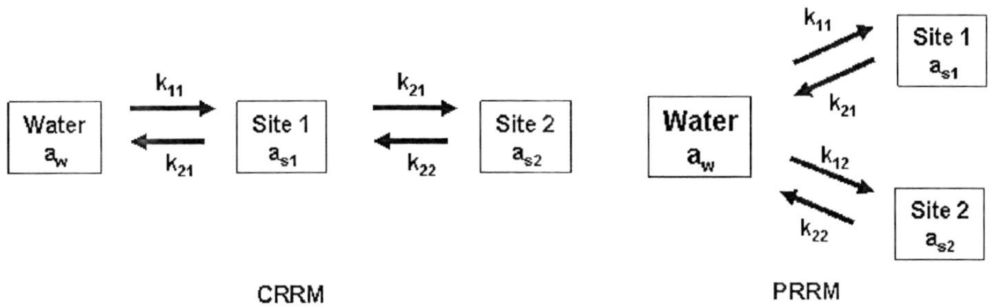

Figure 3. Schematic representation of the two-step models. PRRM is parallel and reversible reaction model and CRRM is consecutive and reversible reaction model.

When taking into account a two consecutive and reversible reaction model (CRRM) for describing interaction of radionuclides at the liquid-solid interface, one supposes the existence of a first mechanism which allows the adsorption of the dissolved ion onto the particles surface. This first reaction is due mainly to electrical forces (such as electrostatic or Van der Waals forces) since dissolved radionuclides occur as charged ions. A second reaction follows which is due to cation exchange or diffusion into the porous structure of the particles. Both reactions are reversible and, therefore, the adsorbed ion can return to the dissolved phase at any step. On the contrary, the two parallel and reversible reaction model (PRRM) assumes that the two reactions involved in the kinetics are independent from each other and are connected only via the cation exchange with the dissolved phase.

The governing equations of both models are as follows:

$$\begin{aligned}
\frac{da_w}{dt} &= -k_{11}.a_w + k_{21}.a_{s1} - k_{12}.a_w + k_{22}.a_{s2} \\
\frac{da_{s1}}{dt} &= k_{11}.a_w - k_{21}.a_{s1} \\
\frac{da_{s2}}{dt} &= k_{12}.a_w - k_{22}.a_{s2}
\end{aligned}$$

Eq. (3)
PRRM

$$\begin{aligned}
\frac{da_w}{dt} &= -k_{11}.a_w + k_{21}.a_{s1} \\
\frac{da_{s1}}{dt} &= +k_{11}.a_w - k_{21}.a_{s1} - k_{12}.a_{s1} + k_{22}.a_{s2} \\
\frac{da_{s2}}{dt} &= +k_{12}.a_{s1} - k_{22}.a_{s2}
\end{aligned}$$

Eq. (4)
CRRM

Both set of equations are analytically solved by applying Laplace transformation and the solution has the following form:

$$a_w(t) = a_w(t=0).\left[A_1.\left(e^{-P_1.t} - 1\right) + A_2.\left(e^{-P_2.t} - 1\right) + 1\right] \quad (5)$$

P_i and A_i (i = 1, 2) are functions of the transfer coefficients k_{ij} and are obtained from the best fit to the experimental uptake curve. The coefficients, in the case of PRRM, can be derived easily from the following set of expressions:

$$\sum_{ij} k_{ij} = -\sum_{i} P_i$$
$$k_{21}k_{22} + k_{11}k_{22} + k_{12}k_{21} = P_1P_2 \qquad (6)$$
$$k_{21} + k_{22} = \sum_{i} P_i(A_i - 1)$$
$$k_{21}k_{22} = P_1P_2 - \sum A_iP_iP_j$$

Although all the parameters are obtained by processing only the time course of concentrations in the dissolved phase, the evolution of the whole system can be determined by solving numerically the set of Equations (3) or (4). Thus, the curves for the material components can be obtained. It may be noted that expression (5) is the analytical solution of both Equations (3) and Equations (4). Consequently, the PRR and CRR models, which are physically different, are mathematically equivalent.

III.3. The Three-Step Model

The three-step model is the most complex description involving four compartments (the dissolved phase and three sites onto the particles) whose application should be well justified by the experimental data.

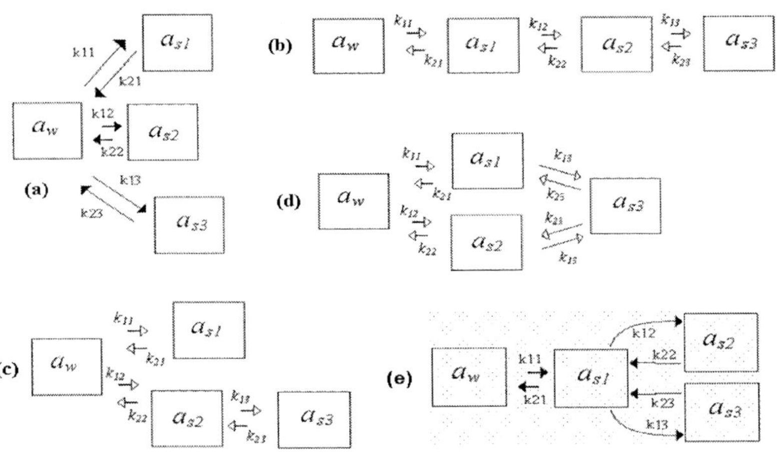

Figure 4. Schematic representation of kinetic box models when four compartments are involved. a_w is the dissolved phase concentration and a_{si} are the concentrations associated to different sites onto the particles.

At least six models could be constructed involving parallel, consecutive reactions and combinations of both of them (Figure 4). It may be noted that there are no selection criteria able to tell about the validity of one model over the others. The set of models are considered as mathematically equivalent since the same multiexponential function could describe the time evolution of concentrations in the dissolved phase independently of the applied model. The question is: what is the physical significance of the kinetic coefficients derived from each model?

A useful illustration of the existence of three-step reaction can be seen in Figure 5 in which the time evolution of ^{133}Ba concentration in the aqueous phase is plotted. The line is the best fit to the sum of three exponential functions reflecting then the superposition of three different adsorption processes in terms of kinetics reaction. The experiment was carried out using natural estuarine waters collected in the Odiel river (southwest of Spain).

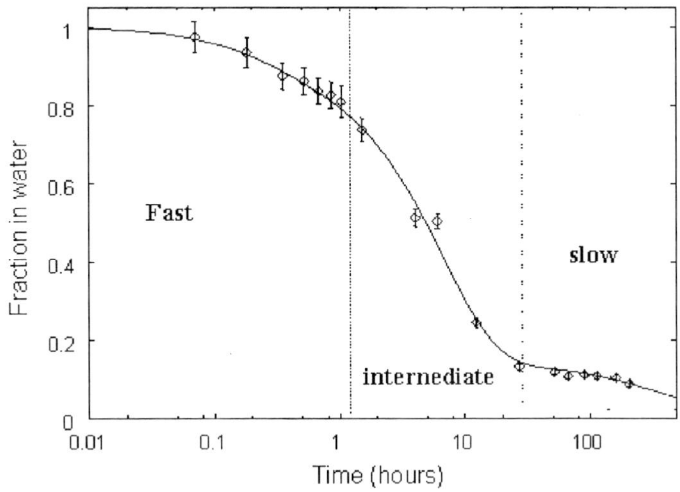

Figure 5. Time evolution of concentrations in the aqueous phase as the result of ^{133}Ba adsorption by suspended particulate matter. The points are the experimental data taken from Barros & Abril, 2004.

Three processes can be clearly distinguished in the uptake curve; a fast adsorption during the first hour followed by a moderated reaction lasting some 20 to 30 hours and finally, an excessively slow reaction which draws the system to a saturation state. The adoption of a model of lower order than the three-step model will be, therefore, a sub-parameterisation of the uptake processes involved in this real case.

Six parameters are obtained from the fitting procedure and are used in the calculation of the kinetic transfer coefficients. By applying the parallel and consecutive models (models a and b in Figure 4), the coefficients k_{ij}(i=1 to 2,j=1 to 3) are obtained and reported in Table 1.

Table 1. Kinetic transfer coefficients k_{ij}(i=1 to 2,j=1 to 3) obtained by applying the three-step parallel and consecutive models to data of Figure 5.

	k_{11}	k_{21}	k_{12}	k_{22}	k_{13}	k_{23}
Parallel (a)	0.365	3.388	0.124	0.019	0.0118	0.0007
Consecutive (b)	0.501	2.472	0.909	0.025	0.0016	0.0007

The computed coefficients were used to solve numerically the governing differential equations in view to determine the evolution of the whole system (the time course of concentrations in the four compartments). Both models are valid in this case and, therefore, the selection of parallel or consecutive description should be arbitrary.

IV. Applicability of Box Models

Kinetically controlled experiments were carried out using radioactive tracers (^{133}Ba, ^{85}Sr, ^{241}Am and ^{239}Pu) to study radionuclides partitioning between natural waters and suspended solid particles. The primordial objective of these studies has been to try to explain the environmental behaviour of radionuclides in natural aquatic environments. Special emphasis has been given to anthropogenic radionuclides which have been present in the environment since 1940s and research into their behaviour, stimulated dominantly by concerns over potential health effects, has been a major area of research in recent decades.

The most commonly used procedure for determining kinetic transfer coefficients is the laboratory batch method using naturally occurring solids and solutions to closely simulate the environmental conditions. It consists of suspending a well characterised sediment of known mass in a known volume of water, or rather using unfiltered water, and adding an identified concentration of an aqueous radionuclide solution. The solution must be continuously stirred with a magnetic stirrer to maintain the particles in suspension. The room temperature should be kept constant during the whole experiment since this physical parameter has a strong effect on the adsorption/desorption. Aliquots of known volume are extracted at controlled elapsed time intervals, transferred to centrifuge vials and immediately centrifuged to separate the solid matter from the liquid. The aqueous phases are finally poured in flasks and the remaining concentration of the radionuclide under study in the supernatant is measured using adequate equipments. It is worth noting that the above procedure provide no certain information regarding the mechanism by which the contaminant are associated to the solid phase.

Strontium

Radiostrontium, particularly its isotope ^{90}Sr, is one of the most intensively studied contaminants in both the terrestrial and marine environments because of its long half life (28.6 years) and its high toxicity (Pröhl et al. 2006, Sysoeva et al. 2005 among others). It is known that it has a relatively weak affinity for solid particles in seawater (Benes & Poliak, 1990) compared to some highly reactive radionuclides such as transuranics. An experiment was carried out by suspending 5.85g of sediment in 1 liter of seawater (the electric conductivity at 20°C was 23 mS) collected from the western Mediterranean Sea. ^{85}Sr was used as a tracer in this experiment (Benkdad et al. 2008).

As can be seen in Figure 6 (a) and (b), the experimental data can be properly fitted to both one step and two step models. The first one (Figure 6a) consists of a single reversible reaction quantified by two kinetic transfer coefficients (k_1 for sorption and k_2 for desorption) where the mathematical solution is an exponential function for each compartment. The second model (Figure 6b) takes into account two reactive sites in the particles (the associated specific activities are a_{s1} and a_{s2} in Bq/kg) which can simultaneously exchange the tracer with the dissolved phase having a specific activity a_w in Bq/l. It may be noted that Sr is a highly hydrated cation at low aqueous concentrations and, therefore, its transfer can be properly described by a simple cation exchange reaction, but this is not always the case. Indeed, in the absence of competitive cations; it has been shown that two reactions of different characteristic times are unambiguously involved in the kinetic evolution of adsorption.

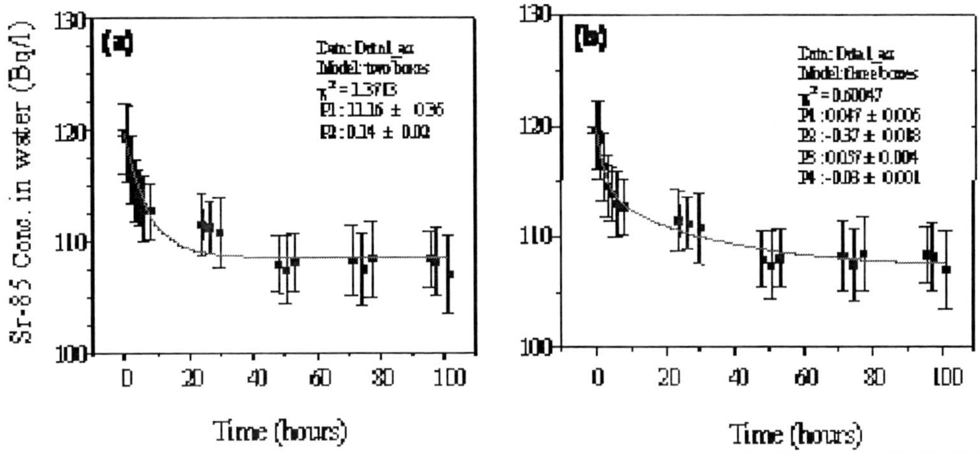

Figure 6. Sorption kinetics of ^{85}Sr in marine aqueous suspension. The points are the experimental data (with 1σ analytical error bars including counting statistics uncertainties) and the lines are the best fit to one-step reversible model (a) and two-step reversible model (PRRM) (b). The P_i values shown in the figure are fitting parameters to Eq. 2 in (a) and to Eq. 5 in (b).

The experimental data (points) were numerically fitted to the theoretical time evolution function (Eq. 2). Table 2 lists the values of the coefficients k_{ij} (in h^{-1}) by using the box model of two-parallel and reversible-reactions. The errors of k_{ij} coefficients were calculated from the uncertainties in fitting parameters P_i (i = 1 to 4).

Table 2. kinetic transfer coefficients, in h^{-1}, for the uptake of ^{85}Sr in aqueous suspension using naturally occurring solids and solution from the western Mediterranean Sea.

k_{11}	0.0018 ± 0.0003
k_{21}	0.028 ± 0.005
k_{12}	0.0172 ± 0.0015
k_{22}	0.35 ± 0.03
k_d (l/kg)	19.9 ± 2.1

All the parameters were obtained by processing the evolution of ^{85}Sr concentration in the dissolved phase. In view of the major difficulty in measuring experimentally the fraction of pollutant associated with different sites in the particles, the evolution of concentrations in all the compartments (water and the two sites in solids) can be determined by a modeling approach using the computed kinetic coefficients. The set of Equations 3, corresponding to the PRR model, was then numerically solved using a finite differences scheme forwarded in time and considering the initial conditions of our experiment. The time step was 0.01h. Thus, predictions of time evolution of the specific activities in water, the two sites in the particles and distribution coefficient were provided by the model and plotted in Figure 7. From this graph, it can be easily noted a sharp increase of activity in site 2 during the first few hours of the experiment, while adsorption onto site 1 is slower and has an asymptotic shape. This emphasizes the fact that site 2 is more accessible to the tracer than site 1. Thus, site 2 corresponds most probably to the active charges located at the external surface of the particles, and site 1 is related to the micropores in the particles where the tracer is strongly bound.

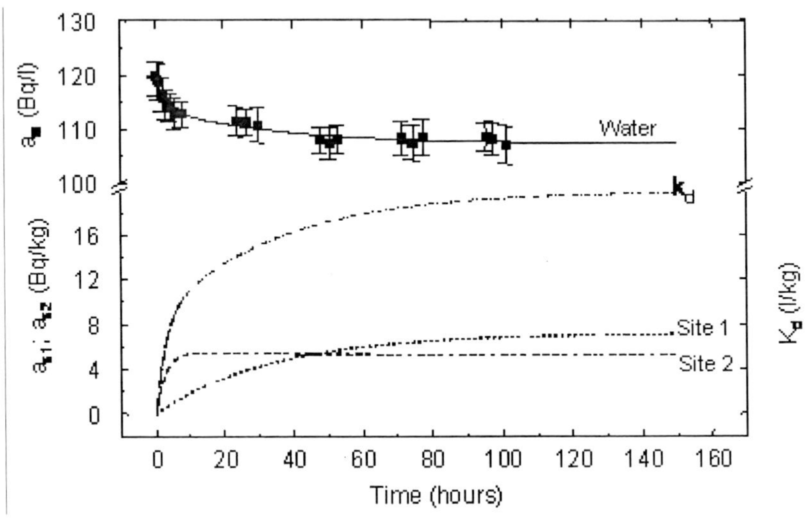

Figure 7. Time evolution of ^{85}Sr concentration in the dissolved phase (line) provided by the two-parallel and reversible-reactions model. The points correspond to the experimental data with 1σ analytical error bars. Predictions of time evolution of specific activities in the two sites and distribution coefficient, kd, are also plotted.

Americium

Americium is an important contributor to radioactivity in nuclear waste with high particle-affinity and complex chemical behaviour in natural systems. The most likely chemical form in the environment is Am(III). Tracer experiments were carried out using unfiltered water samples collected from two different aquatic locations in southern Spain: The Gergal reservoir (NW of Seville – Spain) and the Guadalquivir river (near Seville city).

Figure 8 shows, for both experiments, the time course of the ^{241}Am activity concentration in the dissolved phase. The uptake kinetics of this radionuclide in riverine aqueous suspensions showed three different characteristic times. Consequently, the uptake model must include at least three different reaction channels. In this way, the model (c) of Figure 4 consisting of two parallel and reversible reactions followed by a consecutive adsorption was selected to be applied successfully to the obtained experimental data.

Adsorption is important in the case of the Guadalquivir river experiment because of the high concentration of suspended particulate loads (172 ppm) compared to the experiment corresponding to the Gergal reservoir in which the concentration was only 4 ppm. There is a sharp decrease in the activity of the dissolved phase in the first minutes. In the Gergal experiment the activity in water appears to reach a steady state after few minutes (less than half hour), while in the Guadalquivir experiment a further uptake can be clearly appreciated.

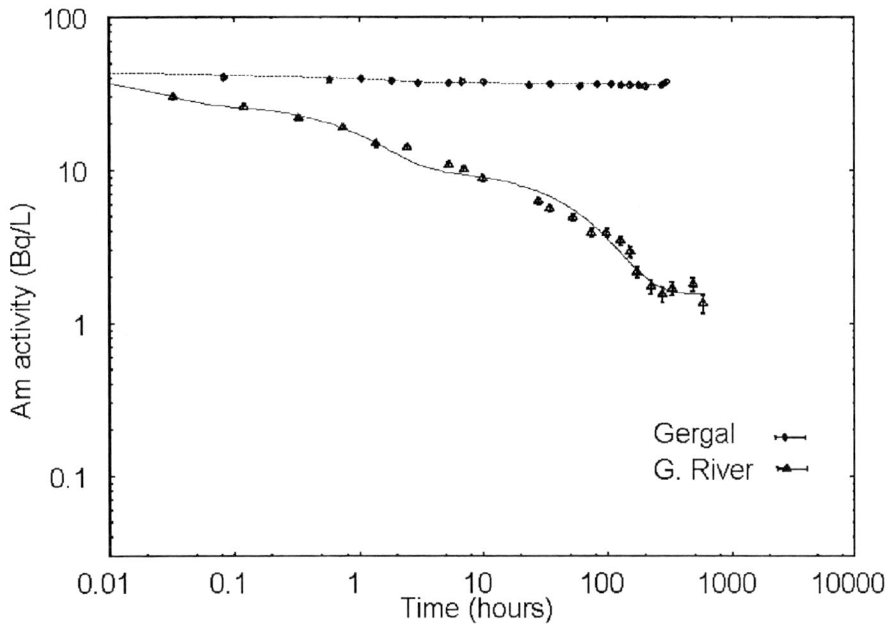

Figure 8. ^{241}Am activity concentrations in the dissolved phase (with 1σ error-bar) corresponding to the tracer experiments using two natural samples (The Gergal reservoir, ms=4 ppm and Guadalquivir river, ms=172 ppm). The lines correspond to the best fits to model (d) (Data from El Mrabet et al. 2004).

Table 3. Parameter values for the model of two parallel reactions plus a consecutive reversible reaction. The coefficients k_{ij} are in h^{-1}. The χ-test values are 0.8 and 1.3 for Gergal reservoir and Guadalquivir river experiments respectively.

	ms (ppm)	$k_{1,1}$	$k_{2,1}$	$k_{1,2}$	$k_{2,2}$	$k_{1,3}$	$k_{3,2}$
Gergal R.	4	0.85 (*)	17.0 ± 1.1	0.07 ± 0.02	0.50 ± 0.15	0.001 (*)	N.I.
G. River	172	40 ± 5	8.5 ± 1.6	1.1± 0.3	0.13 ± 0.05	0.030 ± 0.012	0.0036 ± 0.0022

Values marked with asterisk (*) have errors of the order of 100%. N.I. means that this channel of reaction has not been included in the model.

A two step model, or even a one step reaction model, can describe successfully the experimental data from the Gergal case, while a three-step model is unavoidable to parameterise data from the Guadalquivir experiment. For the sake of homogeneity, the model of two parallel and reversible reaction followed by a consecutive weakly-reversible adsorption (model c in Figure 4) was applied in this study to both experiments for comparison purposes. From Table 3, kinetic coefficients for the forward reaction, $k_{1,j}$ (j=1 to 3), of the first experiment are extremely lower due to the difference in suspended loads concentrations as commented before.

Further experiments, using ^{241}Am as a tracer, were performed to study the effect of suspended load concentration on the adsorption/release rates. The results showed that the kinetic transfer coefficients for sorption increase with the concentration of solids, while desorption coefficients did not show a clear pattern of variation with this parameter.

Plutonium

Plutonium isotopes' behaviour in natural aquatic ecosystems has been widely studied taking account of its oxidation states (Sibley & Sanchez, 1986). At least four oxidation forms Pu(III), Pu(IV), Pu(V) and Pu(VI) exist in aqueous environments, with the ions in hydrated form. The solubility of Pu in natural waters is low and mainly as Pu(V) and Pu(VI). The adsorbed forms are usually in the Pu(III) and Pu(IV) state. Tracer experiments were carried out with ^{239}Pu in unfiltered waters from three aquatic systems (El Mrabet et al. 2001): the Gergal reservoir (NW from Seville), the Guadalquivir river (near the town of Seville) and the Tinto estuary (close to Huelva city). In each experiment, a typical volume of 1 L of water was spiked with a known amount of ^{239}Pu. Results from three experiments are depicted in Figure 9 in logarithmic scale, which shows the time course of the total ^{239}Pu activity, given in counts per minutes (cpm), in the supernatant waters.

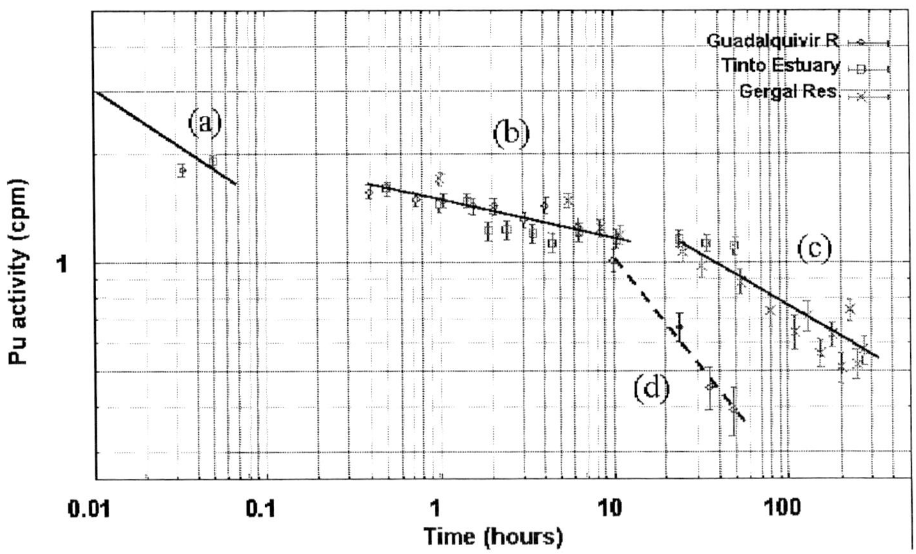

Figure 9. Time evolution of ^{239}Pu activities in the dissolved phase corresponding to the three experiments (points with error bars). Continuous lines indicate the different sequences of the uptake kinetics. Suspended matter concentrations are: 7.37 ppm for the Gergal river, 12.6 ppm for the Guadalquivir river and 7.62 ppm for the Tinto estuary.

The expected initial oxidation state of the tracer is Pu(IV), and an oxidising reaction could follow. The slow increase in concentrations, observed in some cases for long contact times may reflect an oxidation reaction.

Three different regions characterised by different slopes can be distinguished in Figure 9: region (a) corresponds to a fast reaction, lasting for few minutes, during which time up to 30% of dissolved Pu is transferred onto suspended solids; in region (b), a second reaction governs the Pu uptake during 10-20 hours time, transferring up to 30% of remaining Pu in dissolved form. In region (c), the uptake (up to 50% of the remaining Pu in solution) continues during several days. Region (d) corresponds to a faster uptake of Pu from the dissolved phase, which is probably related to flocculation effects. The model of two parallel, reversible and concentration independent reaction followed by a consecutive, non-reversible (with saturation) and concentration independent reaction was successfully applied and the five kinetic coefficients were obtained. In all cases, the first parallel reaction is the faster. It transfers an important fraction of the dissolved Pu to the particulate phase (20% for estuarine waters, 30% for the reservoir and some 42-44% for the river). The period during which this reaction governs the global uptake is about 0.2 h.

The application of a three-step model of single and effective oxidation state in this study allowed reaching a reasonable understanding of the Pu uptake curves. It also provided a set of kinetic parameters useful to describe the most relevant aspects of the electrolytic reactions of ^{239}Pu in the aquatic environment.

V. Conclusion

Kinetic box models are useful approaches to the analysis of time-resolved sorption data and are capable of providing transfer parameters which are important in determining the mobility and rate of removal of radionuclides from the water column. On the other hand, the box models represent an effective way of defining the number of processes involved in the adsorption in aqueous suspensions. A model taking into account a single reaction is generally unable to properly describe both sorption and release kinetics except in some cases such as adsorption of radiostrotium by suspended particulate matter in seawater. Higher order models such as "two-step" and "three-step" are most appropriate since the uptake of radionuclides by solid particles is usually a process where up to three steps can be distinguished: a fast kinetics (minutes), a moderate process (hours) and a slow sorption (days). However, the rationale for the selection of a model instead of another from the same category is still poorly established. This question should be seriously taken into consideration since it is determinant in interpreting the physical meaning of the transfer coefficients. One practical criterion able to discriminate among models is, from the authors' point of view, the statistical analysis of the variability of k_{ij} with physico-chemical parameters. The best model should give a coherent response to changes in these environmental factors.

On the other hand, it may be noted that the use of kinetic transfer coefficients in predictive dispersion models of pollutants has serious limitations which are due to the variability of physico-chemical parameters in natural aquatic systems (conductivity, pH, temperature, etc.) It is well known that rate parameters are sensitive to changes in environmental conditions and, therefore, their inclusion in dispersion models should be in the form of semi-empirical expressions as functions of the most relevant parameters. In this way, many more experimental studies are required to establish quantitatively the variability of transfer coefficients for each radionuclide for a more realistic description of surface-electrolyte interactions.

Finally, investigation on the behaviour of radionuclides in aquatic environments is still of contemporary importance and there will be a continuing requirement for a research capability in this area for as long as nuclear power remains a significant contributor to world energy supply.

REFERENCES

[1] Abril J.M. (1998a). Basic microscopic theory of the uptake kinetics, transfer and distribution of dissolved radionuclides by suspended matter. Part I: Applications. *Journal of Environmental Radioactivity* 41, No3, 325-342.

[2] Abril J.M. (1998(b). Basic microscopic theory of the uptake kinetics, transfer and distribution of dissolved radionuclides by suspended matter. Part II: theory development. *Journal of Environmental Radioactivity* 41, No3, 307-324.

[3] Arnold P.W. (1978). *Surface-electrolyte interactions*. In: The Chemistry of Soil Constituents. (eds.) Greenland and M. H. B. Hayes. Wiley, New York, ISBN 0 471 99619X.

[4] Barros H., Abril J.M. (2005). Constraints in the construction and/or selection of box models for the uptake of radionuclides and metals by suspended particulate matter, *Ecological Modelling*, 185, 371–385.

[5] Barros H., Laissaoui A. & Abril J.M. (2004). Trends of radionuclide sorption by estuarine sediments. Experimental studies using 133Ba as a tracer. *Science of the Total Environment*. 319, 253-267.

[6] Barros H. & Abril J.M. (2004). Experimental and modelling study on the uptake and desorption kinetics of 133Ba by suspended estuarine sediments from southern spain. *Water Research* 38, 749-755.

[7] Benkdad A., Laissaoui A., El Bari H., Benmansour M., IbnMajah M. (2008). Partitioning of radiostrontium in marine aqueous suspensions: laboratory experiments and modeling studies, *Journal of Environmental Radioactivity,* 99, 748-756.

[8] Benes P. & Cernick M. (1992). Kinetics of radionuclide interaction with suspended solids in modeling the migration of radionuclides in rivers. II: Effect of concentration of solids and temperature. *Journal of radioanalytical and Nuclear Chemistry*, 159 No2, 187-200.

[9] Benes P., Cernik M., & Lam Ramos P. (1992). Factors affecting interaction of radiocesium with freshwater solids. II contact time, concentration of the solid and temperature. *Journal of Radioanalytical and Nuclear Chemistry* 159 (2), 201-218.

[10] Benes P. & Poliak R. (1990). Factors affecting interaction of radiostrontium with river sediments. *Journal of Radioanalytical and Nuclear Chemistry* 141, 1, 75-90.

[11] Bird G.A. & Evenden W.G. (1994). Effect of sediment type, temperature and colloids on the transfer of radionuclides from water to sediment. *Journal of Environmental Radioactivity* 22, 219-242.

[12] Bunker D.J., Smith J.T., Livens F.R. and Hilton J. (2001). Kinetics of metal ion sorption on lake sediments – approaches to the analysis of experimental data. *Applied Geochemistry*, 651-658.

[13] Eisma, D. (1993). *Suspended Matter in the Aquatic Environment*. Berlin: Springer.

[14] El Mrabet R., Abril J.M., Manjon G. and Garcia-Tenorio R. (2004). Experimental and modeling study of 241Am uptake by suspended matter in freshwater environment from southern Spain. *Journal of Radioanalytical and Nuclear Chemistry*, 1, 137-144.

[15] El Mrabet R., Abril J.M., Manjon G. & Garcia-Tenorio R. (2001). Experimental and modelling study of plutonium uptake by suspended matter in aquatic environments from southern Spain. *Water Research* 17, 4184-4190.

[16] Hamilton-Taylor J., Kelly M., Kershaw P. and Lambert C.E. Radionuclide aquatic pathways. In: *Radioecology after Chernobyl*, (eds.) Sir Frederick Warner and Roy M. Harrisson. (1993), 177-274.

[17] Jannasch H.W., Honeyman B.D., Balistrieri L.S. and Murray J.W. (1988). Kinetics of trace element uptake by marine particles. *Geochimistry and Cosmochimistry Acta* 52, 567-577.

[18] Laissaoui A. & Abril J.M. (1999). A theoretical technique to predict the distribution of radionuclides bound to particles in surface sediments. *Journal of Environmental radioactivity* 44, 71-84.

[19] Laissaoui A., Abril J.M., Periañez R., Garcia-Leon M. & Garcia Montaño E. (1998). Kinetic transfer coefficients for radionuclides in estuarine waters: Reference values from 133Ba and effect of salinity and suspended load concentration. *Journal of Radioanalytical and Nuclear Chemistry* 237 1-2, 55-61.

[20] MacKenzie A.B. (2000). Environmental radioactivity: experiencen from the 20th century – trends and issues for the 21st century. *The Science of the Total Environment* 249, 313-329.

[21] Nyffler U.P., Li Y.H. & Santschi P.H. (1984). A kinetic approach to describe trace-element distribution between particles and solution in natural aquatic systems. *Geochimica and Cosmochimica Acta* 48, 1513-1522.

[22] Periáñez. R. On the sensitivity of a marine dispersion model to parameters describing the transfers of radionuclides between the liquid and solid phases. *Journal of Environmental Radioactivity* 73 (2004) 101-115.

[23] Pröhl G., Ehlken S., Fiedler I., Kirchner G., Klemt E. and Zibold G. (2006). Ecological half-lives of 90Sr and 137Cs in terrestrial and aquatic ecosystems. *Journal of Environmental Radioactivity*, 91, Issues 1-2, 41-72.

[24] Richter J. (1987). *The Soil as a reactor*. Catena Verlag, ISBN 3-923381-09-03.

[25] Robbins J.A., Linder G., Pfeiffer W., Kleiner J., Stabel H.H. and Frenzel P. (1992). Epilimnetic scavenging of Chernobyl radionuclides in Lake Constance. *Geochimistry and Cosmochemistry Acta* 56, 2339-2361.

[26] Sysoeva A.A, Konopleva I.V. and Sanzharova N.I. (2005). Bioavailability of radiostrontium in soil: Experimental study and modelling. *Journal of Environmental Radioactivity*, 81, Issues 2-3, 269-282.

[27] Sibley T.H., & Sanchez A.L. (1986). Why do Pu(III+IV) and Pu(V+VI) behave similarly in marine experiments. Letter to the Editor. *Journal of Environmental Radioactivity*, 3, 75-78.

[28] Smith J.T., and Comans R.N.J. (1996). Modelling the diffusive transport and remobilisation of Cs-137 in sediments: the effects of sorption kinetics and reversibility. *Geochimistry and Cosmochimistry Acta* 60, 995-1004.

In: Airborne Particulates
Editor: Ming Cheng and Wen Liu
ISBN: 978-1-60692-907-0

Chapter 10

CORRELATIONS BETWEEN DISEASE-SPECIFIC MORTALITIES WITH PARTICULATE AND GASEOUS AIR POLLUTANTS: RISKS FOR CARDIOPULMONARY DISEASE AND FEMALE REPRODUCTIVE ORGAN CANCERS*

Kazuro Iwai[1,†], Kazuhiro Uchimura[1], Shouichi Mizuno[2] and Yoji Miyasaka

[1] Research Institute of Tuberculosis, Japan Anti-tuberculosis Association, Tokyo, Japan
[2] Tokyo Metropolitan Research Institute of Gerontology, Tokyo, Japan
[3] Department of Respiratory Disease, Internal Medicine,
Tokyo University Medical School, Tokyo, Japan

ABSTRACT

Suspended particles, particularly fine particles of less than 2.5 μ m in aerodynamic diameter ($PM_{2.5}$) in the ambient air, has been reported to increase risks of cardiopulmonary diseases on daily death as well as deaths in long-term followed-up cohort studies. We conducted a cross-sectional study in which vital statistics in Japan were used to correlate disease-specific mortalities with air pollution data, including suspended particulate matter (SPM, less than 10 μm in aerodynamic diameter), nitrogen dioxide (NO_2), nitrogen oxide (NO), sulfur dioxide (SO_2) and photochemical oxidants (Ox). A single-pollutant regression analysis revealed a strong correlation of SPM with ischemic heart disease, hypertensive heart disease in both genders, and pneumonia, asthma, chronic bronchitis / emphysema, lung cancer and also breast, uterine and ovarian

---

* A version of this chapter was also published in *Environmental Research Journal Vol. 2 Issue2-3,* edited by Frank Columbus published by Nova Science Publishers, Inc. It was submitted for appropriate modifications in an effort to encourage wider dissemination of research

† Kazuro Iwai: Address: 3-1-24 , Matsuyama, Kiyose-shi, Tokyo, 204-8533, Japan; Tel: 81-424-93-5711, FAX: 81-424-92-4600, E-mail: iwai@jata.or.jp

cancers in females. Cerebrovascular disease, lung infarction and most of cancers did not show any correlation in either gender. NO_2 also showed a high statistically significant correlation with these diseases in a similar manner to SPM, whereas Ox had no association with any of these diseases. Results from the multi-pollutant Poisson regression analysis showed that cardiac diseases remained to correlate with SPM as well as NO_2, while pulmonary diseases correlated only with SPM. The unexpected correlation of female reproductive organ cancer, observed with SPM or NO_2 should further be studied in the future as possible endocrine disrupting-like effects.

Keywords: suspended particles, nitrogen dioxide, disease-specific mortality, cardiopulmonary disease. endocrine-disrupting effects.

INTRODUCTION

Epidemiological studies have shown an increase in cardio-pulmonary diseases in relation with the level of suspended particulate matter in the environmental air, especially for $PM_{2.5}$. Studies on the acute hazardous effects of air pollution have been conducted, and the relationship of pollutant levels to daily mortality [1] [2], frequency of ambulatory outpatient clinic visit and admission [3], increased respiratory symptoms[4] and worsening of lung function [5], have all been reported focusing considerable attention on the effects of particulate pollutants.

Large scale cohort studies have revealed the chronic effects of air pollutants on the mortality of cardio-pulmonary diseases among residents of polluted areas. Studies in 6 large US cities [6] followed up for 14 to 16 years, showed a higher mortality in more polluted cities, and those were correlated with levels of $PM_{2.5}$, sulfate and PM_{10}, but not NO_2 nor ozone, and the cause of deaths was cardiopulmonary disease. Cohort studies with adjustment for age, gender, race, smoking. alcohol use, occupational exposure and other factors [7] [8], demonstrated that PM_{10} and sulfate levels in ambient air correlated with death for all causes and cardiopulmonary diseases at a statistically significant level, but not with lung cancer. In a later study with longer follow up duration [9], cardiopulmonary disease as well as lung cancer showed a significant correlation with $PM_{2.5}$ level, showing an 8 % increase in the risk of lung cancer for each $10\mu g/m^3$ increase in $PM_{2.5}$.

If the entity of "cardiopulmonary" diseases consists not of single but of plural diseases, then questions remained as to what kind of diseases relate to a higher incidence of morbidity or mortality among the inhabitants in polluted areas than in control areas. Secondly, to what extent are the health effects of suspended particles observed in the U.S.A. the same as those in Japan, where the air pollution conditions and mortality patterns may differ between the two.

We used the vital statistics published in Japan, which enumerate age-adjusted mortality rates of various diseases for each prefecture and we compared them with air pollution data estimated at 1,881 points covering the entire country. In our previous report [10], a correlation between yearly disease-specific mortality in the vital statistics and mean yearly concentration of SPM was analyzed, and a statistically significant correlation was found between SPM or calculated $PM_{2.5}$ levels and disease-specific several cardiopulmonary diseases plus, unexpectedly, cancers of female reproductive organs. In the present study, disease-specific single- and multi-pollutant regression analyses were carried out on SPM,

NO_2, NO, SO_2 and Ox. In a single-pollutant analysis, significant correlations were shown for SPM and NO_2 with cardiopulmonary diseases including ischemic/ hypertensive heart diseases, pneumonia, bronchial asthma, chronic bronchitis/emphysema and, unexpectedly, with cancers of the breast, uterus and ovaries. Multivariate Poisson's regression analysis revealed a correlation of pulmonary diseases with SPM, cardiac diseases with SPM as well as nitrogen dioxide, and cancers of female reproductive organs with SPM or nitrogen dioxide.

MATERIALS AND METHODS

Ambient air quality data were obtained from a published data book, the Report on the State of Air Pollution Data, Japan, 2001 [11], which is open to the public. Annual mean concentrations of suspended particulate matter (SPM, 100% of them are 10μm or less in aerodynamic diameter in Japan), nitrogen dioxide (NO_2), nitrogen oxide(NO), sulfur dioxide (SO_2) and photochemical oxidants (Ox, estimated value is close to that of ozone) in the ambient air, estimated at 1,553 residential or commercial sites and 320 road side spots, were used. Carbon monoxide was estimated only at road-side spots but without residential areas, and therefore excluded from the analysis in the present study. The arithmetic annual mean values in 47 prefectures and 13 big cities, a total of 60 districts in Japan, were calculated from the data book.

The statistics of age-adjusted causes of death in 47 prefectures and 13 big cities for the calendar year 2000 were obtained from the Special Report on Vital Statistics of Japan [12]. Classification of disease was in accordance with the International Classification of Disease (ICD-10). Age at time of marriage, age at birth of first child, and number of children were obtained from the Vital Statistics, 2000. Smoking rates by age, gender, prefecture and big city in the whole country for the calendar year 2001 were obtained from another data book, The Nation's Basic Life Survey [13].

For statistical analysis, an estimation of the correlation coefficient and linear regression analysis were done to obtain the slope and a significance value with 95% confidence intervals, by using statistics software programs, The StatMate (ATMS Co., Ltd., Tokyo) and SPSS, Japanese edition. Multivariate Poisson's regression analysis was conducted by using STATA 8 (STATA Inc. TX, U.S.A.).

The Poisson regression model is used for analyzing the count data or rates. The response variable Y has a Poisson distribution and that expected value is depended on predictor variables x_i. The basic model formulation is that the mean of the response variable is a function of predictor variables,

$$E(Y) = \exp\{\beta_0 + \beta_1 x_1 + ... + \beta_n x_n\}$$

because the log of this function produce a linear combination of the predictors, the Poisson regression model is a generalized linear model with a log link function.

Poisson regression is appropriate for rate data. The rate is count of events occurring to a unit of observation, divided by some measure of that unit's *exposure*.

If U is the exposure, the formulation will be

$$E(Y/U) = \exp\{\beta_0 + \beta_1 x_1 + ... + \beta_n x_n\}$$

which implies

$$E(Y) = U\exp\{\beta_0 + \beta_1 x_1 + ... + \beta_n x_n\} = \exp\{\ln(U) + \beta_0 + \beta_1 x_1 + ... + \beta_n x_n\}$$

If all the x_i in the model are fixed constant except one, x_j, the rate raio for a one unit change in x_j is $\exp\{\beta_j\}$.

RESULTS

Mean Yearly Concentration of Each Pollutant, Averaged in 60 Districts of Japan

Mean yearly concentration of SPM, NO_2, NO, SO_2 and Ox, averaged in a total of 60 districts of Japan, and correlation coefficiencies among each pollutant, are shown in table 1. The correlation was strongest between NO_2 and NO; next, SPM and NO_2; SPM and NO, then SPM and SO_2 in decreasing order. Correlations between levels of Ox and other pollutants were mostly negative. The distribution map of pollution levels in each prefecture of Japan (figure 1) indicated that levels of both SPM and NO_2 are high in the prefectures where there are many big cities with high density population and traffic. In contrast, oxidants showed a relatively high level in several northwestern prefectures along the Japan Sea with low population density.

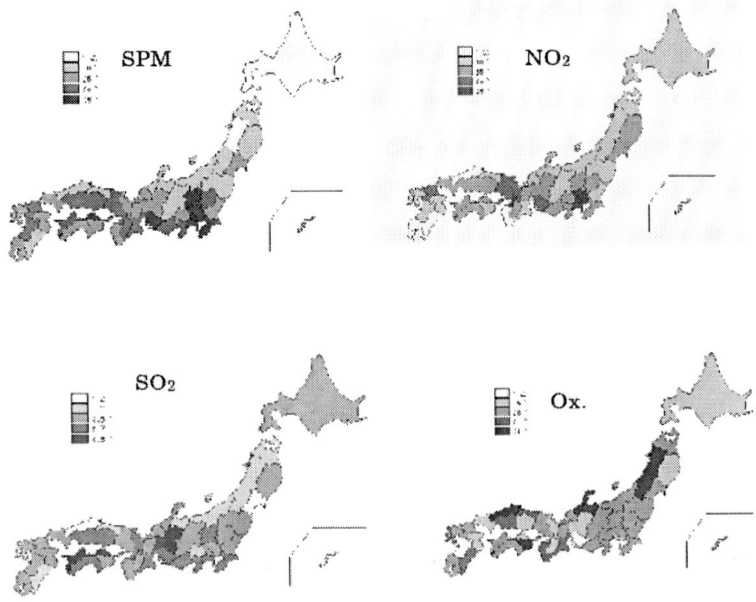

Figure 1. Map of Japan showing distribution of mean annual concentration of each air pollutant (SPM, NO_2, SO_2 and Ox.) by 47 prefectures.

Table 1. Air pollution levels and correlation coefficiencies between each pollutant

	SPM μg/m3	NO_2 ppb	NO ppb	SO_2 ppb	Oxidant ppb
Mean SD	29.8 5.2	18.5 7.2	14.1 7.1	4.4 1.5	28.8 3.6
SPM		0.78	0.64	0.61	-0.39
NO_2			0.94	0.57	-0.51
NO				0.46	-0.49
SO					-0.49

cf: CO was excluded from the table because of small number of estimating spots.

Univariate Correlation Analysis between Age-Adjusted Mortality and Each Pollutant

Mortality in each district varies according to age distribution of the inhabitants. Therefore, age-adjusted disease-specific mortalities were used to analyze the relationship of mortality with air pollution levels in the 60 districts. In males (table 2), both SPM and NO_2 showed a significant correlation with death rates of ischemic heart disease and hypertensive heart disease, NO with ischemic heart disease, and SO_2 with hypertensive heart disease. In females (table 3), SPM had a statistically significant correlation with ischemic and hypertensive heart diseases, pneumonia, bronchial asthma, chronic bronchitis/ emphysema, lung cancer and unexpectedly, breast cancer, uterine cancer and ovarian cancer. NO_2 also had a significant correlation with all these diseases except chronic bronchitis / emphysema. Endogenous NO has been reported from clinical and biological studies to reflect airway inflammation and to be partly involved in the carcinogenic processes, however, ambient NO showed, in this study, a correlation only with a small number of diseases. SO_2 showed a correlation with the least number of diseases. Ox showed completely different results, in which no disease showed a positive significant correlation while 8 of 9 diseases showed a negative correlation, and so was excluded from the following analysis.

Table 2. Age-specific mortality and univariate correlation coefficiency with each pollutant in males

		Correlation coefficiency				
Causes of death	Age-adjusted mortality per 10^5	SPM	NO_2	NO	SO_2	Ox
All death	634.2					
Ischemic heart disease	45	0.41**	0.43***	0.37*	0.24	–0.23
Hypertensive heart disease	1.3	0.33*	0.32*	0.22	0.26*	0.00
Pneumonia	53.1	0.13	0.10	0.08	0.06	0.04
Bronchial asthma	2.8	-0.11	-0.21	-0.14	-0.11	0.10
Chronic bronchitis/ emphysema	9.6	0.03	-0.24	-0.30	0.06	0.05
Lung cancer	46.3	-0.07	-0.09	-0.05	-0.54	0.13

*: $p<0.05$, **: $p<0.01$, ***: $p<0.001$.

Table 3. Age-adjusted mortality and univariate correlation coefficiency with each pollutant in females

		Correlation coefficiency				
Causes of death	Age-adjusted mortality per 10^5	SPM	NO_2	NO	SO_2	Ox
All death	323.9					
Ischemic heart disease	21.7	0.46***	0.53***	0.45***	0.30*	-0.34
Hypertensive heart disease	1.1	0.27*	0.29*	0.23	0.23	0.05
Pneumonia	23.3	043***	0.34**	0.29	0.34	-0.19
Bronchial asthma	1.6	0.34**	0.32*	0.30	0.31	-0.16
Chronic bronchitis/ emphysema	1.6	0.29*	0.19	0.18	0.05	-0.01
Lung cancer	12.3	0.31	0.46	0.49	0.18	-0.28
Breast cancer	10.7	0.37**	0.53***	0.51***	0.12	-0.24
Uterine cancer	5.3	0.45***	0.34**	0.30*	0.40*	-0.37
Ovarian cancer	4.3	0.43	0.50***	0.51***	0.29*	-0.24

*: $p<0.05$, **: $p<0.01$, ***: $p<0.001$.

Poisson's Regression Analysis for the Effects of Co-Pollutants

To analyze the effects of SPM, NO_2, NO and SO_2 on disease-specific mortality, multi-pollutant Poisson's regression model was applied to these data. The results showed, in general, a smaller number of correlating diseases in both genders. In males (table 4), SPM retained its correlation with the mortalities of ischemic and hypertensive heart diseases and chronic bronchitis / emphysema, while NO_2 had a correlation with ischemic heart disease. In this five-pollutant model in females (table 5), SPM preserved correlations with pneumonia, chronic bronchitis/emphysema and uterine cancer, and NO_2 with ischemic heart disease and breast cancer. NO showed a correlation with ischemic heart disease and SO_2 with no disease.

Table 4. Disease-specific mortality with Poisson regression analysis for co-pollutants in males

Causes of death	Incidence rate ratio (95% confidence interval)			
	SPM	NO_2	NO	SO_2
Ischemic heart disease	1.008** (1.005,1.049)	1.011** (0.999,1.049)	0.996 (0.993,1.001)	0.991 (0.981,12.00)
Hypertensive heart disease	1.024* (1.005,1.049)	1.024 (0.999,1.049)	0.985 (0.969,1.001)	1.003 (0.959, 1.050)
Pneumonia	1.003 (0.999,1.007)	1.002 (0.997,1.007)	0.999 (0.995,1.002)	0.998 (0.988,1.007)
Bronchial asthma	1.008 (0.990,1.026)	0.994 (0.971,1.018)	0.998 (0.983,1.014)	0.997 (0.955,1.042)
Chronic bronchitis/ emphysema	1.013 (1.004,1.023)	0.992 (0.980,1.005)	0.995 (0.987,1.004)	1.017 (0.993,1.040)
Lung cancer	0.997 (0.993,1.001)	1.002 (0.997,1.008)	0.999 0.965,1.003)	1.000 (0.990,1.011)

*: $p<0.05$, **: $p<0.01$, ***: $p<0.001$.

Table 5. Disease-specific mortality with Poisson regression analysis for co-pollutants in females

Causes of death	Incidence rate ratio (95% confidence interval)			
	SPM	NO_2	NO	SO_2
Ischemic heart disease	1.006 (0.999,1.012)	1.018** (1.001,1.027)	0.994 (0.988,0.999)	0.992 (0.997,1.008)
Hypertensive heart disease	1.010 (0.990,1.030)	1.025 (0.999,1.052)	0.988 (0.971,1.005)	0.996 (0.948,1.045)
Pneumonia	1.010** (1.004,1.005)	0.999 (0.992,1.007)	1.000 (0.995,1.005)	1.010 (0.995,1.025)
Bronchial asthma	1.010 (0.986,1.034	1.000 (0.970,1.030)	1.004 (0.984,1.024)	1.024 (0.968,1.084)
Chronic bronchitis/ emphysema	1.030 (1.007,1.054)	0.993 (0.964,1.022)	1.004 (0.985,1.025)	0.963 (0.912,1.017)
Lung cancer	0.997 (0.985,1.001)	1.002 (0.997,1.008)	0.999 (0.965,1.003)	1.000 (0.990,1.011)
Breast cancer	1.001 (0.992,1.010)	1..017** (1.005,1.029)	0.998 (0.990,1.005)	0.972 (0.960,0.993)
Uterine cancer	1.018** (1.004,1.031)	0.994 (0.978,1.011)	1.003 (0.992,1.014)	1.030 (0.998,1.063)
Ovarian cancer	1.006 (0.991,1.020)	1.005 (0.986,1.024)	1.005 (0.992,1.017)	0.999 (0.964,1.035)

*: $p<0.05$, **: $p<0.01$, ***: $p<0.001$.

Poisson's Regression Analysis for Confounding Factors Other than Air Pollutants

Poisson's regression model was applied to clarify the possible role of confounding factors other than co-pollutants on the disease-specific mortality. In males, SPM and NO_2 continued to show strong correlation with ischemic and hypertensive heart diseases , while smoking did not show any significant correlation with any of these diseases (data not shown). Poisson's regression analysis for NO_2 and non-pollutant confounders in females revealed a significant correlation of smoking with ischemic heart disease, lung cancer and breast cancer (table 6). However, hormonal factors such as initial child birth age and number of children, which are clinically considered to be important for the onset of female endocrine organ cancers, reflecting long duration of esterogen secretion during life, did not show any significant correlation with these diseases. NO_2 and similarly SPM preserved correlations with many of the diseases, analyzed in this study.

Table 6. Poisson regression analysis on NO2 and other confounding factors in females

		Other confouding factors		
Cause of death	Pollutant NO2	Smoking	Age of first birth	Small number of children
Ischemic heart disease	1.012*** (1.007,1.016)	1.011** (1.005,1.018)	0.986 (0.936.1.039)	0.999 (0.989,1.009)
Hypertensive heart disease	1.024*** (1.011,1.037)	0.980 (0.957,1.003)	0.941 (0.799,1.108)	0.987 (0.958,1.018)

Table 6. (Continued)

		Other confouding factors		
Cause of death	Pollutant NO2	Smoking	Age of first birth	Small number of children
Pneumonia	1.010*** (1.006,1.014)	1.004 (0.997,1.011)	0.928 (0.883,0.975)	0.999 (0.991,1.009)
Asthma	1.029*** (1.014,1.044)	0.993 (0.966,1.019)	0.932 (0.771,1.127)	0.952 (0.921,0.984)
Chronic bronchitis/ emphysema	1.020** (1.005,1.035)	0.987 (0.962,1.014)	1.091 (0.907,1.131)	0.932 (0.905,0.961)
Lung cancer	1.016*** (1.011,1.022)	1.009** (1.000,1.018)	0.987 (0.922,1.056)	0.972 (0.962,0.986)
Breast cancer	1.005 (0.999,1.011)	1.013* (1.003,1.022)	1.028 (0.954,1.108)	1.008 (0.994,1.023)
Uterine cancer	1.013** (1.005,1.021)	1.001 (0.987,1.016)	1.069 (0.963,1.188)	0.962 (0.945,0.980)
Ovarian cancer	1.011* (1.002,1.021)	1.006 (0.991,1.022)	0.989 (0.878,1.003)	1.010 (0.987,1.033)

*: $p<0.05$, **: $p<0.01$, ***: $p<0.001$.

DISCUSSION

The mean annual concentration of air pollutants of 60 districts in Japan, showed to be at a relatively high level of SPM and NO_2 / NO, and the ratio of $PM_{2.5}$ to SPM estimated in Tokyo was to be approximately 0.7 (0.53 – 0.90) [14]. Calculated $PM_{2.5}$ concentration thus corresponded to approximately 20μg/m^3, which exceeds the US Ambient Air-Control Standard of 15 μg/m^3. $PM_{2.5,}$ is considered to originate in most part from oil-combustion exhaust and the high levels of NO_2 / NO are possibly from the same source, including diesel engine emissions.

Correlation among each component of air pollutants showed the highest coefficiency value between NO_2 and NO, suggesting secondary production of NO_2 from NO in the ambient air. Ox showed a very different distribution pattern from those of SPM and NO_2 on the map of Japan, and had almost no correlation with cardiopulmonary diseases, differing from reports from other countries. The reasons for this high Ox level estimated in northern and cooler districts along the Japan Sea is unknown, however, a possible source may be Asian continent from which transportation of air pollutant over the Japan Sea occurred.

There are several epidemiological methods to analyze the health effects of air pollutants. A cohort study, based on the personal data of subjects with followed up health effects, creating possibility for the analysis of confounding factors, is now considered to be the more reliable analytical method, and the previous cohort studies using a huge number of objective persons with a long follow up period, showed an increase of risk in cardiopulmonary disease and lung cancer [9]. The results obtained from our present cross-sectional studies, which is weaker in excluding confounding factors as compared with cohort studies, were in the same line as those of the latter studies, but more disease-specific. In a uni-pollutant regression analysis, ischemic heart disease, hypertensive heart disease, pneumonia, asthma, chronic bronchitis / emphysema and lung cancer were found to have correlation with SPM, but unexpectedly, breast cancer, uterine cancer and ovarian cancer, also showed a highly significant correlation. All cancers of other sites and vascular diseases such as intracranial bleeding, brain or lung infarction, arrhythmia, heart valve disease and myocardial disease, showed no correlation [10].

A single pollutant regression analysis on NO_2 showed a highly significant correlation with ischemic heart diseases in both genders, and cancers of the lung, breast and ovary in females. It is not curious that NO also showed a correlation with the same diseases, as it co-exists with NO_2 in the air. SO_2 did not show a highly significant correlation with any of the diseases, and Ox had none or negative correlation with these diseases. Thus, NO_2 seems to be most important as a co-pollutant of SPM in our environment.

Effects of gaseous co-pollutants in the ambient air have been reported with inconsistent results, varying by location of the studies, temperature or other factors. SPM preserved significant correlation in most of the studies with daily death [15] [16] [17] [18] [19] [20] [21] [22], incidence of emergency admission [23] [24], and asthma attack [25]. CO was shown to be a significant co-pollutant in the studies operated in the US [15] [21] [23] [25]. NO_2 showed positive correlation in the studies in the EU [17], Rome [26], Hong Kong [27] and in Tokyo [24]. Ozone in the US [20], Hong Kong [27], Rome [26], and SO_2 in Korea [16], Canada [19] and Hong Kong [27], correlated with daily deaths. Collecting these results, it can be considered that constitution of air pollutants may differ by places or countries, and the proportion of gasoline to diesel vehicles in these countries may, at least partly, account for these differences of health hazards between the US and the EU / Hong Kong / Tokyo .

Multi-pollutant analysis in our study revealed that both SPM and NO_2 remained in moderate correlation with ischemic heart disease in males. Pneumonia and/or chronic bronchitis / emphysema had a correlation limited to SPM in both genders, breast cancer with NO_2, and uterine cancer with SPM. There may be a trend that heart disease correlates with both SPM and NO_2, pulmonary disease with SPM only and breast cancer with NO_2.

In regards to biological significance of the coexistence of particles and toxic gasses, an important question raised is whether each pollutant acts independently showing additive or synergistic effects. Physically, air-born carbon particles often have a wide surface area, which has a high ability to absorb co-existing gasses and possibly promoting the retention of harmful gasses, and irritating noxious gaseous may damage the clearance system of the inhaled particles by the bronchial epithelium. Biochemically, organic carbons absorbed on the surface of the carbon particles may change to nitrosoamines, e.g. nitrobenzene, a strong carcinogen, and nitrophenol as well-known endocrine disrupting chemicals, which are far stronger in action than original non-nitric compounds. Such biochemical mechanisms of co-existing pollutants may help the understanding of statistical data, however, further analytical studies are required for the role of co-pollutants on health.

Among many possible confounding factors other than co-pollutants, smoking habits have been reported to have a strong association with cardiovascular diseases, lung cancer and recently breast cancer [28]. In our present multi-variate analysis, the smoking rate showed an association with ischemic heart disease, pneumonia, lung cancer and breast cancer, but only in females. As hormonal factors have been clinically considered to be important in breast, endometrial and ovarian cancers as estrogen secretion could play a central role in the carcinogenesis, and anti-esterogenic agents are in fact applied to and effective in breast cancers as therapeutic drugs. However, no correlation was proven in this study for "high initial childbirth age" and "small number of children", while environmental factors, including

SPM or NO_2 and smoking were shown to be more important than endogenous ones in the present data.

Unexpected results in our previous and present study, which we think to be the first observational study in the literature on the subject, is an association of female endocrine organ cancers with SPM and NO_2 in the environmental atmosphere. In our previous study, in which the relationship of air pollution with mortality was analyzed in the three calender years of 1973, 1986 and 2000, ischemic heart disease and also breast and ovarian cancers showed consistent correlations with SPM in these years [29]. The reason why these female organ cancers showed a significant risk ratio related to inhaling pollutants is unknown. However, several recent animal experiments and *in vitro* studies support the possibility of an endocrine disrupting mechanism for diesel exhaust. Male mice exposed long-term to diesel exhaust revealed a marked decrease of spermatogenesis with atrophy of Sertoli cells in the testis [30], and similar effects were observed in rats exposed to both whole and filtered exhausts [31][32]. The morphological and hormonal changes noted in rats exposed to filtered exhaust suggest a presence of hormone-like activity also in gaseous components of diesel exhaust.

Biochemical studies using the extracts of diesel particles revealed the presence of esterogenic and anti-esterogenic agents [33], as well as aryl hydrocarbon receptor-dependent and esterogen receptor-dependent modulation of gene expression in breast cancer cells by the diesel particle extracts [34]. Nitrophenol families which have endocrine disruptive action were detected in diesel particles [35]. Nitrophenol and nitrophenylephenol extracted from diesel particles demonstrated esterogenic activities *in vitro* and in mice [36] [37]. These experimental studies suggest hormone-like activities for the diesel particles as well as for the gaseous components in the diesel exhausts, from which a role of nitro-compounds in the diesel particles emerges.

The possibility of endocrine disturbing action of suspended particles in the ambient air, which consists of dusts from various sources such as waste incineration, factory discharge and diesel exhaust, have been suggested. Nonylphenol, bisphenol, nitorphenols [38] and a trace amount of dioxins including PCDDs and Co-PCBs [39], all of which are known to have esterogenic activities, were found in air-borne dust samples. In human studies, motorway toll-gate young to middle-aged workers were reported to show a decreased mobility and cervical mucous penetration ability of sperm, possibly due to Pb and NO_2 [40].

In conclusion, the present cross-sectional epidemiological study revealed an association of SPM in the ambient air with cardiopulmonary diseases but not with other diseases in disease-specific analysis, which are the same line as in the preceding cohort studies except for cancer of the female endocrine organs. Effects of gaseous co-pollutants, especially of NO_2, seemed to be important for cardiac diseases in certain environmental conditions, while SPM per se seemed so mainly for pulmonary diseases. For breast cancer, NO_2 was considered to have greater importance. However, further many analytical epidemiological studies, as well as biochemical and *in vivo* studies are necessary for this significant issue to obtain concrete evidences.

ACKNOWLEDGMENT

The authors thank Prof Kunio Aoki, former president of Aichi Cancer Center, Nagoya, and Dr. Toru Mori, former Director of the Research Institute of Tuberculosis, JATA. for their adequate comments on the manuscript contents from epidemiological standpoint.

REFERENCES

[1] Schwartz, J. and Dockery, D.W. (1990) Particulate air pollution and daily mortality in Steubenville, Ohio. *Am. J. Epidemiol. 135*, 12-19.

[2] Schwartz, J. and Dockery, D.W. (1992) Increased mortality in Philadelphia associated with daily pollution concentrations. *Am. J. Respir. Dis. 145*, 600-604.

[3] Delfino, R.J., Murphy-Moulton, A.M., Burnet, R.T., Brook, J.R. and Becklane, M.R. (1997) Effects of air pollution on emergency room visits for respiratory illness in Montriol, *Am J Respir Crit Care Med 55*, 568-576

[4] Ware, J.H., Ferris, B.G. Jr., Dockery, D.W., Spengler, J.D., Stram, D.O. and Speitzer, F.E. (1986) Effects of ambient sulfur oxides and suspended particles on respiratory health of preadolescent children. *Am. Rev. Respir. Dis. 133*, 587-594

[5] Ackerman-Liebrich, U., Leuenberger, P., Schwartz, J., Schindler, C., Mon, C., Bolognini, G., et al. (1997) Lung function and long-term exposure to air pollution in Switzerland. *Am. J. Respir. Crit. Care Med. 155*, 122-129.

[6] Dockery, DW., Pope, III C.A., Xu, X., Spengler, J.D., Ware, J.H., Marthia, E.F., et al., (1993) An association between air pollution and mortality in six U.S. cities. *N Engl. J. Med. 329*, 1752-1759.

[7] Pope III, C.A., Then, M.T., Namboodiri, M.M., Dockery, D.W., Evans, J.S., Speizer, F.E., et al. (1995) Particulate air pollution as a predictor of mortality in a prospective study of U.S. adults. *Am. J. Respir. Crit. Care Med. 151*, 669-674.

[8] Samet, J.M., Domonici, F., Curriero, F., Coursac, I. and Zeger, S.L. (2000) Fine particulate air pollution and mortality in 20 U.S. cities, 1987–1994. *N. Engl. J. Med. 343*, 1742-1749

[9] Pope III, C.A., Burnett, R.T., Thun, M.J., Calle, E.E., Krewski, D. and Ito, K. (2002). Lung cancer, cardiopuomonary mortality, and long-term exposure to fine particulate air pollution. *J.A.M.A. 286*, 1132-1141.

[10] Iwai, K., Mizuno, S., Miyasaka, Y. and Mori, T. (2004). Correlation between suspended particles in the environmental air and causes of diseases among inhabitants: Cross-sectional studies using the vital statistics and pollution data in Japan. *Environ. Res. 99*, 106-117

[11] Ministry of Environment. Japan (2003*) The Report on the State of Air Pollution, Japan*, 2001.

[12] Ministry of Health, Labor and Welfare, Japan (2003) Special Report on Vital Statistics, *The Age-adjusted Death Rates by Prefecture.*

[13] Ministry of Health, Labor and Welfare, Japan (2003). The Nation's Basic Life Survey: *Smoking Rate by Age, Gender and Prefecture* 2001.

[14] Sasaki, K. and Sakamoto, K. (2004) Diurnal characteristics of SPM and $PM_{2.5}$ in the urban atmosphere of Tokyo, Japan. *Proceedings of the 13th World Clean Air and Environmental Protection Congress and Exhibition*, IUAPP, London, August 22-27, 2004 , Paper no.433. (Abstract and full papers in compact disk)

[15] Samet, J,M,, Zeger, S.L., Dominici, F., Curriero, F., Coursac, I. and Dockery, D.W. (2000) The national morbidity, mortality, and air pollution study Part II: Morbidity and mortality from air pollution in the United States. *Health Effect Institute Research Report 94*, 1-82.

[16] Lee, J-T., Kim, H., Hong, Y-C., Kwon, H-J., Schwartz, J. and Christiani, D.C. (2000). Air pollution and daily mortality in seven major cities of Korea, 1991-1997. *Environ. Res. Section A. 84*, 247-254.

[17] Katsouyanni, K., Touloumi, G., Samoli, E., Gryparis, A., Tertre, A. L., Monopolis, Y., et al. (2000) Confounding and effect modification in the short-term effects of ambient particles on total mortality: Results from 29 European cities within APHEA2 project. *Epidemilology 12*, 521-531.

[18] Omori, T., Fujimoto, G., Yoshimura, I., Nitta, H. and Ono, M. (2003) Effects of particulate matter on daily mortality in 13 Japanese cities. *J. Epidemiol. 13*, 314-322.

[19] Stieb, D.M., Judek, S. and Burnett, R.T. (2002) Meta-analysis of time-series studies of air pollution and mortality: Effects of gases and particles and the influence of cause of death, age, and season. *J. Air and Waste Manage. Assoc. 52*, 470-484.

[20] Schwartz, J. (2000) Assessing confounding, effect modification, and thresholds in the association between ambient particles and daily death. *Environ. Health Perspect. 108*, 563-568.

[21] Moolgavkar, S.H. (2000). Air pollution and daily mortality in three U.S. counties. *Environ. Health Perspect. 108*, 777- 784.

[22] Braga, A.L.F., Zanobetti, A. and Schwartz, J., 2001. The lag structure between particulate air pollution and respiratory and cardiovascular deaths in 10 US cities. J. *Occup. Environ. Med. 42*, 927-933.

[23] Tolbert, P.E., Klein, M., Metzger, K.B., Peel, J., Flanders, W.D., Tod, K,, et al.(2000) Interim results of the study of particulates and health in Atlanta (SOPHIA). *J. Exposure Analysis Environ. Epidemiol. 10*, 446-460.

[24] Ye, F., Pive, W.T., Ando, M. and Portier, C.J. (2001) Effects of temperature and air pollutants on cardiovascular and respiratory diseases for males and females older than 65 years of age in Tokyo, July and August 1980-1995. *Environ. Health Perspect 109*, 355-359.

[25] Yu, O., Sheppard, L., Lumley, T., Koenig, J.Q. and Shapiro, G.G. (2000) Effects of ambient air pollution on symptoms of asthma in Seattle-area children enrolled in the CAMP study. *Environ. Health Perspect. 108*, 1209-1214.

[26] Fusco, D., Forastiere, P., Michelozzi, P., Spadea, T., Ostro, B., Arca, M., et al. (2001) Air pollution and hospital admissions for respiratory conditions in Rome, Italy. 2001. *Eur. Respir. J. 17*, 1143-1150.

[27] Wong, C-M., Ma, S., Hedley, A.J. and Lam, T-H. (2001) Effect of air pollution on daily mortality in Hong Kong. *Environ. Health Perspect. 109*, 335-340.

[28] Hanaoka, T., Yamamoto, S., Sobue, T., Sasaki, S. and Tsugane, S. (2005) Japan public health center-based prospective study on cancer and cardiovascular disease study.

Active and passive smoking and breast cancer risk in middle-aged Japanese women. *Int. J. Cancer 114*, 317-22.

[29] Miyasaka, Y,, Iwai, K. and Mizuno, S. (2005) Changes of correlation between suspended particles and disease mortality. *Proceedings of the 46th Annual Meeting of Japan Society for Atmospheric Environment*, Nagoya, 2005. p.303

[30] Yoshida, S., Sagai,M,, Oshio, S., Umeda, T., Ihara, T. and Sugamata, M. (1999). Exposure to diesel exhaust affects the male reproductive system of mice. *Int. J. Androl. 22*, 307-315.

[31] Watanabe, N. and Oonuki, Y. (1999) Inhalation of diesel engine exhaust affects spermatogenesis in growing male rats. *Environ. Health Perspect. 107*, 539-544.

[32] Watanabe, N. (2005) Decreased number of sperms and Sertoli cells in mature rats exposed to diesel exhaust as fetus. *Toxicol. Lett. 155*, 51-58.

[33] Taneda, S., Hayashi, H., Sakushima, A., Seki, K., Suzuki, A.K., Kameta, K., et al. (2002) Esterogenic and anti-esterogenic activities of two types of diesel exhaust particles. *Toxicology 170*, 153-161.

[34] Meek, M.D. (1998) *Ah receptor- and esterogen receptor-dependent modulation of gene expression by extracts of diesel exhaust particles*. 79, 114-121.

[35] Mori, Y., Kamata, K., Toda, N., Hayashi, H., Seki, K., Taneda, S., et al., 2003. Isolation of nitrophenols from diesel exhaust particles (DEP) as vasodilatation components. *Biol. Pharm. Bull. 26*, 384-385

[36] Taneda, S., Mori, Y., Kamata, K., Hayashi, H., Furuta, C., Li, C., et al. (2004) Esterogenic and anti-androgenic activities of nitorphenols in diesel exhaust particles (DEP). *Biol. Pharm. Bull. 27*, 835-837.

[37] Furuta, C., Suzuki, A.K., Taneda, S., Haayashi, H., Mori, Y., Li, C., et al. (2004) Esterogenic activities of nitrophenols in diesel exhaust particles. *Biol. Reprod. 70*, 1527-1533

[38] Ruddel, R.A., Broady, J., Spengler, J.D., Vallarino, J., Geno, P.W., Sun, G., et al. (2001) Identification of selected hormonally active agents and animal mammary carcinogenesis in commercial and residential air and dust samples. *J. Air Waste Manage. Assoc. 51*, 499-513.

[39] Yoshioka, H., Sasaki, Y., Tsukui, T., Iimura, F. and Higashino, K. (2004) Gas-particel partitioning of PCDDs and Co-PCBs in the atmosphere. *J. Jap. Soc. Atoms. Environ,. 39*, 1-11.

[40] DeRosa, M., Zarrilli, M., Paesano, L., Carbone, U., Boggia, B., Petretta, M., et al. (2003) *Traffic pollutants affects fertility in men. Human Reprod.* 18, 1055-1060.

Reviewer's Comments

General Comments

In this manuscript the authors conducted a cross-sectional study in which vital statistics in Japan were used to correlate disease-specific mortalities with air pollution data, including suspended particulate matter (SPM), nitrogen dioxide (NO_2), nitrogen oxide (NO), sulfur dioxide (SO_2) and photochemical oxidants (Ox).

They showed that a strong correlation of SPM with ischemic heart disease, hypertensive heart disease in both genders, and pneumonia, asthma, chronic bronchitis / emphysema, lung cancer and also breast, uterine and ovarian cancers in females.

Results from the multi-pollutant Poisson regression analysis showed that cardiac diseases remained to correlate with SPM as well as NO_2, while pulmonary diseases correlated only with SPM. The unexpected correlation of female reproductive organ cancer, observed with SPM or NO_2 should further be studied in the future as possible endocrine disrupting-like effects.

The results presented by authors in this manuscript were important, because analyses of air pollution data were carried out at 1,881 points covering all of Japan. The air pollution conditions and mortality patterns might differ from those of U.S.A.

It is very important finding that a strong correlation of SPM with breast, uterine and ovarian cancers in females was obtained. The reason is not clear yet, however, the fact that these three cancers require estrogen for their development is interesting to consider that SPM contain diesel exhaust particles (DEP) and DEP have various endocrine disrupting materials. Furthermore inhalation of airborne particles into the lungs leads to presumptive mutations in mouse germ cells that can be passed on to the next generation.

Suggested Comments

1) DEPs contain many substances that stimulate Ah receptors, such as the polycyclic aromatic hydrocarbon containing benzo[*a*]pyrene.
   (Westerholm, R., and Egeback, K. E. (1994): Exhaust emissions from light- and heavy-duty vehicles: chemical composition, impact of exhaust after treatment, and fuel parameters. *Environ. Health Perspect.* 102 Suppl 4:13-23.)
2) Dioxins are fat-soluble environmental pollutants that disrupt the endocrine system of humans (see Harper). The aryl hydrocarbon receptor (AhR) binds dioxins; interacts with the AhR nuclear-translocator protein (ARNT) and sex hormone steroid receptors, such as estrogen receptor α(ERα) and androgen receptor (AR); and promotes hormone-independent transcriptional activation.
   (J. W. Harper, A degrading solution to pollution. *Nature* 446, 499-500 (2007). [PubMed])
3) In addition, dioxins also inhibit sex steroid signaling, and Ohtake *et al.* show that one mechanism for this inhibition is by promoting the proteasome-mediated degradation of the steroid hormone receptors.
   (F. Ohtake, A. Baba, I. Takada, M. Okada, K. Iwasaki, H. Miki, S. Takahashi, A. Kouzmenko, K. Nohara, T. Chiba, Y. Fujii-Kuriyama, S. Kato, Dioxin receptor is a ligand-dependent E3 ubiquitin ligase. *Nature* 446, 562-566 (2007). [PubMed])
   Combined 1), 2) and 3), polycyclic aromatic hydrocarbons in DEP cause the endocrine disrupting activity through Ah receptors.
4) Monohydroxy derivatives of PAHs in DEP have estrogenic and antiestrogenic activities.
   (Ken Takeda, Naomi Tsukue and Seiichi Yoshida (2004): Endocrine-Disrupting Activity of Chemicals in Diesel Exhaust and Diesel Exhaust Particles (Review). Environmental Sciences, 11,33-45.)
   "Hayakawa et al. reported that DEPs contain both PAHs and their monohydroxy derivatives (OHPAHs). It can be presumed that OHPAHs have estrogenic and

antiestrogenic activities because the structure of some OHPAHs is similar to that of estrogen [1].

Estrogenic and antiestrogenic activities of 80 kinds of PAHs and OHPAHs having 2 to 6 rings were measured by the yeast two-hybrid assay. Estrogenic and antiestrogenic activities were indicated by their effects relative to the effects of 17β-estradiol and 4-hydroxytamoxifen (4-OHT), respectively [2].

A study of the relationship between the activities obtained from the experiments and structural or physicochemical properties calculated using computer-aided chemistry (CAChe) is in progress. They found three important relationships. First, OHPAHs have estrogenic and antiestrogenic activities while PAHs do not. The activities of OHPAHs were dependent on the ring number or position of the hydroxyl group. Second, OHPAHs having 2 to 5 rings, particularly 2-hydroxyfluorene, 4-hydroxybenz[*a*]anthracene, and 3-hydroxybenzo[*a*]pyrene, have strong estrogenic activity. Finally, OHPAHs having 4 rings have strong antiestrogenic activity. The activity of 3-benzo[*c*]phenanthrene was stronger than that of 4-OHT, which is a typical antagonist[2]. "

1. Hirose, T., Morito, K., Kizu, R., Toriba, A., Hayakawa, K., Ogawa, S., Inoue, S., Muramatsu, M., and Masamune, Y. (2001): Estrogenic/antiestrogenic activities of benzo[*a*]pyrene monohydroxy derivatives. *J. Health Sci.* 47(6) 552–558
2. Hayakawa, K., Onoda, Y., Kamiya, M., Hirose, T., Toriba, A., and Kizu R. (2002): Structure-activity relationship for estrogenic/antiestrogenic activities of monohydroxylated polycyclic aromatic hydrocarbons. *The 5th Annual Meeting of Japan Society of Endocrine Disrupters Research.* 25-26.

5) Prenatal exposure to diesel exhaust (DE) elevated serum testosterone levels of the exposed group significantly at 12 weeks.
(Naoka Ono, et al. (2007): Prenatal Exposure to Diesel Exhaust Impairs Mouse Spermatogenesis. Inhalation Toxicology, 19:275–281)
6) Inhalation of airborne particles into the lungs leads to presumptive mutations in mouse germ cells that can be passed on to the next generation.
(Somers, CM et al. (2004): Reduction of particulate air pollution lowers the risk of heritable mutations in mice. Science 304, 1008-1010)

Review by Ken Takeda,PhD, Professor , Department of Hygiene-Chemistry, Faculty of Pharmaceutical Sciences, Tokyo University of Science.

SHORT COMMUNICATIONS

In: Airborne Particulates
Editor: Ming Cheng and Wen Liu

ISBN: 978-1-60692-907-0

Short Communication A

PHOTOCHEMICAL DEGRADATION OF SELECTED NITRO- AND OXY-POLYCYCLIC AROMATIC HYDROCARBONS ON AIRBORNE PARTICLES UNDER SIMULATED SOLAR UV-IRRADIATION

Takayuki Kameda*[a], Yuki Nakayama[a], Tomoko Goto[a], Takeshi Koyanagi[b], Hiroshi Bandow[c], Keiichi Fujimori[b], Akira Toriba[a], Ning Tang[a] and Kazuichi Hayakawa[a]

[a]Graduate School of Natural Science and Technology, Kanazawa University, Kakuma-machi, Kanazawa, Ishikawa 920-1192, Japan

[b]College of Engineering, Osaka Institute of Technology

[c]Graduate School of Engineering, Osaka Prefecture University

ABSTRACT

Polycyclic aromatic hydrocarbons (PAHs) and their nitrated/oxygenated derivatives (NPAHs, OPAHs) are atmospheric pollutants which exhibit various biological activities such as mutagenic and carcinogenic activities. These compounds are generally generated via incomplete combustion of organic materials or chemical conversion of the parent compounds and released into the atmosphere with airborne particles. In this study, photolysis rate constants and quantum yields of seven OPAHs and eighteen NPAHs including 3-nitrobenzanthrone (3-NBA), a nitrated aromatic ketone showing extremely strong mutagenicity were newly determined to simulate the atmospheric fate of the compounds on airborne particles under UV irradiation. NPAHs were more easily photodecomposed than the corresponding oxygenated derivatives. 9-Nitroanthracene had the largest photolysis rate constant, while 4-nitropyrene and 3-NBA were found to be the most photoresistant of the nitrated aromatic compounds tested in this study under UV irradiation. Benzo[*c*]phenanthrene-5,6-quinone which has photolysis rate constant similar to that of 9-nitrophenanthrene degraded the fastest of the seven OPAHs. 1,2-

* E-mail: kameda@p.kanazawa-u.ac.jp
Tel: +81-76-234-4458 Fax: +81-76-234-4456

Benzanthraquinone was the most photoresistant of all the substituted PAHs examined in this study. Atmospheric lifetime of the compounds due to photolysis was successfully estimated using the actinic flux on the earth's surface, photolysis rate constants, and quantum yields obtained in this study.

INTRODUCTION

Polycyclic aromatic hydrocarbons (PAHs) and their nitrated/oxygenated derivatives (NPAHs, OPAHs) are ubiquitous atmospheric pollutants with toxic, mutagenic, and carcinogenic properties [1, 2]. These compounds are predominantly produced from a wide variety of anthropogenic sources such as the incomplete combustion of fossil fuels for industrial plants, heating, as well as diesel powered vehicles [3–5]. NPAHs, which generally exhibit higher mutagenicity and carcinogenicity than the parent PAHs, are also generated from atmospheric reactions of PAHs released into the gas-phase with radical species such as OH- and NO_3-radicals and nitrogen oxides [6–10]. OPAHs are also known to be secondarily formed via oxidation of PAHs and NPAHs in the atmosphere [11, 12]. These derivatized PAHs generally have lower volatility than the parent PAHs and are likely to be deposited on airborne particles, especially fine particles, which are easily taken into the inside of human body, and diffuse into the atmosphere.

Photoinduced degradation is a dominant pathway for the degradation of particle-associated PAH derivatives [13, 14]. Solar radiation in UV region can modify PAH derivatives to form new compounds, which may exhibit different types of biological effects such as disruption of endocrine systems and production of reactive oxygen species (ROS) in human body [15]. Thus, it is necessary to understand the photodecomposition of PAH derivatives in order to understand their effects on humans. However, the rate constants and the quantum yields regarding the photolysis of the derivatized PAHs, which are the most significant factors for the photodecomposition, have not been studied well.

In this study, we conducted photodegradation experiments for selected OPAHs and NPAHs including 3-nitrobenzanthrone (3-NBA), a nitrated aromatic ketone that is strongly mutagenic [16], on a glass surface as a simple model of airborne particles and determined the photolysis rate constants and the quantum yields for the PAHs derivatives in the system. Furthermore, the atmospheric lifetime of the compounds due to photodecomposition was estimated using the actinic flux on the earth's surface, photolysis rate constants, and quantum yields obtained in this study.

EXPERIMENTAL SECTION

Photoreaction Experiments

Seven OPAHs, eighteen NPAHs, two PAHs, and 1-hydroxypyrene (1-OHPy) (Fig. 1) were applied to the photoreaction experiments, which were employed with a Pyrex photo-reactor (5 L, 10 cm ID × 70 cm length) using ten black light lamps (20 W, Toshiba, FL20S-BLB) as the radiation source. The inside surface of the reactor was uniformly coated with 150 nmol of the reactants and then irradiated by the black light lamps placed around the reactor. The total incident photon flux reaching to the reactor surface (I, photons cm^{-2} s^{-1}), measured with a Hatchard-Parker actinometer using potassium ferrioxalate photoreduction [17], was 7.0×10^{16} photons cm^{-2} s^{-1}. The incident photon flux at the wavelength λ (I_λ) was not observed by the chemical actinometer employed in this study. Therefore, I_λ was determined by distributing I among the UV range from 310 to 400 nm of the wavelength (maximum intensity was obtained at 350 nm, Fig. 2) based on the relative spectral intensity of the irradiation by the black light lamp.

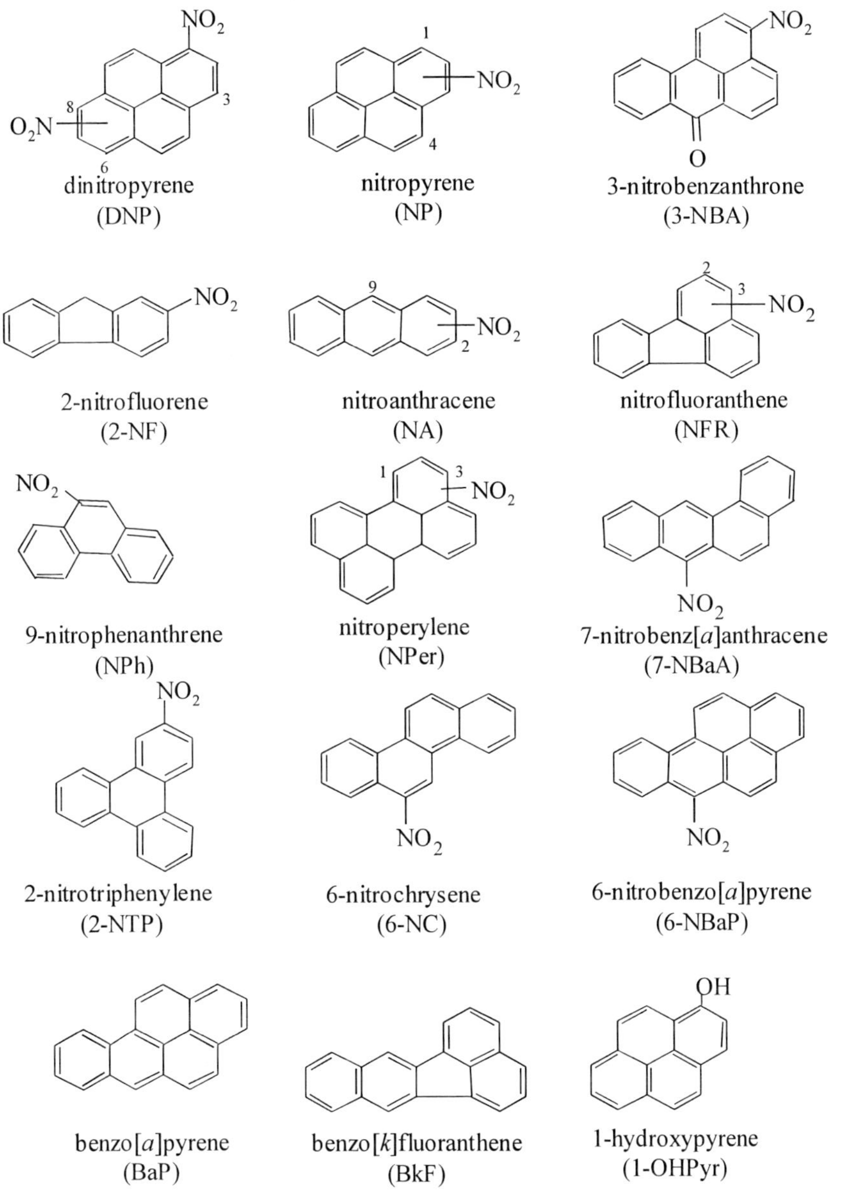

Figure 1. (Continued on next page)

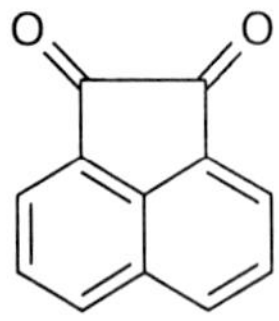

Acenaphthenequinone

(AcQ)

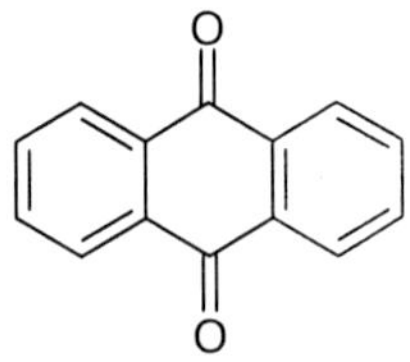

Anthraquinone

(AQ)

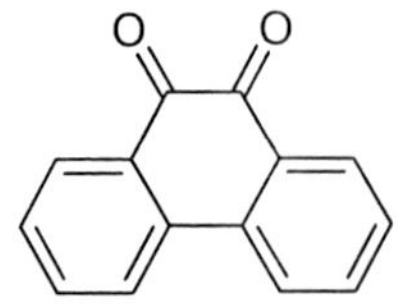

9,10-Phenanthrenequinone

(9,10-PQ)

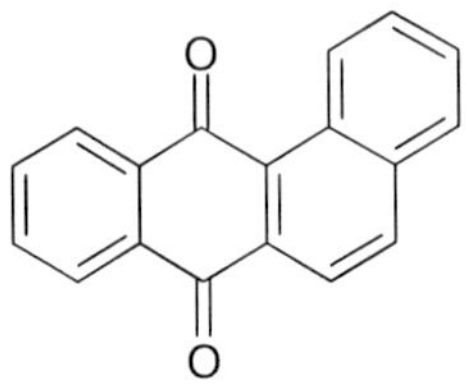

1,2-Benzanthraquinone

(1,2-BAQ)

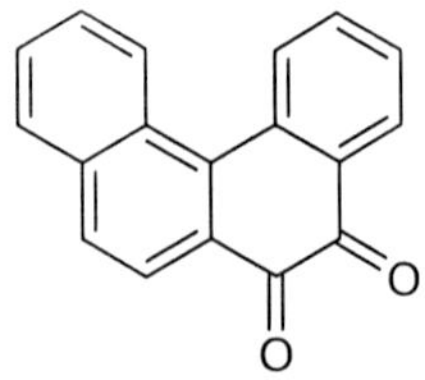

Benzo[*c*]phenanthrene-5,6-quinone

(B[*c*]P-5,6-Q)

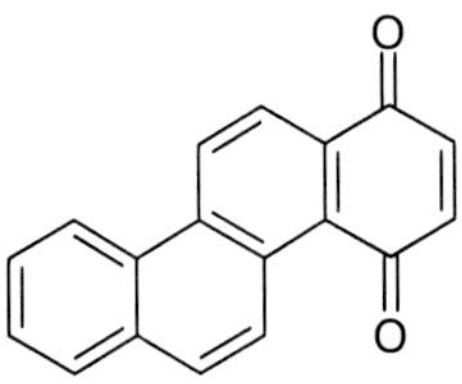

1,4-Chrysenequinone

(1,4-CQ)

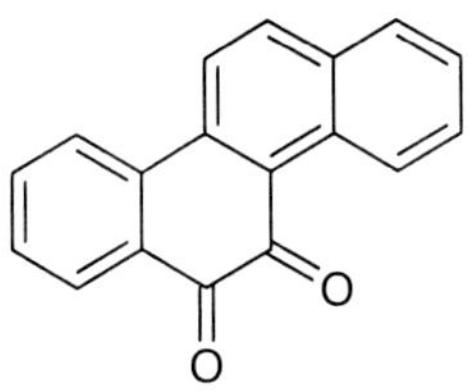

5,6-Chrysenequinone

(5,6-CQ)

Figure 1. Structures of the compounds examined in this study.

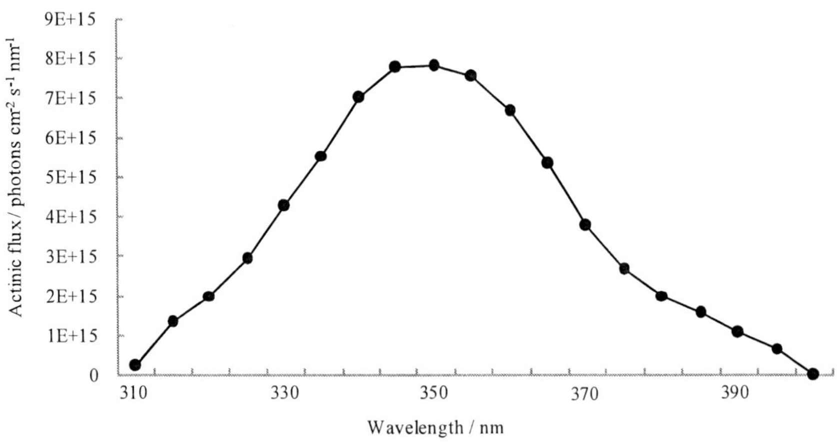

Figure 2. Actinic flux of ten black-light lamps employed in this study. Total flux was determined by the Hatchard-Parker actinometer to be 7.0×10^{16} photons cm^{-2} s^{-1}. See text for the detail.

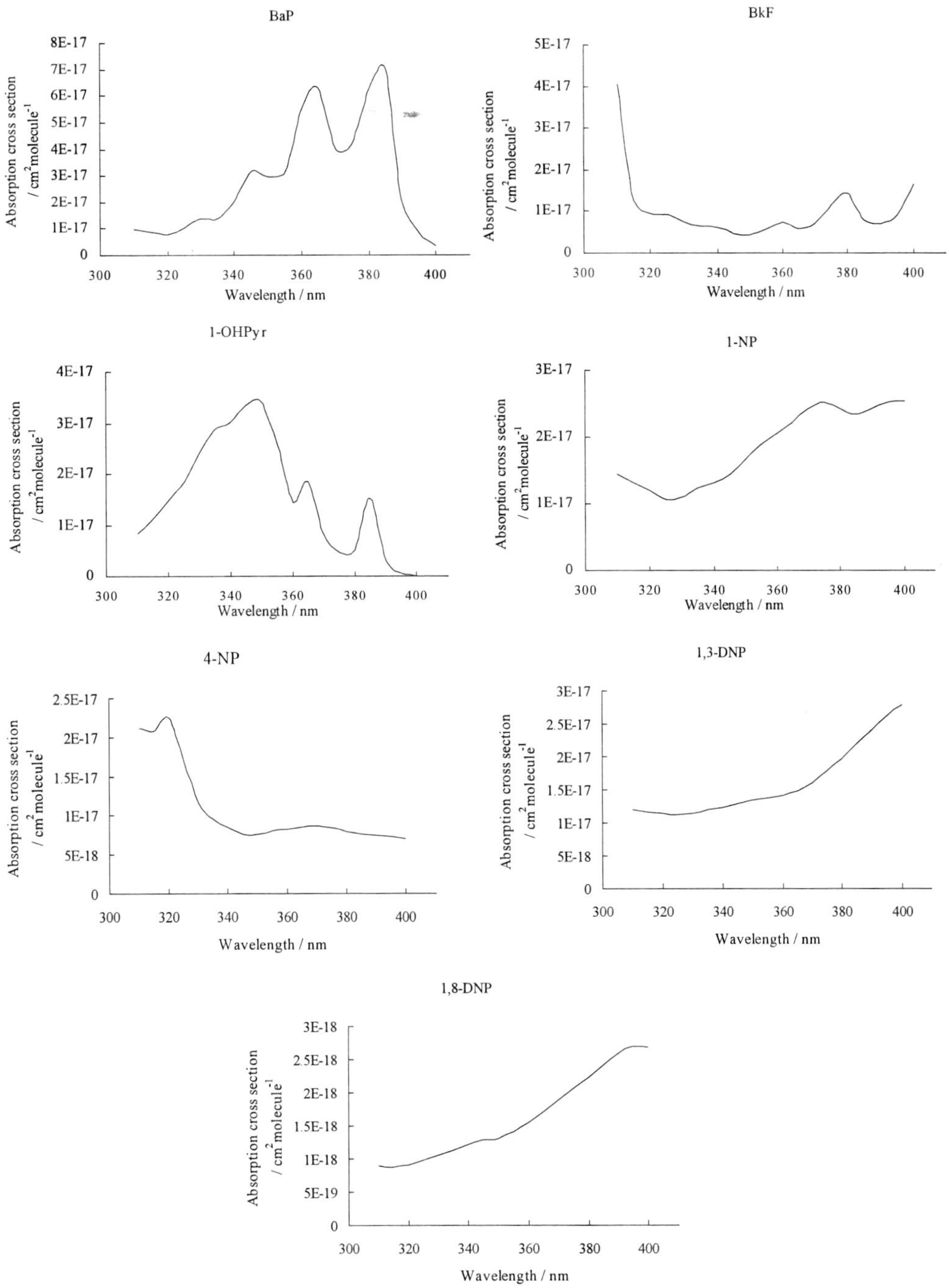

Figure 3. (Continued on next page)

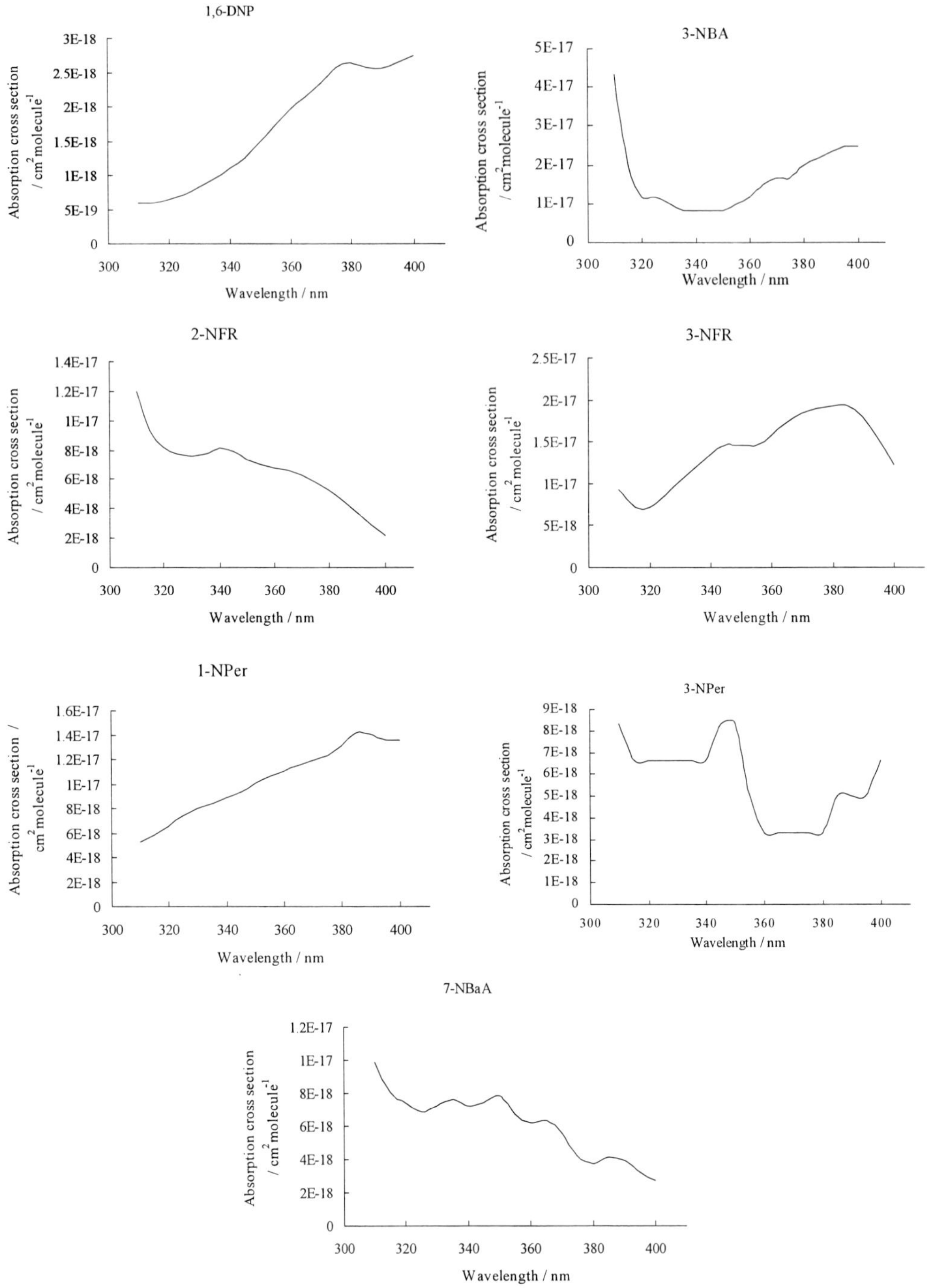

Figure 3. (Continued on next page)

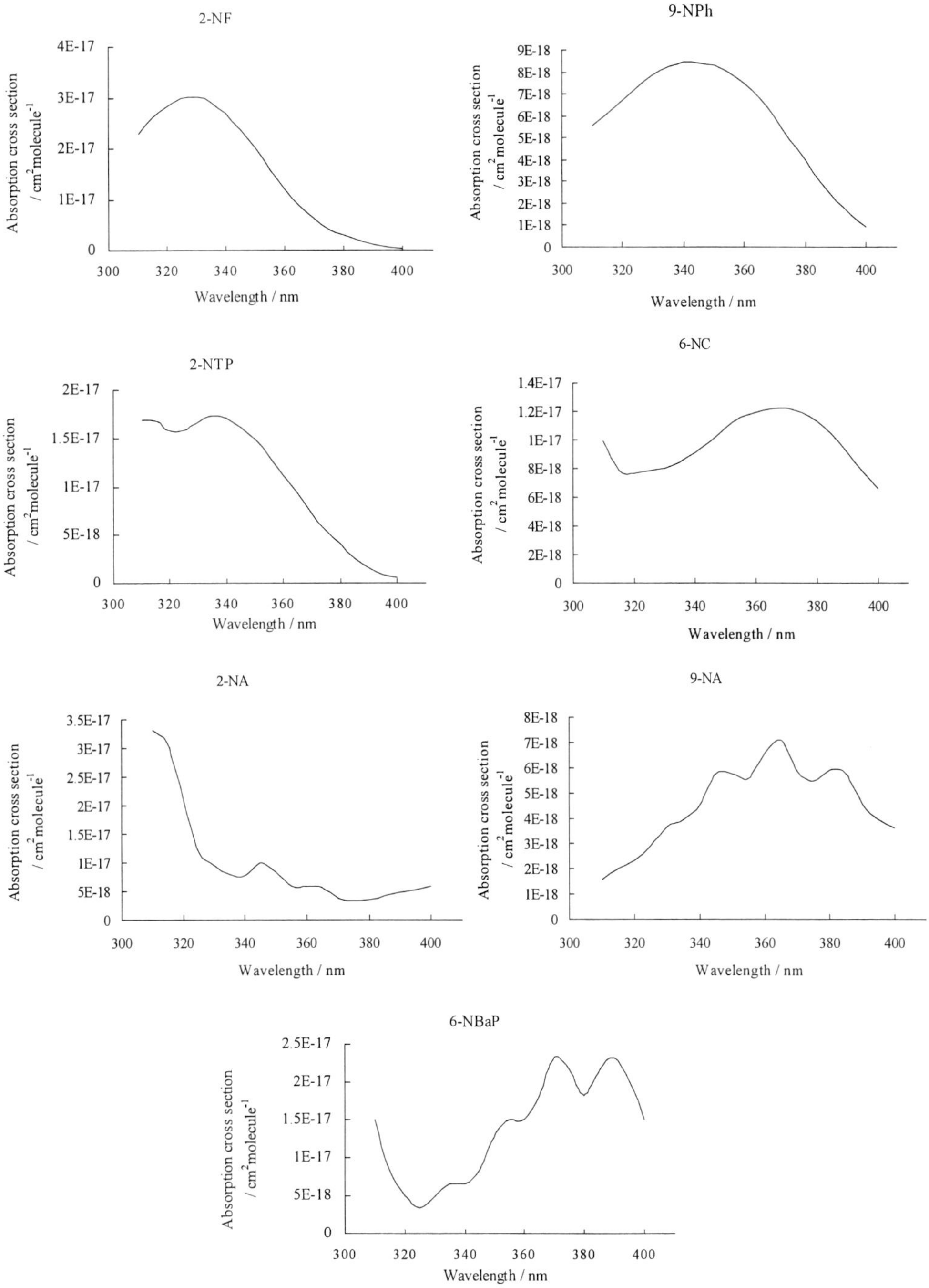

Figure 3. (Continued on next page)

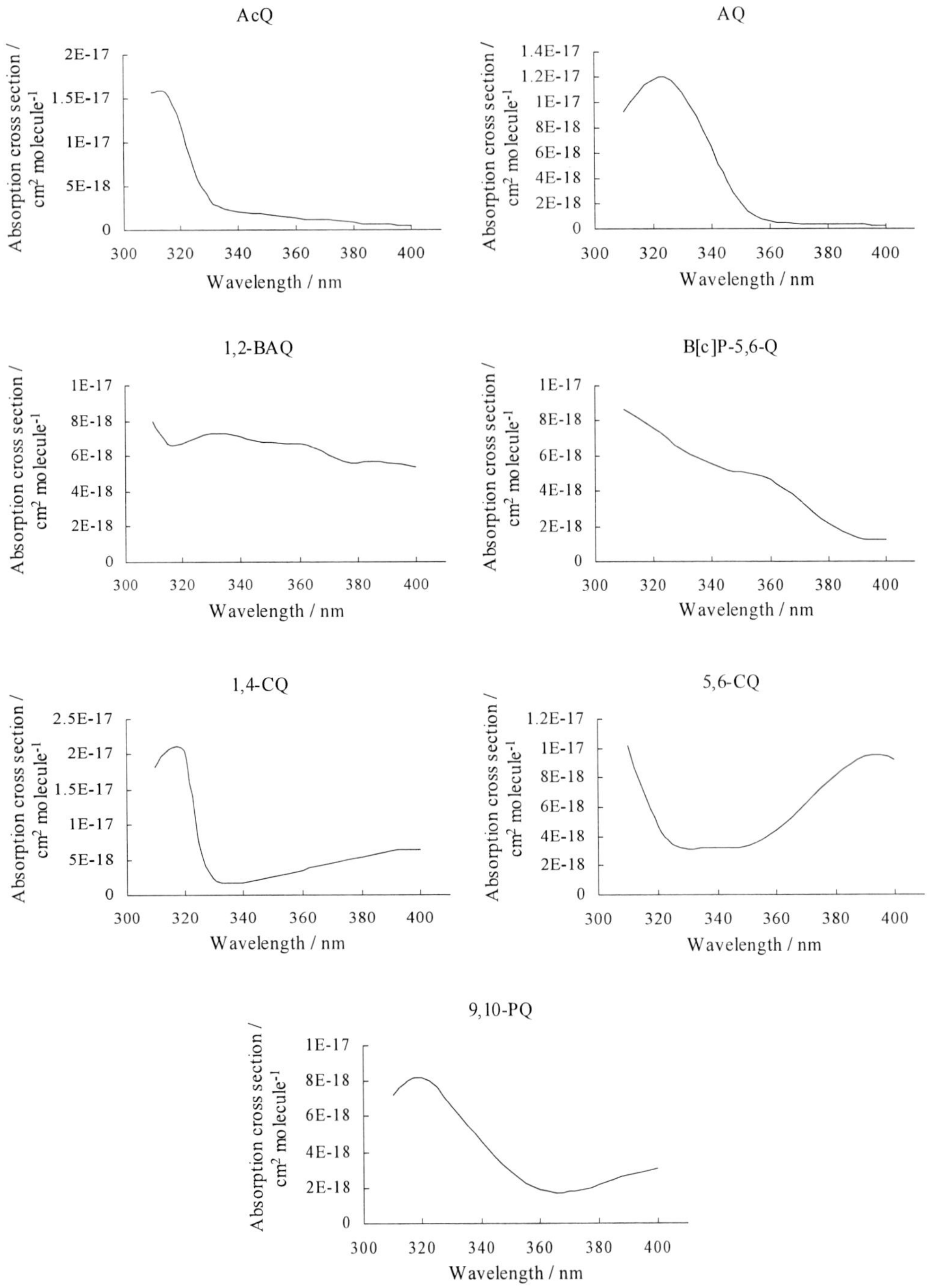

Figure 3. UV absorption spectrum of the compounds examined in this study.

The average quantum yields (ϕ) for the compounds tested in this study were determined as follows.

ϕ = number of molecules reacted via the photoreaction/number of photons absorbed = $kC_0/\Sigma\, I_\lambda\, \sigma_\lambda C_0$

where k (sec^{-1}) is the initial photolysis rate constant of the compounds measured experimentally, C_0 (molecules cm^{-3}) is the initial concentration of the compounds, and σ_λ (cm^2 $molecule^{-1}$) is the absorption cross section of the compounds at the wavelength λ, which was determined by UV-visible spectrophotometer in ethanol (Fig. 3). Since the quantum yield for the direct photolysis induced by the emission line λ (ϕ_λ) could not be obtained in this study, the average quantum yield ϕ was substituted for ϕ_λ.

Chemical Analysis of Nitro-PAH By HPLC

The reactant compounds on the inside surface of the photo-reactor after irradiation were extracted with 15 mL of acetonitrile. A sufficiently high recovery ratio of more than 90% was confirmed for the compounds from the reactor. An aliquot of the sample solution was subjected to HPLC analysis. A column-switching and chemiluminescence detection HPLC system was employed for NPAH analysis, with several modifications of the previously reported system [*18–22*]. Briefly, the system consists of four HPLC pumps, a 6-port switching valve, two separation ODS columns (Wako Pure Chemicals Ind., Wakosil-II 5C18AR and Imtakt, Cadenza CD-C18, 3.0 mm ID × 250 mm), a reducer column (Jasco, NPpak-RS, 4.0 mm ID × 10 mm), a concentration column (GL science, ODS-3, 4.0 mm ID × 30 mm), and a chemiluminescence detector (Soma Optics, S-3400). The chemiluminescence reagent solution was an acetonitrile solution containing 0.04 mmol L^{-1} bis(2,4,6-trichlorophenyl)oxalate and 15 mmol L^{-1} H_2O_2. The mobile phase was methanol/water (3/1, v/v) for the first separation and reduction of nitro-PAH, and acetonitrile/imidazole-perchloric acid buffer (1/1, v/v) for the second separation. HPLC analysis for OPAHs, PAHs, and 1-OHPy was performed with a Hitachi L-6200 pump with an F-1050 spectrofluorometric detector or an L-4200 spectrophotometric detector and an Inertsil ODS-3 column (GL science, 3.0 mm ID × 250 mm).

Chemicals

1-Nitropyrene (1-NP), 1,3-, 1,6- 1,8-dinitropyrenes (DNPs), 2-nitrofluorene (2-NF), 3-nitrofluoranthene (3-NFR), 3-nitroperylene (3-NPer), 6-nitrochrysene (6-NC), 7-nitrobenz[a]anthracene (7-NBaA), 9-nitroanthracene (9-NA), and 1-OHPy were obtained from Sigma-Aldrich Co. 1-Nitroperylene (1-NPer), 2-nitroanthracene (2-NA), 2-nitrofluoranthene (2-NFR), 3-NBA, 4-nitropyrene (4-NP), 6-nitrobenzo[a]pyrene (6-NBaP), 9-nitrophenanthrene (9-NPh), benzo[c]phenanthrene-5,6-quinone (BcP-5,6-Q), and 5,6-chrysenequinone (5,6-CQ) were purchased from Chiron AS. 2-Nitrotriphenylene (2-NTP) was obtained from ChemBridge Co. Acenaphthenequinone (AcQ) and 9,10-phenanthraquinone (9,10-PQ) were purchased from Fluka. 1,4-Chrysenequinone (1,4-CQ), 1,2-benzanthraquinone (1,2-BAQ), anthraquinone (AQ), benzo[a]pyrene (BaP), and benzo[*k*]fluoranthene (BkF) were obtained from Wako Pure Chemical Ind., Ltd. All solvents and other chemicals used were analytical grades from Wako Pure Chemical Ind., Ltd.

Results And Discussion

The concentration of the reactant compounds exposed to UV light showed an exponential decrease with irradiation time (Fig. 4). Decay of the compounds during the initial irradiation can be described by the first-order kinetics as follows:

$$\ln (C_0/C_t) = kt$$

where C_0 and C_t are the concentrations of the compounds at time 0 and t, respectively, and k is the initial photolysis rate constant. Table 1 shows k and the lifetime (τ) of the compounds obtained from the photoreaction employed in this study. The largest photolysis rate constant, 3.2×10^{-3} sec^{-1}, was observed for 9-NA, while 4-NP and 3-NBA were found the most stable of the nitrated compounds under UV-irradiation. It is hypothesized that photoreactivity of NPAHs is governed by the orientation of the nitro group, i.e. NPAHs having perpendicular nitro groups are more easily photodecomposed than those having parallel ones. [*12, 23*] The fast photodegradation observed for 9-NA, 6-NBaP, and 7-NBaA, which have a perpendicular nitro group, was consistent with the hypothesis. Although the photoreaction products of NPAHs in this system were not identified, the main products were reported as quinoid PAHs in the literature [*12*], as well as photoreaction products of PAHs. BcP-5,6-Q which has a similar photolysis rate constant to 9-NPh degraded the fastest of the seven OPAHs. 1,2-BAQ was the most resistant to photodecomposition of all the substituted PAHs examined in this study. Although previous studies for the photostability of OPAHs were quite limited, Cvrčková and Ciganek reported that 9,10-PQ was less stable than AQ under UV irradiation [*24*], that is consistent with our results.

Table 1. Quantum yield, photolysis rate constant, and lifetime of the PAH derivatives: (a) Irradiated by black light lamps; (b) Irradiated by sunlight on the earth's surface

		(a) Black light lamp		(b) Sunlight [a]	
	Quantum yield ϕ	Photolysis rate constant k (s^{-1})	Lifetime τ (h) [b]	Photolysis rate constant k_{sun} (s^{-1})	Lifetime τ_{sun} (h) [b]
BaP	1.96×10^{-6}	2.22×10^{-5}	12.5	3.72×10^{-6}	74.6
BkF	4.10×10^{-6}	9.89×10^{-6}	28.1	2.22×10^{-6}	125.3
1-OHPyr	3.05×10^{-5}	2.38×10^{-4}	1.2	2.98×10^{-5}	9.3
1,3-DNP	1.67×10^{-5}	8.15×10^{-5}	3.4	1.78×10^{-5}	15.7
1,6-DNP	3.21×10^{-4}	1.75×10^{-4}	1.6	3.70×10^{-5}	7.5
1,8-DNP	4.43×10^{-4}	2.23×10^{-4}	1.2	4.86×10^{-5}	5.7
1-NP	1.31×10^{-5}	7.85×10^{-5}	3.5	1.56×10^{-5}	17.8

		(a) Black light lamp		(b) Sunlight [a]	
	Quantum yield ϕ	Photolysis rate constant k (s^{-1})	Lifetime τ (h) [b]	Photolysis rate constant k_{sun} (s^{-1})	Lifetime τ_{sun} (h) [b]
1-NPer	4.07×10^{-4}	1.42×10^{-3}	0.2	2.78×10^{-4}	1.0
2-NA	3.61×10^{-4}	1.01×10^{-3}	0.3	1.71×10^{-4}	1.6
2-NF	5.07×10^{-5}	3.26×10^{-4}	0.9	4.21×10^{-5}	6.6
2-NFR	4.24×10^{-5}	1.05×10^{-4}	2.6	1.59×10^{-5}	17.4
2-NTP	2.25×10^{-5}	1.02×10^{-4}	2.7	1.32×10^{-5}	21.1
3-NBA	1.28×10^{-5}	5.23×10^{-5}	5.3	1.25×10^{-5}	22.2
3-NFR	1.70×10^{-5}	8.56×10^{-5}	3.2	1.56×10^{-5}	17.8
3-NPer	8.13×10^{-5}	1.65×10^{-4}	1.7	2.81×10^{-5}	9.9
4-NP	2.21×10^{-5}	7.31×10^{-5}	3.8	1.29×10^{-5}	21.5
6-NBaP	3.18×10^{-4}	1.38×10^{-3}	0.2	2.84×10^{-4}	1.0
6-NC	1.65×10^{-4}	5.93×10^{-4}	0.5	9.93×10^{-5}	2.8
7-NBaA	9.09×10^{-4}	2.12×10^{-3}	0.1	3.19×10^{-4}	0.9
9-NA	1.73×10^{-3}	3.15×10^{-3}	0.1	5.21×10^{-4}	0.5
9-NPh	1.88×10^{-4}	4.79×10^{-4}	0.6	6.49×10^{-5}	4.3
1,2-BAQ	2.02×10^{-5}	4.74×10^{-5}	5.9	7.96×10^{-6}	34.9
1,4-CQ	5.33×10^{-5}	7.63×10^{-5}	3.6	1.76×10^{-5}	15.8
5,6-CQ	6.38×10^{-5}	9.86×10^{-5}	2.8	2.39×10^{-5}	11.6
9,10-PQ	1.75×10^{-5}	2.23×10^{-4}	1.2	3.90×10^{-5}	7.1
AcQ	2.68×10^{-4}	2.35×10^{-4}	1.2	3.98×10^{-5}	7.0
AQ	6.02×10^{-5}	8.38×10^{-5}	3.3	1.23×10^{-5}	22.7
B[c]P-5,6-Q	2.51×10^{-4}	4.33×10^{-4}	0.6	6.16×10^{-5}	4.5

[a] Calculated based on the quantum yields and the absorption cross sections for the compounds and the actinic flux on the earth's surface. See text for the detail. [b] Defined as the inverse of the photolysis rate constant.

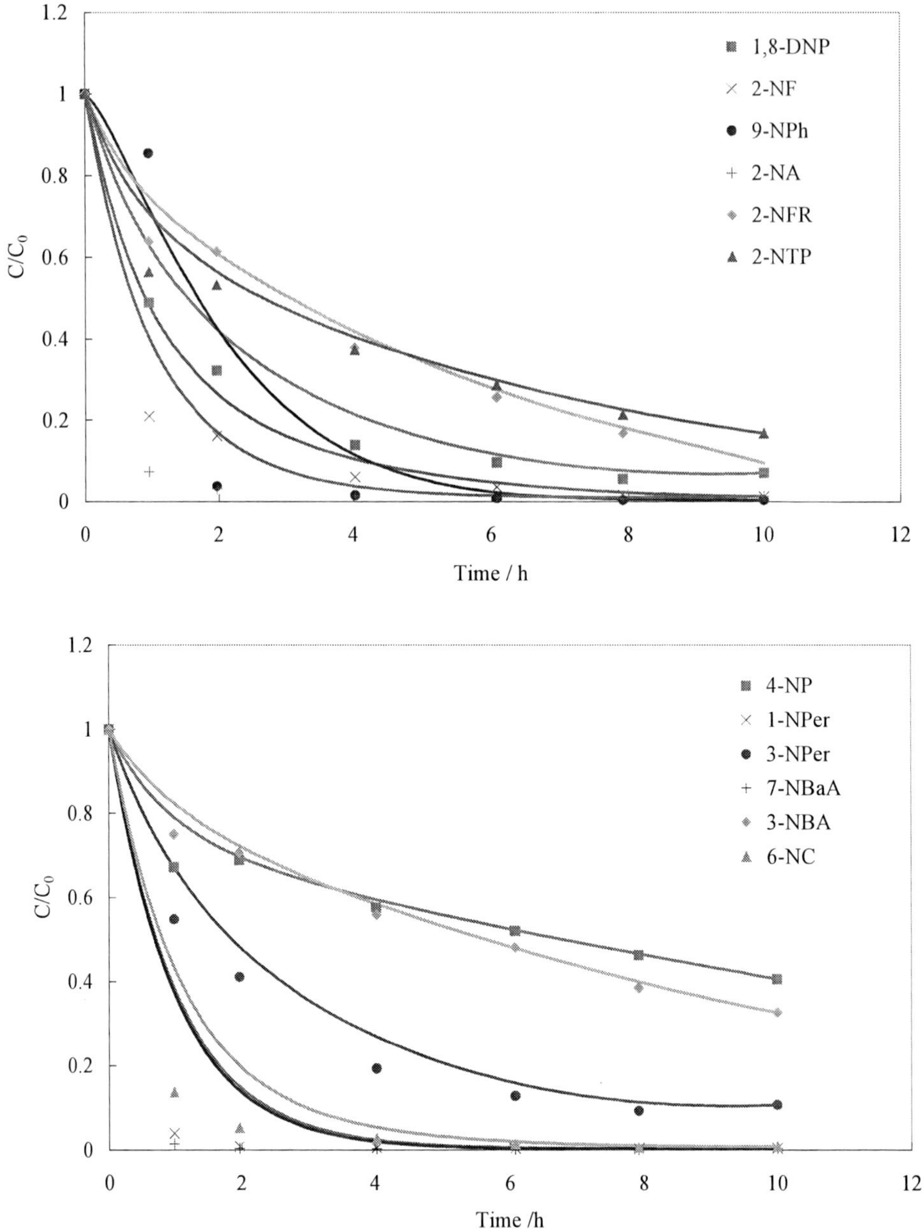

Figure 4. (Continued on next page)

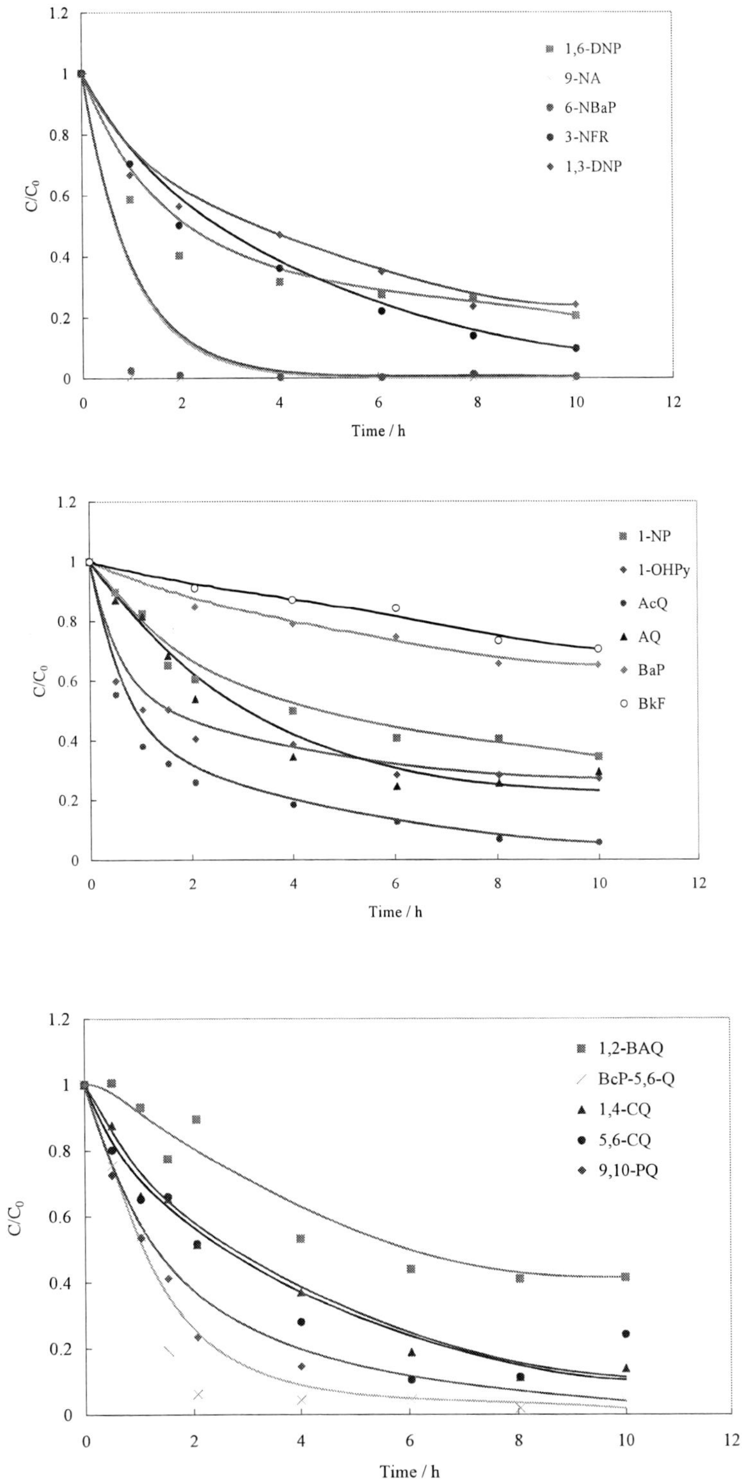

Figure 4. Photodegradation curves for the compounds examined in this study. Each data point is the mean value of the results obtained by three independent experiments.

Table 2: Calculated Actinic Flux at the Earth's Surface as a Function of Wavelength and Zenith Angle [25]

Wavelength		Actinic flux (photons $cm^{-2}sec^{-1}$)									
(nm)	Exp [a]	0° [b]	10°	20°	30°	40°	50°	60°	70°	78°	86°
290-295	14	0.001	0.001	--	--	--	--	--	--	--	--
295-300	14	0.037	0.035	0.027	0.017	0.008	0.003	--	--	--	--
300-305	14	0.362	0.347	0.302	0.234	0.154	0.078	0.025	0.004	0.001	--
305-310	14	1.280	1.240	1.140	0.965	0.737	0.476	0.227	0.060	0.010	0.001
310-315	14	2.850	2.810	2.650	2.360	1.960	1.440	0.858	0.332	0.083	0.008
315-320	14	3.940	3.910	3.730	3.420	2.960	2.330	1.560	0.739	0.245	0.027
320-325	14	4.960	4.900	4.710	4.380	3.880	3.180	2.260	1.200	0.466	0.067
325-330	14	7.140	7.060	6.820	6.400	5.760	4.830	3.570	2.020	0.864	0.155
330-335	14	7.540	7.470	7.250	6.840	6.220	5.300	4.010	2.380	1.070	0.221
335-340	14	7.450	7.380	7.180	6.810	6.230	5.370	4.140	2.520	1.170	0.262
340-345	14	8.030	7.970	7.760	7.380	6.790	5.900	4.600	2.850	1.340	0.310
345-350	14	8.020	7.950	7.750	7.400	6.840	5.980	4.710	2.970	1.410	0.327
350-355	14	9.080	9.010	8.800	8.420	7.800	6.860	5.450	3.470	1.660	0.385
355-360	14	8.360	8.300	8.120	7.780	7.230	6.390	5.120	3.300	1.590	0.364
360-365	14	9.250	9.190	9.000	8.640	8.060	7.160	5.780	3.770	1.830	0.413
365-370	15	1.140	1.130	1.110	1.070	0.997	0.890	0.724	0.477	0.233	0.052
370-375	15	1.030	1.020	1.000	0.965	0.906	0.811	0.664	0.442	0.217	0.048
375-380	15	1.150	1.140	1.120	1.070	1.020	0.914	0.753	0.506	0.251	0.054
380-385	15	0.948	0.942	0.926	0.895	0.845	0.763	0.633	0.429	0.214	0.045
385-390	15	1.040	1.040	1.020	0.987	0.933	0.846	0.705	0.484	0.244	0.051
390-395	15	1.070	1.070	1.050	1.020	0.965	0.878	0.737	0.510	0.259	0.053
395-400	15	1.300	1.290	1.280	1.240	1.180	1.070	0.905	0.632	0.324	0.066
400-405	15	1.540	1.530	1.510	1.460	1.390	1.280	1.080	0.761	0.394	0.079
405-410	15	1.720	1.710	1.690	1.640	1.570	1.440	1.230	0.872	0.455	0.090
410-415	15	1.820	1.810	1.780	1.740	1.660	1.530	1.310	0.937	0.497	0.097

Table 2. (Continued)

415-420	15	1.840	1.830	1.810	1.770	1.690	1.560	1.340	0.968	0.514	0.100
420-430	15	3.620	3.600	3.560	3.480	3.340	3.090	2.680	1.950	1.050	0.201
430-440	15	3.720	3.710	3.670	3.590	3.450	3.220	2.800	2.070	1.130	0.214
440-450	15	4.320	4.300	4.260	4.180	4.030	3.770	3.310	2.470	1.380	0.255
450-460	15	4.730	4.720	4.670	4.590	4.430	4.160	3.680	2.780	1.570	0.278
460-470	15	4.850	4.840	4.800	4.710	4.560	4.300	3.820	2.910	1.660	0.299
470-480	15	4.960	4.950	4.910	4.830	4.690	4.430	3.960	3.040	1.760	0.314
480-490	15	4.930	4.930	4.880	4.810	4.680	4.430	3.970	3.070	1.790	0.341
490-500	15	4.940	4.930	4.890	4.820	4.690	4.460	4.010	3.120	1.840	0.319
500-510	15	4.930	4.920	4.880	4.820	4.690	4.460	4.020	3.140	1.850	0.310
510-520	15	4.850	4.830	4.800	4.740	4.620	4.410	3.980	3.130	1.860	0.309
520-530	15	4.930	4.920	4.880	4.820	4.710	4.490	4.060	3.200	1.910	0.524
530-540	15	4.920	4.920	4.890	4.830	4.710	4.490	4.070	3.210	1.920	0.299
540-550	15	4.840	4.830	4.800	4.750	4.640	4.430	4.020	3.170	1.900	0.291
550-560	15	4.810	4.800	4.780	4.720	4.610	4.410	4.000	3.170	1.900	0.285
560-570	15	4.860	4.850	4.820	4.760	4.660	4.450	4.040	3.200	1.920	0.279
570-580	15	4.960	4.950	4.920	4.870	4.760	4.550	4.130	3.270	1.960	0.276
580-600	16	1.010	1.010	1.000	0.992	0.971	0.931	0.848	0.676	0.409	0.059
600-620	16	1.010	1.000	1.000	0.990	0.970	0.931	0.852	0.682	0.417	0.060
620-640	16	1.010	1.010	1.000	0.995	0.978	0.943	0.867	0.703	0.439	0.068
640-660	16	1.010	1.010	1.010	1.000	0.986	0.954	0.883	0.724	0.467	0.078
660-680	16	1.010	1.010	1.010	1.000	0.990	0.960	0.893	0.740	0.482	0.087
680-700	16	1.010	1.010	1.010	1.000	0.988	0.961	0.897	0.749	0.496	0.094

[a]The second column lists power of 10 by which all entries should be multiplied. [b]Zenith angles.

Photolysis rate constants of the tested compounds under irradiation of sunlight (k_{sun}) were estimated based on the actinic flux on the earth's surface [*25*] ($I_{\lambda\text{-sun}}$, zenith angle φ = 30 °, Table 2), σ_λ, and ϕ, i.e. $k_{sun} = \Sigma\, I_{\lambda\text{-sun}}\sigma_\lambda\phi$. Table 1 also shows k_{sun} and the estimated lifetime of the compounds under irradiation of sunlight (τ_{sun}). Atmospheric lifetimes of the compounds due to photodecomposition were estimated to be 0.5 – 22 h for NPAHs and 4.5 – 35 h for OPAHs. We found that OPAHs were generally more stable against photo irradiation than NPAHs. This finding indicates that the risk induced by OPAHs is critical for human health, because we may be continuously exposed by OPAHs due to their longer residence time in the atmosphere. Recently, it has been found that ROS, which can cause severe oxidative stress connected with inflammatory processing, are produced in larger amounts via chain reactions induced by quinoid PAHs in the human body [*15*]. Further studies on the biological effects and the atmospheric formation and decomposition mechanisms for OPAHs are required.

REFERENCES

[1] Schuetzle, D. Sampling of vehicle emissions for chemical analysis and biological testing. *Environ. Health Perspect.* 1983, *47*, 65-80.

[2] Salmeen, I.T.; Pero, A.M.; Zator, R.; Schuetzle D.; Riley, T.L. Ames assay chromatograms and the identification of mutagens in diesel particle extracts. *Environ. Sci. Technol.* 1984, *18*, 375-382.

[3] Schuetzle, D.; Riley, T.L.; Prater, T.J.; Harvey, T.M.; Hunt, D.F. Analysis of nitrated polycyclic aromatic hydrocarbons in diesel particulates. *Anal. Chem.* 1982, *54*, 265-271.

[4] Bamford, H.A.; Bezabeh, D.Z.; Schantz, M.M.; Wise, S.A.; Baker, J.E. Determination and comparison of nitrated-polycyclic hydrocarbons measured in air and diesel particulate reference materials. *Chemosphere* 2003, *50*, 575-587.

[5] Bezabeh, D.Z.; Bamford H.A.; Schantz, M.M.; Wise, S.A. Determination of nitrated polycyclic aromatic hydrocarbons in diesel particulate-related standard reference materials by using gas chromatography/mass spectrometry with negative ion chemical ionization. *Anal. Bioanal. Chem.* 2003, *375*, 381-388.

[6] Zielinska, B.; Arey, J.; Atkinson, R.; Ramdahl, T.; Winer, A.M. The nitroarenes of molecular weight 247 in ambient particulate samples collected in southern California. *Atmos. Environ.* 1989, *23*, 223-229.

[7] Ciccioli, P.; Cecinato, A.; Brancaleoni, E.; Frattoni, M.; Zacchei, P.; Miguel, A.H.; de Castro Vasconcellos, P. Formation and transport of 2-nitrofluoranthene and 2-nitropyrene of photochemical origin in the troposphere. *J. Geophys. Res.* 1996, *101*, 19567-19581.

[8] Reisen, F.; Arey, J. Atmospheric reactions influence seasonal PAH and nitro-PAH concentrations in the Los Angeles basin. *Environ. Sci. Technol.* 2005, *39*, 64-73.

[9] Atkinson, R.; Arey, J. Atmospheric chemistry of gas-phase polycyclic aromatic hydrocarbons: formation of atmospheric mutagens. *Environ. Health Perspect.* 1994, *102*, 117-126.

[10] Phousongphouang, P.T.; Arey, J. Sources of the atmospheric contaminants, 2-nitrobenzanthrone and 3-nitrobenzanthrone. *Atmos. Environ.* 2003, *37*, 3189-3199.

[11] Cvrčková, O.; Ciganek, M.; Šimek, Z. Anthracene, chrysene, their nitro- and methyl-derivatives photostability in isooctane. *Polycycl. Aromat. Compd.* 2006, *26*, 331-344.

[12] Warner, S.D.; Farant, J.-P.; Butler, I.S. Photochemical degradation of selected nitropolycyclic aromatic hydrocarbons in solution and adsorbed to solid particles. *Chemosphere* 2004, *54*, 1207-1215.

[13] Finlayson-Pitts, B.J.; Pitts, Jr, J.N. Chemistry of the upper and lower atmosphere; Academic Press: San Diego, CA, 2000; pp 514.

[14] Kamens, R.M.; Guo, Z.; Fulcher, J.; Bell, D. Influence of humidity, sunlight, and temperature on the daytime decay of polyaromatic hydrocarbons on atmospheric soot particles. *Environ. Sci. Technol.* 1988, *22*, 103-108.

[15] Chung, S.W.; Chung, H.Y.; Toriba, A.; Kameda, T.; Tang, N.; Kizu, R.; Hayakawa, K. An environmental quinoid polycyclic aromatic hydrocarbon, acenaphthenequinone, modulates cyclooxygenase-2 expression through reactive oxygen species generation and nuclear factor kappa B activation in A549 cells, *Toxicol. Sci.* 2007, *95*, 348-355.

[16] Enya, T.; Suzuki, H.; Watanabe, T.; Hirayama, T.; Hisamatsu, Y. 3-nitrobenzanthrone, a powerful bacterial mutagen and suspected human carcinogen found in diesel exhaust and airborne particulates. *Environ. Sci. Technol.* 1997, *31*, 2772-2776.

[17] Hatchard, C.G.; Parker, C.A. A new sensitive chemical actinometer: II. Potassium ferrioxalate as a standard chemical actinometer. *Proc. R. Soc. London* 1956, *A235*, 518–536.

[18] Murahashi, T.; Kizu, R.; Kakimoto, H.; Toriba, A.; Hayakawa, K. 2-Nitrofluoranthene, 1-, 2- and 4-nitropyrenes and 6-nitrochrysene in diesel-engine exhaust and airborne particulates. *J. Health Sci.* 1999, *45*, 244-250.

[19] Kameda, T.; Inazu, K.; Hisamatsu, Y.; Takenaka, N.; Bandow, H. Determination of atmospheric nitro-polycyclic aromatic hydrocarbons and their precursors at a heavy traffic roadside and at a residential area in Osaka, Japan, *Polycycl. Aromat. Compd.* 2004, *24*, 657-666.

[20] Hayakawa, K.; Kitamura, R.; Butoh, M.; Imaizumi, N.; Miyazaki, M. Determination of diamino- and aminopyrenes by high performance liquid chromatography with chemiluminescence detection. *Anal. Sci.* **1991**, *7*, 573-577.

[21] Murahashi, T.; Hayakawa, K. A sensitive method for the determination of 6-nitrochrysene, 2-nitrofluoranthene and 1-, 2- and 4-nitropyrenes in airborne particulates using high-performance liquid chromatography with chemiluminescence detection. *Anal. Chim. Acta* 1997, *343*, 251-257.

[22] Hayakawa, K.; Noji, K.; Tang, N.; Toriba, A.; Kizu, R.; Sakai, S.; Matsumoto, Y. A high-performance liquid chromatographic system equipped with on-line reducer, clean-up and concentrator columns for determination of trace levels of nitropolycyclic aromatic hydrocarbons in airborne particulates. *Anal. Chim. Acta* 2001, *445*, 205-212.

[23] Yang, D.T.C.; Chou, A.; Chen, E.; Chiu, L.-H. Photodecomposition of environmental nitro-polycyclic aromatic hydrocarbons. *Polycycl. Aromat. Compd.* 1994, *5*, 201-208.

[24] Cvrčková, O.; Ciganek, M. Photostability of polycyclic aromatic hydrocarbons (PAHs) and nitrated polycyclic aromatic hydrocarbons (NPAHs) in dichloromethane and isooctane solutions. *Polycycl. Aromat. Compd.* 2005, *25*, 141-156.

[25] Demerjian, K.L.; Schere, K.L.; Peterson, J.T. Theoretical estimates of actinic (spherically integrated) flux and photolytic rate constants of atmospheric species in the lower troposphere. *Adv. Environ. Sci. Technol.* 1980, *10*, 369–459.

In: Airborne Particulates
Editor: Ming Cheng and Wen Liu

ISBN: 978-1-60692-907-0

Short Communication B

GENOTOXIC AND ULTRASTRUCTURAL EVIDENCES OF DAMAGE INDUCED BY FINE ($PM_{2.5}$) AND COARSE (PM_{10}) AIR PARTICULATE MATTER ON MURINE MACROPHAGES

***Anna Poma**[*a], **Maria Laura Di Giorgio**[a], **Augusta Galano**[a], **Pietro Picozzi**[b], **Pierpaolo Aimola**[a] **and Anna Maria Ragnelli**[a]*

[a] Department of Basic and Applied Biology, University of L'Aquila, Department of Physics[b], University of L'Aquila, Via Vetoio 1, I-67010 Coppito-L'Aquila. Italy

ABSTRACT

Urban air, particularly in high traffic areas, contains mutagenic and carcinogenic pollutants, including both organic compounds, such as benzene and polycyclic aromatic hydrocarbons, and inorganic substances, as for instance nickel, chromium and other heavy metals. Extracts of urban air particulates induce cancer in animals and are mutagenic in bacteria and mammalian cells. In recent years, increasing levels of PM_{10} (coarse) and $PM_{2.5}$ (fine) particulate matter, i.e. particles with equivalent diameter less than, respectively, 10 μm and 2.5 μm, have been associated with an increment in cardiorespiratory mortality and respiratory-related emergency department visits. Furthermore, a correlation between mutagenitricy data, airborne micropollutants and meteorological parameters has been established. In this work we have evaluated: a) the genotoxic effects of carbon black, coarse, fine airborne particles and particulate ranging from and 0.65-0.43 μm collected in L'Aquila (a town in Central Italy) on the murine macrophage cell line RAW 264.7; b) the ultrastructural evidences of damage induced by fine and coarse air particulate matter. The results obtained give evidence of higher cytotoxicity of fine PM on cultured murine macrophages, as compared with coarse PM.

Keywords: carbon black, air particulate matter; $PM_{2.5}$; PM_{10}; genotoxicity; ultrastructural damages

---

* poma@univaq.it

INTRODUCTION

Atmospheric dust or particulate matter indicate a mixture of solid and liquid particles, suspended in air, with specific chemical, morphological or dimensional characteristics due to emission source or processing. Atmospheric dust can be defined as total suspended particles (TSP) or particulate matter (PM). Total suspended particles are an heterogeneous mixture of solid and liquid particles which are suspended in air because of their dimensions ranging from few nanometers to dozens of micrometers. On the other hand, different methods are employed for the classification of particulate matter, the most used of which is based on the distribution of dust in "classes" as a function of particle size (measured in μm), while their presence in the air is quantified in terms of concentration (expressed in $\mu g/m^3$ or in micrograms of particles per cubic meter air environment).

In recent years the interest of investigators has shifted in particular on a fraction of TSP, PM_{10}, which includes particles with aerodynamic diameter less than 10 μm, and, more recently, on smaller particles, $PM_{2.5}$ (with an aerodynamic diameter less than 2.5 μm) and "ultrafine" particulate matter (with aerodynamic diameter less than 0.1 μm). Although there may be some dimensional overlap between PM_{10} and $PM_{2.5}$, the two classes are generally distinct both in terms of emission sources, and for the chemical composition and the behavior in the atmosphere. Atmospheric dust levels have typical values (annual average of PM_{10}) of 5 $\mu g/m^3$ in remote oceanic areas, 40 $\mu g/m^3$ in rural areas and 50 to 100 $\mu g/m^3$ in urban areas [1]. These data clearly show that municipal levels are two-fold higher than particulate values in rural areas, and these in turn are only about eight-fold higher than those in remote areas: this is due to the long residence time of particles in the atmosphere and to the important contribution of natural sources to the emission of particulate matter.

In order to give a physical characterization of atmospheric dusts, we have to consider particle dimensions as the most important parameter. Important characteristics are also shape, chemical composition, vapor pressure, hygroscopy, deliquescence and refraction index. The particles can act as condensation nuclei for steam under conditions of high humidity and then undergo changes through processes that occur during wet phase, first of all the absorption of SOx and subsequent acid attack. Moreover, oxidizing agent and photodissociation phenomena may lead to the generation of hydroxyl radicals (HO•) which are necessary for the beginning and the propagation of many typical chain reactions involving different components of the atmosphere. The particles are generally hygroscopic and, consequently, their weight may increase because of water absorption; however, hygroscopy may be reduced due to the presence of hydrophobic organic substances. Therefore, water can affect mass concentration and size distribution of particulate matter, producing significant variability even in harmful effects: for example, an acid fog has a higher penetrability in the respiratory tract than acid rain. Moreover high levels of humidity can locally promote removing of particles.

Even though particulate matter has a very heterogeneous composition, especially as regards trace elements, it usually consists of few components, commonly grouped in the following three classes [2]:

- Inorganic ions, from secondary aerosols (NO_3, NH_3, SO_4);
- Organic carbon (OC) and elemental carbon (EC or Black Carbon, BC);
- Crustal elements (Crustal, Soil or Mineral), marine aerosols (Sea Spray);

These components are in different proportions in particles of different sizes: crustal elements are more concentrated in the coarse fraction, while carbon is particularly abundant in the fine fraction.There is a fourth class of suspended particulate matter that differs from the other three components in a substantial manner, being represented by biological material such as pollen, bacteria, spores and tiny plant fragments. The typical size of these dust particles is usually higher than one micrometer, so that they belong to the coarse fraction.

Carbonaceous particulate matter is produced mainly in combustion processes and emitted directly in the atmosphere. "Black carbon" was one of the first pollutants to be monitored and used as a tracer of aerosols from primary combustion processes, causing loss of visibility in industrial districts; currently, it is also used in specific studies on vehicular traffic [3]. Organic carbon rises, most of all, from oxidation of combustion products, such as VOCs (Volatile Organic Compounds) and their subsequent condensation, decomposition in aqueous phase, absorption (mainly on the surface of elemental carbon particles) or adsorption [4]. The organic carbon found on emitted particulate matter contains more than 100 different compounds, including alkanes, benzaldehyde and polycyclic aromatic hydrocarbons (PAHs), particularly hazardous to human health [5]. The best known PAH is Benzo(a)pyrene (BaP), which becomes highly carcinogenic after metabolic activation. High concentrations of PAHs are present in soot generated from the burning of biomass and coal and by discharges of diesel and petrol vehicles.

Particulate matter contains a variety of elements with a very different percentage distribution among particles of different origins, due to processes of genesis and transformation which influence composition of atmospheric dust. Indeed, certain elements are considered "tracers" of the sources from which they were emitted:

- Al, Fe, Ca, Si, Ti: erosion of rocks and soil (crustal elements); coal combustion
- Na, Cl, Mg: sea spray; chloride from waste incineration
- C: incomplete combustion of fossil fuel
- V: combustion of bitumen oil
- Pb: burning of fuels rich in lead or waste containing it
- Zn: combustion of wood and tires

Sodium and magnesium compounds are present in substantial percentages and originate mainly from marine aerosols (which can be transported over long distances inland by the wind) and salts used for the dissolution of ice during winter road maintenance. Calcium and potassium compounds, present in rocks and soil, can be found in the atmosphere because of phenomena of resuspension of dust from the earth surface. Other metal compounds are usually present in trace quantities. As for chlorides, apart from being generally associated with sodium, they may also be released into the atmosphere in the form of hydrogen chloride from the burning of coal and incineration processes.

The dosimetric classification is based on the capability of particles to penetrate the respiratory tract:

- Inhalable fraction, that enters only the upper respiratory tract;
- Thoracic fraction, that reaches lungs;
- Respirable fraction, that is able to reach alveoli.

Fine particles can be absorbed by the human body mainly through the skin or by inhalation. The absorption depends chiefly on solubility in water and lipids of the pollutant (usually the potential for dermal absorption is inversely proportional to the molecular weight of the substance) and to a lesser extent on temperature, contact surface area and duration of adhesion. Absorption of particulate matter is faster when it occurs by inhalation and the main target organ is, then, the respiratory system. The deposition of particles in the respiratory tract is mainly caused by a radial spread, by gravitational sedimentation and by particular properties of the particles (chemical reactivity, hygroscopy, aerodynamic diameter, shape and size) and is also influenced by respiratory frequency and route of entry (nose or mouth). The particle size, in particular, is a factor closely related to the effects of particulate matter on health: small particles have greater surface area for the absorption of carcinogenic organic pollutants, like polycyclic aromatic hydrocarbons, that are carried into the respiratory system by suspended particles. Furthermore, particles with aerodynamic diameter smaller than one micrometer can easily reach pulmonary alveoli. This is particularly significant, if we consider that direct and extended contact with the alveoli strongly facilitates the carcinogenic effect. Many epidemiological studies conducted on workers exposed have highlighted the role of inhalation of polycyclic aromatic hydrocarbons in the development of lung cancer. In addition, the particulate matter causes a "mechanical" irritation on the lungs and bronchi. It was also shown that the ultrafine fraction is more harmful than the coarse and fine fractions of identical composition. This may be due, in part, to its lower solubility, but especially to its smaller dimensions that prevent it from a complete phagocytosis.

In the alveolar region the first cellular defense mechanism towards the particulate material is carried out by resident macrophages that phagocytate and carry particles in the mucosal layer, thus ensuring the sterility of the alveoli and preventing the entry of smaller particulates in interstitial spaces. Macrophages maintain the integrity and sterility of the respiratory system, both by the process of phagocytosis and through a brief or prolonged inflammatory response depending on the concentration of particulate matter to remove. During this inflammatory response process, the macrophages may secrete and release many substances, among which reactive oxygen species (ROS: superoxide anion, hydrogen peroxide, hydroxyl radical), with their high bactericidal activity, are of considerable importance. Moreover, during phagocytosis, activated alveolar macrophages, respond with an oxidative "burst" characterized by a rapid consumption of molecular oxygen that leads to further production of ROS. These are usually neutralized by antioxidant defenses of the tissue which, in this way, maintains its structural integrity. However, a high concentration of particulate matter deposited in the lung causes a strong stimulation of phagocytic activity and inflammatory response with consequent notable increase in the production of ROS. The high concentration of ROS produced can exceed the cellular antioxidant capabilities, with consequent onset of "oxidative stress" and severe lung damage. The highest risk of oxidative stress can be attributed to ultrafine particulate and particulate containing transition metals, especially iron. Finally, it has been shown that large particles cause a worsening of asthma, while oedema and changes in heart rhythm are connected with the deposition of small particles in the lungs [6]. Epidemiological studies, conducted in the U.S. and Europe on healthy adults, suggest an association between long-term exposure to PM_{10} and the increase of respiratory and cardiovascular diseases and even hospital admissions, especially in urban areas with intense traffic. Moreover, clinical and epidemiological studies have shown that acute or chronic exposure to particulate matter causes numerous disorders and cardiopulmonary diseases, such as bronchial asthma attacks, chronic obstructive pulmonary disease and loss of lung function. The harmful effects on human health depend, as mentioned above, both on the ability of particles to overcome the defense barriers and on the ability to remove and/or neutralize the toxic substances carried by them. In this work we studied the genotoxic effects and the ultrastructural damage induced by urban airborne particulate matter on mammalian macrophages (RAW 264.7), the first line of defense in the alveolar region.

MATERIALS AND METHODS

Airborne particulates from L'Aquila were collected with a multistage particle sampler (Andersen sampler); carbon black, from Degussa, is composed by particles with average diameter of 20 nm. Both particulate matter and carbon black were visualized by SEM and subjected to SEM-EDX analysis to determine the chemical inventory of aerosol particle surface. The micronucleus test was performed according to the procedure of Ying et al. (2003) [7], the Comet test according to Ravanat et al. (2004) [8], and the MTS assay according to Takako et al. (2001)[9]. Cells were fixed, dehydrated and embedded in Durcupan ACM resin and then thin sections were examined at transmission electron microscope [10].

RESULTS AND DISCUSSION

Sem Analysis

In order to evaluate the superficial chemical composition of particulate matter and carbon black, collected samples were subjected to SEM-EDX analysis. Particulate matter was collected on a polycarbonate filter and analysed with a Philips SEM505 scanning electron microscope equipped with Microanalysis LINK AN10000, employing a Si(Li) detector with ultra-thin window (elemental analysis $Z \geq 6$) (Fig. 1). Microanalysis has shown that particulate matter is mainly composed of carbon especially in fractions with smaller particles.

Moreover, microanalysis of carbon black, coarse and fine particulate revealed the presence of chlorine, due to culture media (DMEM) in which particulate and carbon black samples were resuspended during storage. However data show a characteristic chemical composition both for environmental particulate and carbon black (Fig. 1). Indeed, as shown in figure 1, (A) and (C) (carbon black and fine particulate), the highest peak corresponds to the carbon peak. Coarse fraction shows typical peaks corresponding to calcium and sodium. Notable is the absence of cytotoxic elements such as heavy metals (Pb, Cd, Fe) both in fine and coarse particulate and in carbon black. Overall data show no significant differences in chemical composition of different samples; this allows us to consider particle dimension as the main parameter to evaluate cytotoxicity and genotoxicity of these substances. Moreover, collected samples of carbon black and particulate were further observed by scanning electron microscopy in order to conduct morphological analysis on different fractions and to confirm the dimensions of particulate matter sampled with a multistage Andersen ambient particle sizing sampler. Micrographies in figure 2 show carbon black, coarse (PM_{10}), fine ($PM_{2.5}$), and

0.65-0.43 μm PM collected on polycarbonate filter put on 0-3rd, 4th -6th and 7th stage of the particle sampler.

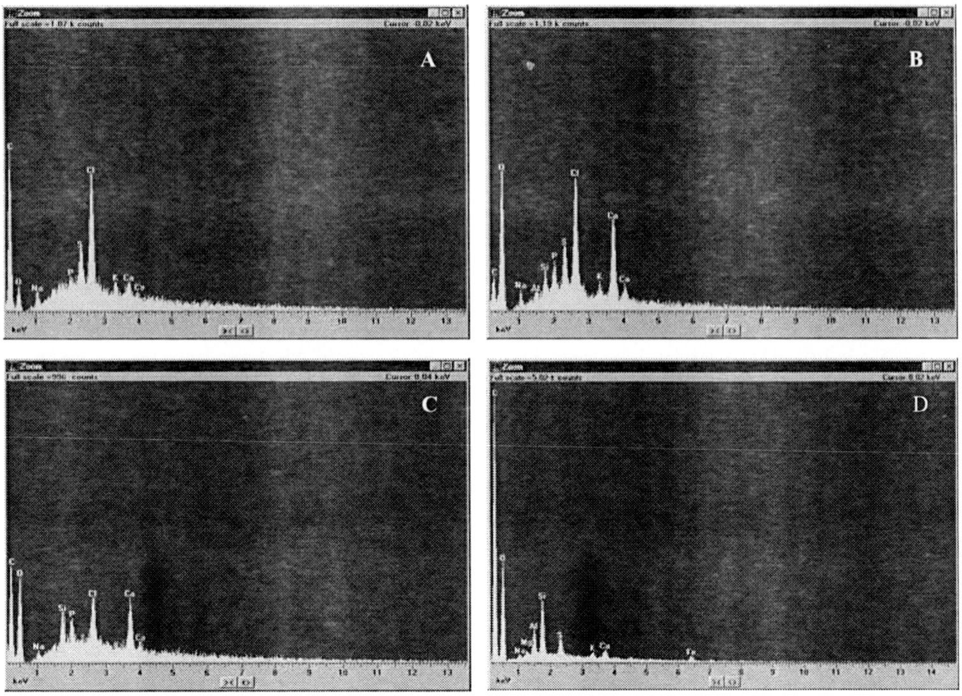

Figure 1

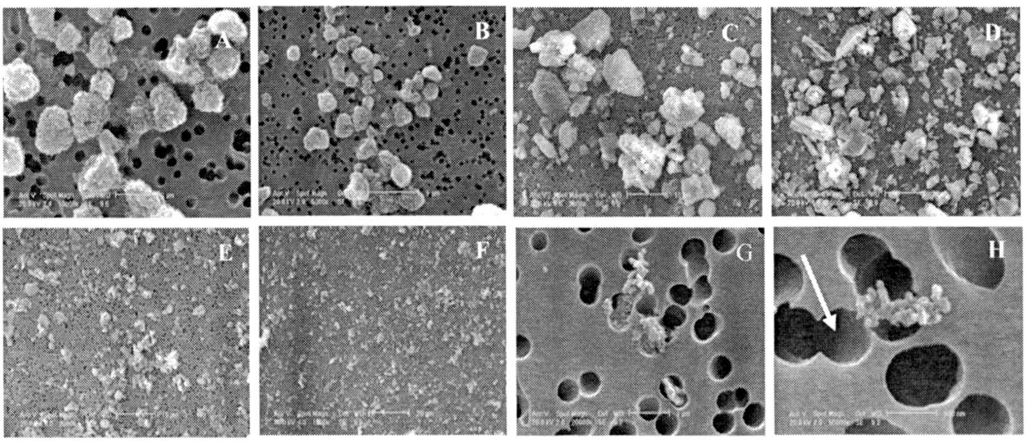

Figure 2. Airborne particles collected on polycarbonate membranes and observed at SEM. A) and B) carbon black; C) and D) coarse air particulate matter (PM10); E) and F) fine air particulate matter (PM 2.5); G) and H): 0.65-0.43 μm particulate matter. White arrow, filter pore (0.4μm).

PARTICULATE CYTOTOXICITY

Cytotoxiciy data show that particulate matter and carbon black treated cells have lower proliferation rate if compared with control cells (Fig. 3) both after 24 and 48 hour. In particular, fine fraction and fraction collected on 7th stage (particles with aerodynamic diameter ranging from 0.65-0.43μm) induce reduction of cell proliferation in a dose dependent manner (Fig. 3).

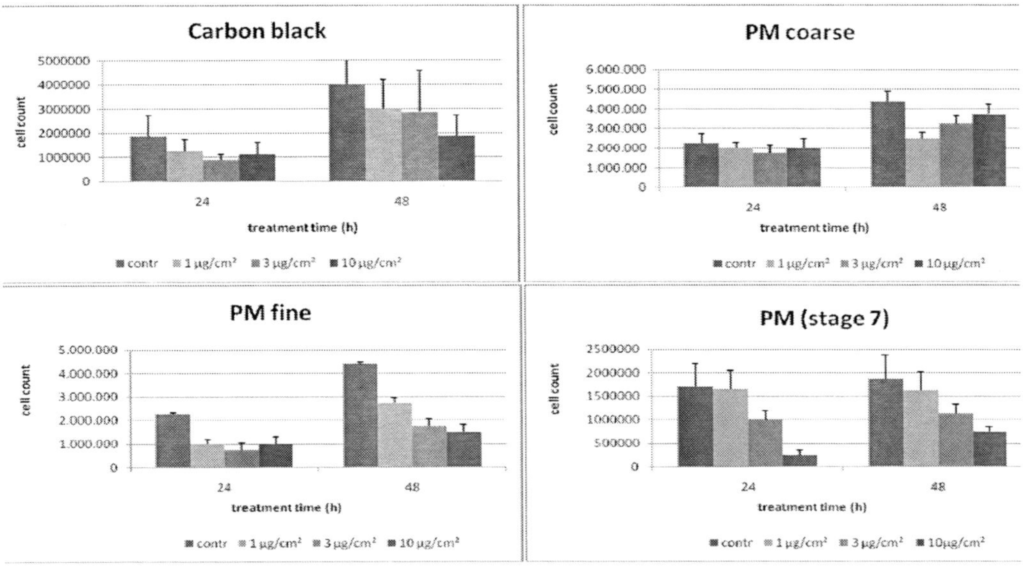

Figure 3. Cell proliferation of fine and coarse PM-treated RAW 264.7; viable cells as measured by trypan blue dye exclusion test (mean values + standard deviation). Cell viability was also determined by measuring MTS [3-(4,5-dimethylthiazol-2-yl)-5-(3-carboxymethoxyphenyl)-2-(4-sulfophenyl)-2H-tetrazolium] reduction to formazan soluble in culture medium. Table 1 shows average values of optical density obtained at 490 nm.

Table 1. Effects of fine and coarse airborne particulate matter on the viability of RAW 264.7 macrophage cell line. O. D (mean values + S.D.) are reported.

Treatment time (h)	Concentrations	PM coarse	PM fine	PM (0.65-0.43μm)	Carbon black (20 nm)
24	contr	0,96±0,21	0,96±0,21	0,88±0,25	0,49±0,10
	1 μg/cm²	0,85±0,02	1,08±0.01	0,5±0,13	0,58±0,11
	3 μg/cm²	0,76±0.01**	1,25±0.01**	0,46±0,10	0,62±0,02*
	10 μg/cm²	0,78±0,02**	1,03±0.01**	0,62±0,16	0,61±0,10 **
48	contr	0,91 ±0,08	0,91±0,08	1,21±0,06	1,27±0,11
	1 μg/cm²	0,74±0,03*	0,75±0.05*	1,32±0.05**	1,42±0,17 **
	3 μg/cm²	1,26±0.01**	0,74±0.01	1,55±0,34	1,41±0,19
	10 μg/cm²	1,07±0.01**	0,77±0,00**	1,23±0,21+	0,96±0,06 **

**p< 0.01; *p<0.05

Cells treated for 24 hours with coarse particulate matter or with particles collected on stage 7 (0.65-0.43μm) show a decreased MTS-reducing activity if compared with control cells, while those treated with fine particulate matter or carbon black have increased activity. After 48 hours of treatment with coarse particulate matter, carbon black or stage 7 particles, cells show a higher MTS-reducing activity than control cells, whereas this activity is lower for fine particulate matter-treated cells. On average, metabolic activity increases during the treatment period. These data are consistent with those relating to the rate of proliferation, showing an activation of macrophages in contact with particulate matter. Indeed, *in vivo* macrophages, of which RAW 264.7 cells are an excellent *in vitro* model, when stimulated by particulate matter (dust, bacteria) change from an inactive into an active phase. Activation is accompanied by a significant increase in the consumption of oxygen, glycolysis, release of cytokines such as interleukin-6 and tumor necrosis factor - α and a simultaneous decrease in the rate of proliferation [11].

Particulate Genotoxicity

Histograms presented in figure 4 illustrate results concerning the induction of micronuclei after 72-hour treatment with coarse, fine, or stage 7 particulate matter, or with carbon black, at three different concentrations: 1, 3, 10 μg/cm^2 .

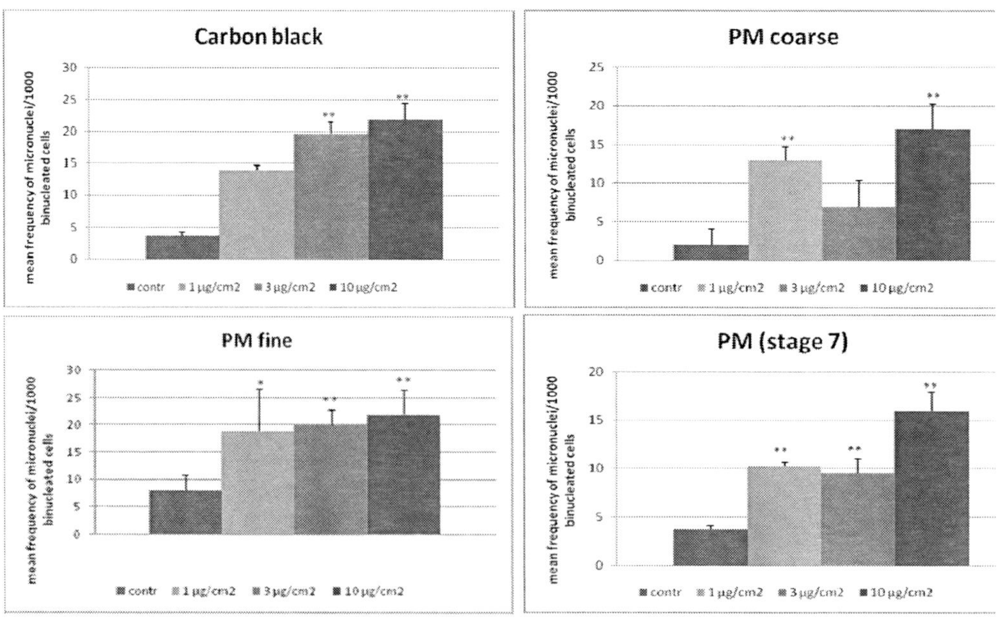

Figure 4. Frequency of micronuclei (referred to the total number of cells with micronuclei in 1000 cells) in RAW 264.7 cultured cells exposed for 72h to different concentrations of fine, coarse or stage 7 particulate matter, or carbon black. Mean values + S. D. ; *p<0.05, **p<0.01.

Values show that particulate matter and carbon black treatment induce micronuclei in a dose dependent manner and that the percentages of micronuclei are higher in treated cells than in controls. Treatment with fractions containing larger particles causes induction of a greater number of micronuclei if compared to those treated with fine particles and carbon black. This is probably due to an incomplete phagocytosis by macrophages. Nevertheless, smaller particulate can induce an indirect genotoxic effect because of the great surface area, on which toxic substances that catalyze extended synthesis of ROS can be absorbed [12]. These, in fact, can interact with organic molecules, deeply damaging tissues and causing degradation and fragmentation of DNA. Treatment with carbon black, at the highest concentrations tested, causes a significantly lower increment in micronuclei frequency than what observed after treatment with particulate matter; this may be due to the formation of aggregates of carbon black nanoparticles when this substance is administered to cultured cells. Genotoxicity was further evaluated with the comet assay. This assay is relatively easy and quick and allows one to evaluate DNA damage cell by cell, even if subsequent data interpretation may result rather difficult. To quantify DNA damage several different comet parameters can be analyzed such as: % DNA in tail, tail length, tail moment or Olive moment. In this regard, it has to be said that there is no general consent about the validity and interpretation of these parameters in different laboratories. Moreover, we have to consider a number of other factors that may influence the net outcome of this assay, such as the type of DNA damage caused by the mutagenic substance (clastogenic or aneugenic), as well as the different DNA repair mechanisms that can be activated in response to the genotoxic lesion, with their different efficiency and quickness. To quantify DNA damage, we have analyzed, by TriTek CometScore Freeware, the tail moment (percentage of tail DNA divided by 100 and multiplied by the tail length) (Fig. 5). The frequencies of comets in treated cells were higher than in control cells for any concentration and any particulate fraction (data not shown), although data do not highlight a dose-dependent increase of DNA damage.

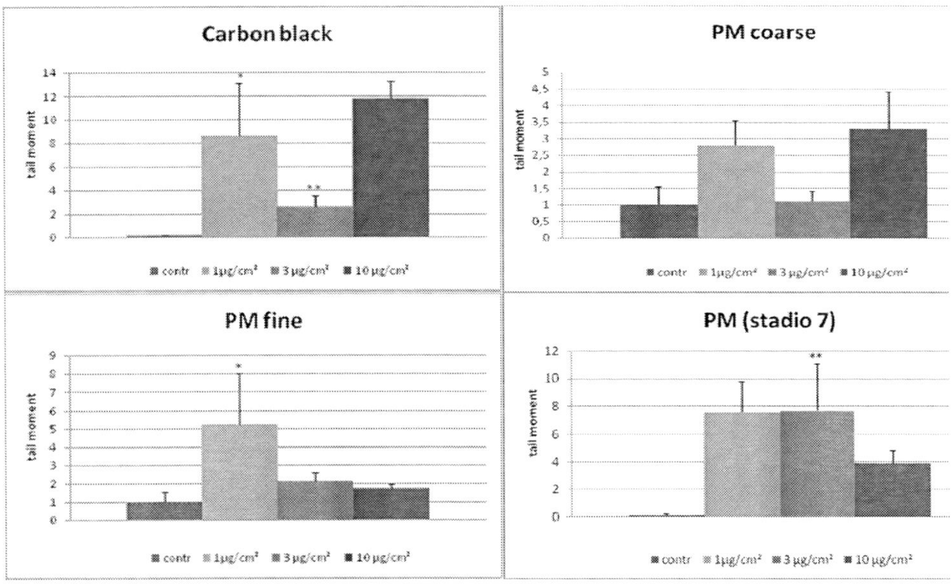

Figure 5. DNA damage, expressed by tail moment (TM) in RAW 264.7 cultured cells exposed for 15h to fine or coarse particulate matter, stage 7 PM or carbon black at different concentrations. The tail moment represents the percentage of tail DNA multiplied by the tail length divided by 100. Mean values + S.D. ; * $p < 0.05$, ** $p<0.01$

TEM ANALYSIS

Untreated (control) RAW 264.7 cells, when observed by transmission electron microscopy, exhibited typical features of normal cultured murine macrophages, with large rounded and predominantly euchromatic nuclei, in which one or two nucleoli were generally present. The cell surface was covered with sparse microvilli-like structures, while the cytoplasmic organelles appeared normal. Occasionally, some cells contained well-evident phagolysosomes, indicative of phagocytic activity (Fig. 6A). Treatment with either fine or coarse PM induced a strong increase in the phagocytic activity of these cells, as revealed by the change in nuclear morphology (with irregular contour), the appearance of rather numerous cytoplasmic protrusions and the abundant phagolysosomes (Fig.s 6B and 7A, B, D, E).

Furthermore, necrotic cells were frequently observed (Fig. 6B). Similar results were obtained after exposure of RAW 264.7 cells to carbon black (Fig. 6C and 7C, F). For all the experimental conditions no particular ultrastructural alterations could be detected, whereas internalized particles were clearly recognizable within phagolysosomes (Fig 7). In the case of carbon black treatment, particle agglomerates of different size were phagocytized (Fig. 7, C and F).

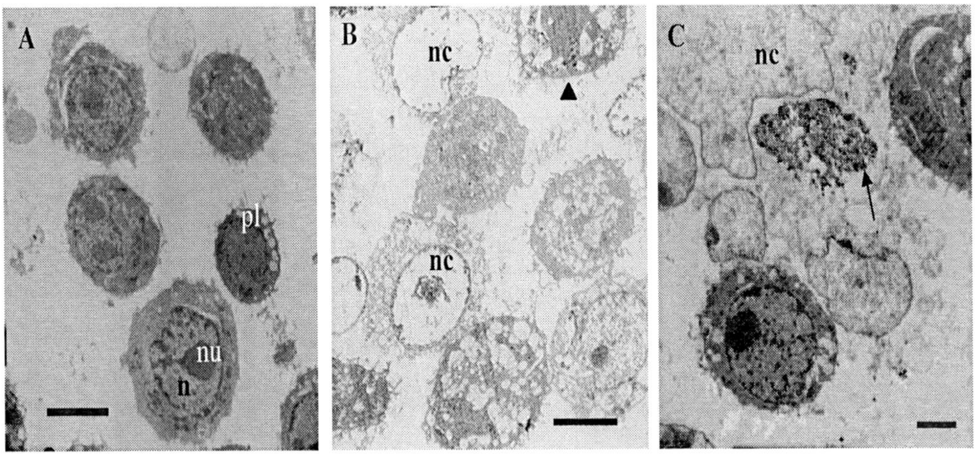

Figure 6. transmission electron micrographs of ultrathin sections of control (A), fine PM (B) or carbon black (C) -treated RAW264.7 cells. Black arrow: aggregate of carbon black; arrow head: fine particulate; nc: necrotic cells; n: nucleus; nu: nucleolus; p: particulate matter; pl: phagolysosome. Bars: A, B, 5μm; C, 2 μm

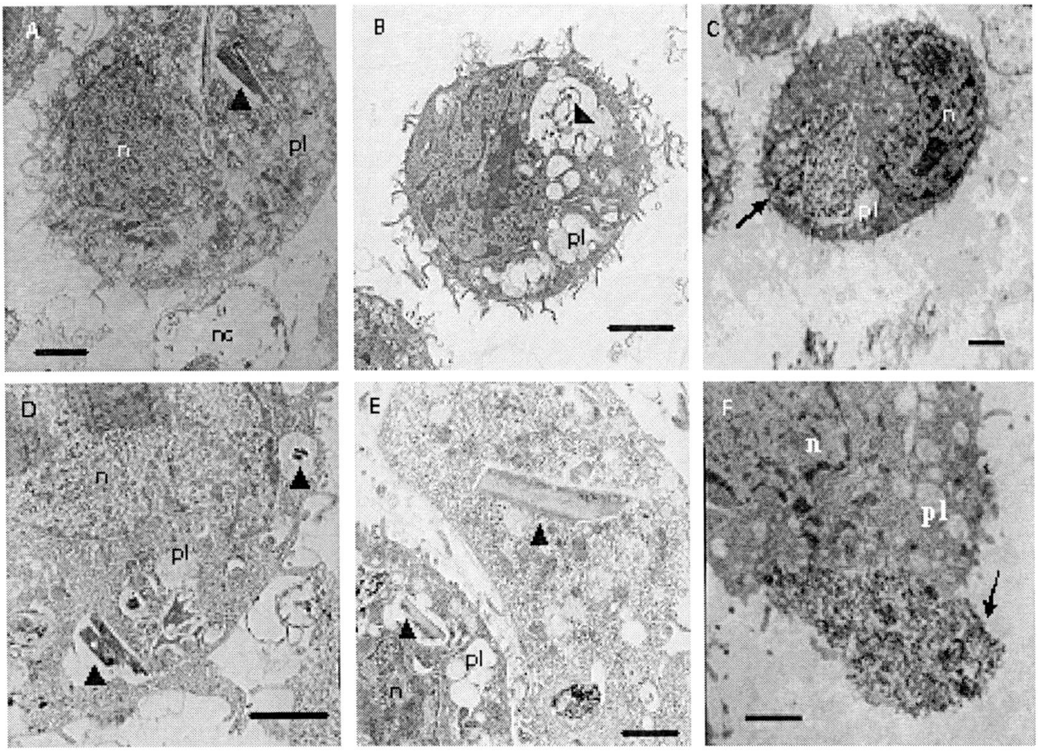

Figure.7. transmission electron micrographs of ultrathin sections of RAW 264.7 cells treated with fine PM (A and D), coarse PM (B and E) or carbon black (C and F). Arrows head: particulate matter within phagolysosomes; black arrow: agglomerate of carbon black within phagolysosome (C) and during phagocytosis (F); cb: agglomerate of carbon black; n: nucleus; nc: necrotic cells; pl: phagolysosome. Bars: A, B and C, 2 µm; D, E and F, 1 µm.

CONCLUSION

On the whole, the results indicate that fine particulate matter contains a high percentage of C-rich particles. The characteristic functional groups of most frequent organic pollutants have been reported, and it is evident that particulate matter surface chemistry represents an important factor for biological and cyto/genotoxic effects on involved cells, especially macrophages. Moreover, the results presented here demonstrate the utility of in vitro tests in animal cells for assessing toxicity and genotoxicity of urban air particulate matter. The murine macrophage RAW-264.7 cell line has proved a reliable and sensitive in vitro model for fine and coarse particulate genotoxicity testing, not only useful for the study of the release of inflammatory mediators induced by coarse and fine urban air particles, as recent survey of studies indicates.

The results of the treatments with fine, as compared with those with coarse PM, give further evidence for the higher citotoxicity of fine particulates. In addition, fine particulate, particulate collected on the last stage of particle sampler and carbon black are more genotoxic than coarse PM. These results are consistent with data shown in other works which underline that fine and, in particular unltrafine particulate matter is more harmful to human health than the coarse fraction. These fractions include the largest number of particles per volume unit, presenting a wide surface area on which there may be absorption of toxic substances that catalyze the synthesis of reactive oxygen species (ROS) [13].

So, even if the smaller particles can escape a complete phagocytosis by macrophages, they may induce an indirect genotoxic effect and can also overcome the barrier of pulmonary alveoli and enter the bloodstream, reaching other target organs like the brain.

References

[1] WHO, World Health Organization. WHO Strategy on Air Quality and Health. *Revised final draft.*, 2001, Ginevra.

[2] Lenshow, P; Abraham, J; Kutzner, K; Lutz, M; Preuß, JD; Reichenbächer, W. Some ideas about the sources of PM10. *Atmospheric Environ.*, 2001, 35: 23-33.

[3] Zhu, Y; Hinds, WC; Kim, S; Sioutas, C. Concentration and size distribution of ultrafine particles near a major highway. *J. Air & Waste Manage. Assoc.*, 2002, 52: 1032-1042.

[4] Seigneur, C. Current status of Air Quality models for particulate matter. *J. Air & Waste Manage. Assoc.*, 2001, 51(11): 1508-1521.

[5] Rogge, WF; Hildemann, LM; Mazurek, MA; Cass, GR; Simoneit, BRT. Sources of fine organic aerosol, 2. Non catalyst and catalyst equipped automobiles and heavy duty diesel trucks. *Environ. Sci. Technol.*, 1993, 27: 636-651.

[6] Van Bree, L. & Cassee, F. R. (2000). Toxicity of Ambient Air PM 10. A critical rewiew of potentially causative PM properties and mechanisms associated with health effects. *Netherlands*, 9-16.

[7] Ying, T; Ishikawa, H; Piao, FY; Yamamoto, H; Yamauchi, T; Duan, ZW; Zhang, YM; Ma, MY; Cui, JS. Micronucleus assay of human lynphocytes: a comparison of cytokinesis-block and human capillary blood lymphocytes methods. *J. Occup. Health*, 2003, 45: 408-409.

[8] Ravanat, JL; Sauvaigo, S; Caillat, S; Martinez, GR; Medeiros, MHG; Di Mascio, P; Favier, A; Cadet, J. Singlet oxygen-mediated damage to cellular DNA determined by the comet assay associated with DNA repair enzymes. *Bio. Chem.* 2004, 385: 17-20.

[9] Yamaguchi, T. & Yamazaki, H. (2001). Cytotoxicity of airborne particulates sampled roadside in rodent and human lung fibroblasts. *Journal of Health Science*, 47: 272-277

[10] Amicarelli, F; Bucciarelli, T; Poma, A; Aimola, P; Di Ilio, C; Ragnelli, AM; Miranda, M. Adaptive response of human melanoma cells to methylglyoxal injury. *Carcinogenosis*, 1998, 19 (3): 519-523.

[11] Santini, M. T. & Rainaldi, G. (2004). Environmental Fine Particulate Matter (PM 2.5) Activates the RAW 264.7 Macrophage Cell Line Even at Very Low Concentrations as Revealed by 1H NMR. *Chem. Res. Toxicol.*,17: 63-74.

[12] Renwick, L. C. & Donaldson, K. (2000). Impairment of Alveolar Macrophages Phagocitosis by Ultrafine Particles. *School of Life Sciences, Napier University, Edinburgh, EH10, Scotland.*

INDEX

A

E

H

I

J

K

L

N

O

P

Q

R

T

U